Sensor Data Analytics: Challenges and Methods for Data-Intensive Applications

Sensor Data Analytics: Challenges and Methods for Data-Intensive Applications

Editors

Felipe Ortega
Emilio López Cano

MDPI • Basel • Beijing • Wuhan • Barcelona • Belgrade • Manchester • Tokyo • Cluj • Tianjin

Editors
Felipe Ortega
Department of Signal Theory
and Communications and
Telematics Systems and
Computing; Data Science
Laboratory (DSLAB), Centre
for Intelligent Information
Technologies, University Rey
Juan Carlos, Madrid, Spain

Emilio López Cano
Department of Computer
Science and Statistics, Data
Science Laboratory (DSLAB),
Research Centre for
Intelligent Information Technologies,
Rey Juan Carlos University,
Madrid, Spain

Editorial Office
MDPI
St. Alban-Anlage 66
4052 Basel, Switzerland

This is a reprint of articles from the Special Issue published online in the open access journal *Entropy* (ISSN 1099-4300) (available at: https://www.mdpi.com/journal/entropy/special_issues/ Data_Analytic).

For citation purposes, cite each article independently as indicated on the article page online and as indicated below:

LastName, A.A.; LastName, B.B.; LastName, C.C. Article Title. *Journal Name* **Year**, *Volume Number*, Page Range.

ISBN 978-3-0365-4851-7 (Hbk)
ISBN 978-3-0365-4852-4 (PDF)

Contents

About the Editors

Felipe Ortega

Felipe Ortega (Associate Professor) works in the Dept. of Signal Theory and Communications, Telematics Systems and Computing, at the Higher School of Telecommunications Engineering, Rey Juan Carlos University (Madrid, Spain). He coordinates the data engineering line in the Data Science Lab (DSLAB) research group at this univeristy. His main interests include data science processes, data engineering, machine learning operations, data visualization, large-scale data indexing, and distributed computing. He is the author of more than 30 research works, including journal articles, conference papers, and 3 books. Dr. Ortega has participated in more than 35 national and international research projects in many different areas such as software engineering, energy management, risk assessment, health sciences, cybersecurity, and lifestock farming, among others. He also has more than 18 years of teaching experience at the graduate, postgraduate, and doctorate levels.

Emilio López Cano

Emilio López Cano (Associate Professor) works in the field of Computer Science and Statistics at Rey Juan Carlos University. His research interests include applied statistics, statistical learning, and methodologies for quality. He has given more than 1,000 hours of in-company training. He is the author of the SixSigma R package, published in the CRAN repository with an average of 1500 downloads per month, and of two monographs on quality methodologies with R in Springer. He is constantly transferring research results with companies via technology transfer contracts. He is also the president of the technical subcommittee of standardization UNE (member of ISO) CTN 66/SC 3 (Statistical Methods), a collaborating teacher in the Spanish Association for Quality (AEC), and the president of the R Hispano association (Spanish R users group).

Editorial

Sensor Data Analytics: Challenges and Methods for Data-Intensive Applications

Felipe Ortega * and Emilio L. Cano *

Data Science Laboratory, Research Centre for Intelligent Information Technologies, Rey Juan Carlos University, 28933 Madrid, Spain
* Correspondence: felipe.ortega@urjc.es (F.O.); emilio.lopez@urjc.es (E.L.C.)

Citation: Ortega, F.; Cano, E.L. Sensor Data Analytics: Challenges and Methods for Data-Intensive Applications. *Entropy* **2022**, *24*, 850. https://doi.org/10.3390/e24070850

Received: 2 June 2022
Accepted: 17 June 2022
Published: 21 June 2022

Publisher's Note: MDPI stays neutral with regard to jurisdictional claims in published maps and institutional affiliations.

Sensors have become a key element for the development of the Information Society. An ever-increasing number of improved sensor devices capture information for decision-making tools, either to be interpreted by humans or to be plugged back into the system for autonomous operation, self-diagnostics and resilience. It is possible to find sensor applications spanning almost any area, including healthcare and medicine, retail and logistics, smart agriculture and animal farming, industry digitalisation, smart cities, energy grids, transport or security, among many others [1]. Analytics is a term connected to the practice of data science that refers to the analysis of data using statistical tools and techniques, machine learning, information theory, pattern recognition and other methods. Outcomes stemming from this task constitute essential inputs for data-driven decision-making [2,3].

The current overabundance of data, generated in many cases by sensors, together with the refinement of standard methodologies for data science and engineering [4] has led to the rise of a fourth scientific paradigm, the so-called *data-intensive scientific discovery* [5]. Indeed, one of the most challenging aspects for the development of data-intensive applications has been how to cope with massive and complex datasets effectively, especially in situations in which real-time requirements arise [6,7]. In this regard, sensors provide an unrivalled data source to match these needs, as they can provide timestamped information with enough level of detail to characterise observed phenomena adequately.

Information theory [8] plays a central role for knowledge extraction in sensor data analytics, such as the analysis of data in the frequency domain [9], the essential concept of *entropy* [10] and efficient data representation and compression [11]. As a result, many new methods based on information theory have been developed in modern data science [12]. This Special Issue presents nine original contributions encompassing a wide variety of sensor data analytics applications, in which information theory is used to obtain knowledge from data in different domains.

Gajowniczek et al. [13] develop a novel method for data streams clustering, applicable to complete time series representing customer electricity consumption. This method leverages new Fast Fourier Transform (FFT) [9] features to improve its performance, showing the importance of information theory principles in this type of analysis. Wearable sensors tracking human activity and behaviour are at the core of several works, including applications in rehabilitation of visually impaired people [14], automated human activity recognition [15] and walking behaviour detection for elderly people [16]. The last two works attach importance to the application of information gain and neural networks to detect activity profiles accurately. Alfaro et al. [17] propose a new method to distribute the training process using the SVM algorithm, which can be applicable to Wireless Sensor Networks (WSN), aggregating the local contributions from individual sensors using Voronoi regions. Once again, this demonstrates the critical role of information aggregation in this kind of energy and location-aware sensor application. Sensor placement optimisation is the topic of another work [18], using Gaussian priors and the Fisher Information Matrix (FIM) to show important properties that can enhance recommendations on the best possible location for a given device set. Sun et al. present an interesting application of sensor data analytics to estimate vehicles accident risk [19]. This is an emerging topic that raises significant interest among insurance companies, taking advantage of the more precise tracking capabilities enabled by built-in sensors installed in vehicles. In turn, Esteban-Escaño et al. present an interesting application of sensor data analysis to predict acidemia in electronic fetal monitoring [20], using machine learning algorithms, stressing the use of cross-entropy optimisation function along this process, to adjust the best possible predictive model. Finally, the last work [21] presents a novel methodology for cattle behaviour profiling and classification that, again, uses both time-domain and frequency-domain features to improve the accuracy of this classification task.

In summary, these contributions offer a diverse and representative portfolio of sensor data analytics applications in different scenarios, in which information theory and data science methods perform a central role in order to successfully accomplish the proposed challenges in each case.

Funding: This research received no external funding.

Conflicts of Interest: The authors declare no conflict of interest.

References

1. Perros, H. *An Introduction to IoT Analytics*; Chapman & Hall/CRC Data Science Series; CRC Press: Boca Raton, FL, USA, 2021.
2. Anderson, C. *Creating a Data-Driven Organization*; O'Reilly Media Inc.: Sebastopol, CA, USA, 2015.
3. Provost, F.; Fawcett, T. *Data Science for Business: What You Need to Know about Data Mining and Data-Analytic Thinking*; O'Reilly Media Inc.: Sebastopol, CA, USA, 2013.
4. Wirth, R.; Hipp, J. CRISP-DM: Towards a Standard Process Model for Data Mining. In Proceedings of the 4th International Conference on the Practical Applications of Knowledge Discovery and Data Mining, Manchester, UK, 11–13 April 2000; Volume 1, pp. 29–40.
5. Hey, A.J.; Tansley, S.; Tolle, K.M. *The Fourth Paradigm: Data-Intensive Scientific Discovery*; Microsoft Research Redmond: Redmond, WA, USA, 2009; Volume 1.
6. Kelleher, J.D.; Tierney, B. *Data Science*; MIT Press: Cambridge, MA, USA, 2018.
7. Bifet, A.; Gavalda, R.; Holmes, G.; Pfahringer, B. *Machine Learning for Data Streams: With Practical Examples in MOA*; MIT Press: Cambridge, MA, USA, 2018.
8. Shannon, C.E. A Mathematical Theory of Communication. *Bell Syst. Tech. J.* **1948**, *27*, 379–423. [CrossRef]
9. Proakis, J.G.; Manolakis, D.G. *Digital Signal Processing: Principles, Algorithms and Applications*; Pearson: London, UK, 2006.
10. Bishop, C.M. *Pattern Recognition and Machine Learning*; Information Science and Statistics; Springer: Berlin/Heidelberg, Germany, 2006.
11. MacKay, D.J. *Information Theory, Inference and Learning Algorithms*; Cambridge University Press: Cambridge, UK, 2003.
12. Rodrigues, M.R.D.; Eldar, Y.C. (Eds.) *Information-Theoretic Methods in Data Science*; Cambridge University Press: Cambridge, UK, 2021. [CrossRef]
13. Gajowniczek, K.; Bator, M.; Ząbkowski, T. Whole Time Series Data Streams Clustering: Dynamic Profiling of the Electricity Consumption. *Entropy* **2020**, *22*, 1414. [CrossRef] [PubMed]
14. Reyes Leiva, K.M.; Jaén-Vargas, M.; Cuba, M.Á.; Sánchez Lara, S.; Serrano Olmedo, J.J. A Proposal of a Motion Measurement System to Support Visually Impaired People in Rehabilitation Using Low-Cost Inertial Sensors. *Entropy* **2021**, *23*, 848. [CrossRef] [PubMed]
15. Liu, L.; He, J.; Ren, K.; Lungu, J.; Hou, Y.; Dong, R. An Information Gain-Based Model and an Attention-Based RNN for Wearable Human Activity Recognition. *Entropy* **2021**, *23*, 1635. [CrossRef] [PubMed]
16. Aznar-Gimeno, R.; Labata-Lezaun, G.; Adell-Lamora, A.; Abadía-Gallego, D.; del Hoyo-Alonso, R.; González-Muñoz, C. Deep Learning for Walking Behaviour Detection in Elderly People Using Smart Footwear. *Entropy* **2021**, *23*, 777. [CrossRef] [PubMed]
17. Alfaro, C.; Gomez, J.; Moguerza, J.M.; Castillo, J.; Martinez, J.I. Toward Accelerated Training of Parallel Support Vector Machines Based on Voronoi Diagrams. *Entropy* **2021**, *23*, 1605. [CrossRef] [PubMed]
18. Zhou, R.; Chen, J.; Tan, W.; Yan, Q.; Cai, C. Optimal 3D Angle of Arrival Sensor Placement with Gaussian Priors. *Entropy* **2021**, *23*, 1379. [CrossRef] [PubMed]
19. Sun, S.; Bi, J.; Guillen, M.; Pérez-Marín, A.M. Driving Risk Assessment Using Near-Miss Events Based on Panel Poisson Regression and Panel Negative Binomial Regression. *Entropy* **2021**, *23*, 829. [CrossRef] [PubMed]
20. Esteban-Escaño, J.; Castán, B.; Castán, S.; Chóliz-Ezquerro, M.; Asensio, C.; Laliena, A.R.; Sanz-Enguita, G.; Sanz, G.; Esteban, L.M.; Savirón, R. Machine Learning Algorithm to Predict Acidemia Using Electronic Fetal Monitoring Recording Parameters. *Entropy* **2022**, *24*, 68. [CrossRef] [PubMed]
21. Cabezas, J.; Yubero, R.; Visitación, B.; Navarro-García, J.; Algar, M.J.; Cano, E.L.; Ortega, F. Analysis of Accelerometer and GPS Data for Cattle Behaviour Identification and Anomalous Events Detection. *Entropy* **2022**, *24*, 336. [CrossRef] [PubMed]

Article

Whole Time Series Data Streams Clustering: Dynamic Profiling of the Electricity Consumption

Krzysztof Gajowniczek *, Marcin Bator and Tomasz Ząbkowski

Department of Artificial Intelligence, Institute of Information Technology, Warsaw University of Life Sciences-SGGW, 02-776 Warsaw, Poland; marcin_bator@sggw.edu.pl (M.B.); tomasz_zabkowski@sggw.edu.pl (T.Z.)
* Correspondence: krzysztof_gajowniczek@sggw.edu.pl

Received: 9 November 2020; Accepted: 11 December 2020; Published: 15 December 2020

Abstract: Data from smart grids are challenging to analyze due to their very large size, high dimensionality, skewness, sparsity, and number of seasonal fluctuations, including daily and weekly effects. With the data arriving in a sequential form the underlying distribution is subject to changes over the time intervals. Time series data streams have their own specifics in terms of the data processing and data analysis because, usually, it is not possible to process the whole data in memory as the large data volumes are generated fast so the processing and the analysis should be done incrementally using sliding windows. Despite the proposal of many clustering techniques applicable for grouping the observations of a single data stream, only a few of them are focused on splitting the whole data streams into the clusters. In this article we aim to explore individual characteristics of electricity usage and recommend the most suitable tariff to the customer so they can benefit from lower prices. This work investigates various algorithms (and their improvements) what allows us to formulate the clusters, in real time, based on smart meter data.

Keywords: clustering; data stream; machine learning; smart metering; time series

---

1. Introduction

The advances in smart metering solutions have enabled that gathering information about customer power consumption in real time is feasible and it can be successfully used for data exploration to bring actionable recommendations. The data (in the form of a time series) from the smart grid still makes challenges to analyze it due to the very large size, high dimensionality, skewness, sparsity, and number of seasonal fluctuations, including daily and weekly effects. Although the analysis requires a lot of effort to discover the segmentation of entities based on their electricity consumption data, the benefits, as the result of the data insights, would be very appealing to the electricity providers [1]. By supplying providers with demand response predictions on aggregated level, due to segmentation (other terms such as clustering and grouping are used interchangeably), and revealing the real economic structure of the entities (e.g., individual users, households, small business) the goal is to fit into the integrated planning system, where the appropriate real-time actions could be proposed to meet the system demands effectively [2]. Well recognized consumption patterns itself are also a source of valuable insight to determine optimal tariff rates for the users and to deal with the spikes in electricity demand.

The analysis of the data streams (in this article we deal with time series and therefore we will use term time series data streams as well) coming from the grid over consecutive time windows allows for a better understanding of the usage characteristics. With the data arriving in a sequential form the underlying distribution is subject to changes over the time intervals what is referred to as concept drift [3,4]. For example, the changes in smart meter streaming data may be the result of many factors, including those related to weather conditions, to week days or those related to price incentives [5]. It is often observed that smart meter readings received at an instant intervals may have a dynamic distribution or may contain a large number of sparse and missing values. Therefore, traditional algorithms are not applicable directly nor suitable for these type of data as they extract patterns from data by assuming the global properties (what requires the complete training data set), rather than capturing the local ones.

Time series data streams have their own specifics in terms of data exploration and processing, because, usually, it is not possible to process the whole history in memory. The reason for that is that data are coming very fast so the processing and the analysis should be done incrementally using sliding windows (overlapping or non-overlapping) or using other approaches like the stochastic learning weak estimators [6]. Classical clustering algorithms aim to divide a set of objects (observations) into groups so that objects in the same group are more similar to each other than objects in other groups. The literature on time series data stream clustering makes a distinction in terms of what is the subject of grouping [3]. The first approach tries to cluster observations from a single univariate or multivariate time series data stream through lots of promising tools and methods [7]. On the other hand, second approach tries to analyze multiple time series data streams, generated by several sources (e.g., smart meters), in order to find a division of sources. In literature the latter problem is also known as attribute clustering [8]. Despite the proposal of many clustering techniques dedicated for the first approach, only a few of them are dedicated to the second approach. Due to that in this article we focus on multiple time series data streams clustering, as this is one of the most important challenge in data stream mining.

In many countries, all over the world, the retail electricity demand side of the market consists of several groups of end users. In Poland, for instance, the vast majority of consumers belong to the so-called tariff group G (mostly households). Other end users belong to so-called tariff groups A (top, strategic customers), B (large, key customers) which are supplied from the high and medium voltage grid, while group C consists of customers connected to the low voltage grid, consuming electricity for business purposes and they are called commercial customers [2]. For low-voltage households, operators have set up several different tariff groups which differ in the time zone (single or two time zone meters) and whether or not electricity is used for heating. The most general tariff group for households is G11, i.e., customers with single time zone meters and flat price per kWh. The other tariff groups, G12, G12r, and G12w, are time and weekdays. G12 is effective between 10 p.m. and 6 a.m. and between 1 p.m. and 3 p.m., while G12w is additionally effective during the weekends (between 10 p.m. on Friday and 7 a.m. on Monday). G12r is effective seven days a week between 10 p.m. and 7 a.m. and between 1 p.m. and 4 p.m.

The main goal of this article is to investigate technical aspects of the existing clustering algorithms for time series data streams. The secondary goal is to explore individual characteristics of electricity usage and to recommend the most suitable tariff to the customers so they can benefit from lower prices, thus optimize the expenses. The research shall be conducted on the basis of a dataset provided by the Irish Commission for Energy Regulation (CER; detailed analysis) [5] and two other datasets, which are described later. We investigate various algorithms (and their improvements) what allows us to formulate the clusters in real time based on smart meter data. Basically, we develop a clustering approach applicable for data streams with the primary motivation to create well defined user profiles what may further allow to create more predictable groups of customers. The contribution of this article can be summarized as follows:

- We have created the framework and measures to compare and to evaluate time series data streams clustering algorithms;
- New Fast Fourier Transformation based features were created (calculated in liner time) to compress and to represent time series using the business context;
- Comparative study between the state-of-the-art time series data streams clustering algorithms was prepared;
- Comparative study between overlapping and non-overlapping windows and their impact on the choice of an optimal tariff was prepared; and

- Finally, an approach for dynamic consumer segmentation and prediction of an optimal tariff was proposed.

We believe that our contribution would address the gap related to those aspects of dynamic profiling where there was no clear conclusion with regards to the benefit of using overlapping vs. non-overlapping windows and the impact of those on the results of clustering algorithms.

The remainder of this paper is organized as follows: Section 2 provides an overview of the similar research problems for data stream time series clustering and electricity consumption segmentation. In Section 3, the theoretical framework of the proposed algorithm is presented. In Section 4, the research framework is outlined, including the details of numerical implementation, evaluation measure description, and algorithm parameter settings. Section 5 outlines the experiments and presents the discussion of the results. The paper ends with concluding remarks in Section 6.

2. Literature Review

Whilst the vast majority of customers belong to a single tariff with high volatility within the group, it creates a number of challenges, including short-term and long-term forecasting to meet the demand side response (DSR) of electricity operators, not to mention the stability of the whole network [9]. Obviously, daily energy consumption does not depend only on the composition of the customer's tariffs, but also it depends on many external factors related to specific days, atmospheric phenomena, and weather conditions [10]. Due to that, there is a need for an objective approach to increase the effectiveness and efficiency of network management and operations by dividing mass markets into consumer groups with clearly similar patterns of behavior. This can be supported by statistical clustering methods what helps to formulate valid and meaningful clusters based on the available measurements data e.g., hourly.

Given the huge number of low-voltage customers, especially households, hourly measuring and recording equipment are a serious shortage. Both, the future demand and the initial settlement of customers are determined based on the load shape associated with specific tariff group. In that case, a similar energy demand structure determines the number of groups. Statistical and engineering techniques [11–14], time series [15–17], and neural networks [16,18,19] are used for load profiling. Based on the literature review, there is a clear and increasingly recognizable research trend that addresses the challenges of segmentation of electricity end-users. For example, the application of the k-means algorithm for clustering of the daily load profiles of individual users was described in [17,20–22]. A comparison of clustering algorithms for classifying household electricity consumers Kohonen's self-organization map (SOM), and including hierarchical clustering, was analyzed by [2,23].

The literature on data streams clustering is quite extensive and includes the methods (1) aiming at grouping of the observations of a single data stream; and (2) proposals that monitor the proximity between multiple data streams in order to find the division of streams into clusters. The state-of-the-art survey of a multivariate or single univariate data stream clustering methods is available in [3]. Authors have presented a comprehensive survey on this phenomenon which discusses various types of data stream clustering techniques and the corresponding challenges. So far, most of the attention has been devoted to observations-based data streams clustering, which focuses on clustering of the observations from the single data stream. Reference is made to several categories of methods, including: Grid-based stream methods, partitioning stream methods, density-based stream methods, hierarchical stream methods, and growing neural gas-based methods. The flagship methods in those categories are: Str-FSFDP [24], MuDi [25], D-Stream [26], ClusStream [27], DBSTREAM [28], BIRCH [29], E-Stream [30], and StreamKM++ [31].

A more detailed analysis of the literature on grouping of multiple data streams (or time series stream), which is the subject of this article, is desired. For example, the recent methods are constructed in a way to ensure the division of streams over time [32–39]. All of them monitor the proximity of data streams using a record flow and introduce some strategies to obtain partitioning of streams into a set of clusters. Other interesting methods, such as [40–43], are focused on monitoring proximity between streams, but these do not include a grouping stage.

In the broader context of the techniques used for electricity consumption data driven by explosive growth of time-series data and the capability of the methods there are interesting attempts which propose a cohort of dominant data set selection algorithms for electricity consumption time series with a focus on discriminating the dominant data set that is a small data set but capable of representing the key information carried by time series with an arbitrarily small error rate [44].

Authors in [34] discussed the clustering on-demand framework (COD) involving a single data scan to derive online statistics. The COD consists of two stages, namely the online maintenance (providing an effective mechanism for maintaining hierarchical summaries of data streams) and offline clustering (finding approximations of desired sub-streams from the summary hierarchy according to cluster queries). Based on this algorithm Chen [39] introduced the CORREL-cluster algorithm offering a time horizon segmentation scheme and statistical information storage for each time segment.

A tree-like hierarchy of clusters evolving with the data and using a top-down strategy has been introduced in [38]. The Online Divisive-Agglomerative Clustering algorithm (ODAC), incorporates correlation-based measure of similarity between time series, dividing each node by the furthest pair of streams. Due to the splitting and merging, operators algorithm is able to detect and to adapt to the data in the presence of the concept drift. The performance of the ODAC algorithm has been next improved by TS-Stream algorithm which calculates several descriptive time series measures and builds a decision tree [37]. Adequate measures are selected on the basis of the criterion of minimizing variance. As previously, the algorithm can gradually expand or reduce the tree according to changes in the stream that change the node variance. Finally, in [45] authors have presented an extended version of the TS-Stream algorithm, that overcomes some base algorithm drawbacks. After those modifications the final tree structure reaches its full size immediately and it can have leaves with the number of time series above a certain threshold (otherwise the tree would be very complex and deep).

Algorithm called IDEStream has been introduced by [39]. In this approach an autoregressive modelling (AR) is used to measure the correlation between data streams and it uses the estimated frequency spectrum to extract the relevant data stream characteristics such as attenuation rate, phase, and amplitude. Authors in [36] presented a two phase algorithm which uses a gamma mixture model to identify dense units of incoming data in the first phase. Aim of the second phase is to cluster the time series from one time window, while third phase performs incremental clustering between received groups of two consecutive time windows.

In [32] authors have developed a powerful online version of the fuzzy C-means algorithm (FCM-DS), allowing to quickly calculate the approximate distance between the streams, thanks to the scalable online transformation of the original data. In [35] authors have presented an algorithm called ClipStream where time-series are compressed and represented by interpretable features separated from clipped representation. Next, based on such data transformation the K-medoids method with the Partition Around Medoids (PAM) algorithm cluster the data streams.

Finally, paper [8] presents a strategy which is based on the independent processing of incoming data batches, through a preliminary summarization using histograms, followed by local clustering carried out on histograms, which ensures further summarization of the data. To track the proximity of data between data streams over time they used local clustering outputs to update the proximity matrix.

3. Time Series Data Streams Clustering Algorithms

3.1. Notations and Data Representation

A time series is an ordered sequence of values of a variable at equally spaced time intervals (e.g., 30 min electricity load readings). Let us assume that, $s_j = \left\{s_{j,1}, s_{j,t}, \ldots, s_{j,n}\right\}^T$ is a partial realizations from a j-th ($j = 1, \ldots, m$) real-valued processes $S_j = \left\{S_{j,t}, t \in \mathbb{Z}\right\}$. Formally, the problem of grouping the time series data streams can be defined as follows. Let $S = \left\{s_1^T, s_j^T, \ldots, s_m^T\right\}$ be the data stream composed of m time series each of length n (S is a matrix with m rows and n columns). For a l-th ($l = 1,\ldots,k$) overlapping or non-overlapping time windows (blocks) with w time slots (intervals), B_l (with partial realization $b_j = \{b_{j,1}, b_{j,t}, \ldots, b_{j,w}\}^T$) is a subset (of columns) of S, i.e., a matrix of dimension $m \times w$ (each block consists of a subset of times series from the same time interval. For a given block, $L_l = \{L_{l_1}, L_{l_o}, \ldots, L_{l_p}\}$ represents a partition (of rows) of B_l such that L_{l_o} is the o-th cluster of L_l, $L_{l_o} \cap L_{l_p} = \varnothing$, $\forall o \neq p$ and $\cup_{o=1}^{p} L_{l_o} = B_l$ [37].

An exemplary data representations for overlapping (bottom part) and non-overlapping (upper part) windows with final clustering are depicted in Figure 1. On both figures on the left-hand side, there are m time series data streams, S, divided into k blocks each of length w (here $w = 5$). The right part of this figure illustrates an exemplary partition of the m time series from the l-th window (B_l) into L_{l_p} cluster.

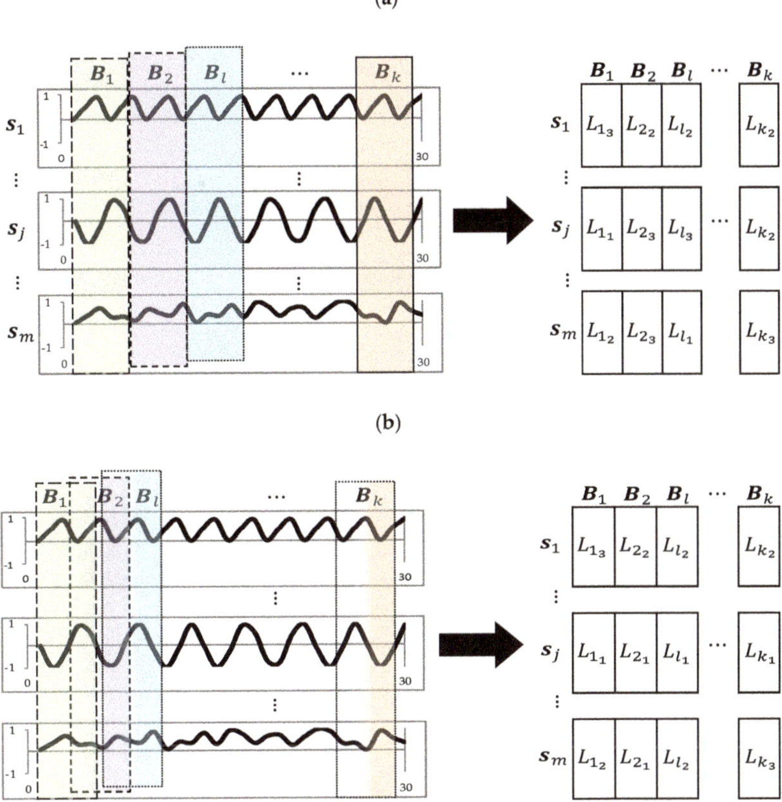

Figure 1. An exemplary data representation model with clustering: (**a**) Non-overlapping windows, (**b**) overlapping windows.

3.2. Histogram-Based Clustering Algorithm

The algorithm presented by [8] is composed of 4 main phases, where phases 1-3 are done online, while phase 4 is done offline. The goal of the phase 1 is to represent each time series data stream as a series of histograms by dividing the incoming data into (by default) non-overlapping time windows (this assumption will be further extended) and calculating the histogram of each l-th window:

$$H_j^l = \left\{ (I_1, \pi_1), \ldots, \left(I_p, \pi_p\right), \ldots, (I_P, \pi_P) \right\}, \tag{1}$$

where I_p denotes P successive bins/intervals associated with the relative frequencies π_p (weights), which sum up to 1. In this way, one can obtain, for each time window, a set of histograms which become the input for the local clustering procedure.

The purpose of the phase 2 is to get a local data partition (using BIRCH algorithm [29]) on a set of histograms that summarize the data behavior in each window. In order to do that the L_2 Wasserstein metric (distance) should be introduced, which simply calculate the distance between any two histograms H_k^l and H_j^l. As shown in [46] this metric requires an initial homogenization step to ensure consistency of distance calculations, which is based on the histogram configurations. Since all histograms are uniformly dense in each I_p interval, their quantile functions Q_j^l are piecewise linear. Aforementioned homogenization step consists in dividing Q_j^l functions in such a way that piecewise linear functions are defined on the same set of h cumulative probability values $q_v = \sum_{p=1}^{v} \pi_p$, $(v = 1, \ldots, h)$ [8]. To make the computation faster, according to the authors [46], each bin $I_v = \left[\overline{I_v}; \underline{I_v}\right]$ in the histogram can be represented as a function of a radius and a center, i.e., $I_v = [c_v - r_v; c_v + r_v]$, where $c_v = \left(\overline{I_v} + \underline{I_v}\right)/2$ is the centre of each interval and $r_v = \left(\overline{I_v} - \underline{I_v}\right)/2$ is the radius. Finally, using this representation the L_2 Wasserstein distance is as follows:

$$d_W^2\left(H_k^l, H_j^l\right) = \sum_{v=1}^{h} \pi_v \left[\left(c_v^k - c_v^j\right)^2 + \frac{1}{3}\left(r_v^k - r_v^j\right)^2\right]. \tag{2}$$

The formula allows to take into account the features of two histograms being compared in terms of shape, range and location.

To perform a local clustering on l-th batch, aforementioned BIRCH algorithm requires two information about each o-th group ($o = 1, \ldots, p$), i.e., histogram centroid (average) $\overline{H_o^l}$ and L_2 Wasserstein-based variance σ_o^{2l}. According to the [47] and based on the Formula (2), the mean of a set of histograms of equal frequency is obtained by the average of the centers and the average of the radii of the corresponding h intervals:

$$\overline{H^l} = \{([\bar{c}_1 - \bar{r}_1; \bar{c}_1 + \bar{r}_1], \pi_1) \ldots ([\bar{c}_v - \bar{r}_v; \bar{c}_v + \bar{r}_v], \pi_v) \ldots ([\bar{c}_h - \bar{r}_h; \bar{c}_h + \bar{r}_h], \pi_h)\}, \tag{3}$$

where:

$$\bar{c}_v = m^{-1} \sum_{j=1}^{m} c_v^j; \ \bar{r}_v = m^{-1} \sum_{j=1}^{m} r_v^j. \tag{4}$$

On the other hand, a volatility measure for a set of histograms is the average of the L_2 Wasserstein measure between each j-histogram and the average histogram defined in Formula (3):

$$\sigma^{2l} = \frac{1}{m} \sum_{j=1}^{m} d_W^2\left(H_j^l, \overline{H^l}\right). \tag{5}$$

The rationale in favor of this phase is to perform a single scan of the input data in order to obtain a division into a large number of clusters with low variability. To do that authors in [8] adopted the basic BIRCH algorithm to histogram-based data structures. Whenever a new time window is introduced, the algorithm allocates each H_j^l histogram to existing micro-clusters or generates new micro-clusters according to a fixed threshold u that controls the growth of heterogeneity in micro-clusters. In other words, if the L_2 Wasserstein distance to the nearest micro-cluster centroid is smaller than the predefined threshold $d_W^2\left(H_j^l, \overline{H_o^l}\right) < u$ then H_j^l histogram (representation of the time series data stream) is assigned to this cluster, otherwise it creates entirely new cluster, with the initialized variance σ_o^{2l} set to at the L_2 Wasserstein distance to the nearest cluster.

In phase 3 an update of the proximity matrix $A^l = \left[a^l(k,j)\right]$ is performed, which registers the dissimilarities between the streams. The proximity matrix is updated incrementally (each cell $a^l(k,j)$) each time a new data window is processed in phase 2, therefore, it tracks the proximities over time, using information only from the local partitions. If two histograms H^l_k and H^l_j fall into the same micro-cluster the proximity matrix is updated by adding the value of the variance σ^{2l}_o of this cluster:

$$a^l(k,j) = a^l(k,j) + \sigma^{2l}_o. \tag{6}$$

On the other hand, if these two histograms fall into different micro-clusters, the cell is updated by adding the mean of two distances:

$$a^l(k,j) = a^l(k,j) + \frac{d^2_W\left(H^l_k, \overline{H^l_o}\right) + d^2_W\left(H^l_j, \overline{H^l_p}\right)}{2}, \tag{7}$$

i.e., L_2 Wasserstein distances to the nearest micro-cluster centroids for both histograms. This update strategy allows to use only information from the micro-clusters, thus it requires only $m^2/2$ operations.

Finally, phase 4 provides an ultimate global clustering of the time series data streams from B_l block by grouping the updated proximity matrix into L_l. In order to obtain such partition DCLUST algorithm [48] is employed which minimizes intra-cluster variability, expressed by the sum of the dissimilarities between all pairs of elements within a cluster:

$$\sum_{o=1}^{p} \sum_{k,j \in L_{l_o}} a^l(k,j) \rightarrow min. \tag{8}$$

According to the authors [8] histograms are fast to compute with the time complexity $O(wP)$. The generation and the update of histogram micro-clusters, through a single scan of the histograms in a window, induces the time complexity of the algorithm is linear in m and p.

3.3. ClipStream Algorithm

The ClipStream algorithm is composed of two main phases [35], i.e., online data abstraction (representation) and an offline clustering. The first data representation phase includes a fast and incremental method of calculating feature vector from each B_l block named FeaClip and automatic detection of outliers. The second offline phase aims at grouping of a new data abstraction, aggregation of time series data streams in the cluster and the change detection process.

The feature extraction approach from the first phase is based on a so called clipped representation. Let us first define a short window b^{short} as a subsequence of an original time series data stream s of length z (z is shorter than window length w, and it could represent e.g., each day having 24 or 48 recordings; see also Section 3.1.) and a long window b^{long} which consists of last d consecutive short windows (therefore it is of length $d*z$). Next, a new representation (with reduced dimensionality $p < z$) of b^{short} is repr^{short} defined as below, first:

$$\hat{b}^{short}_t = f(x) = \begin{cases} 1 & \text{if } b^{short}_t > \mu \\ 0 & \text{otherwise} \end{cases}, \tag{9}$$

$\hat{b}^{short}$ is a clipped (bit-level) abstraction of the original block, where μ denotes a mean value of b^{short}. Then, the compression method called Run Length Encoding (RLE) [49] is applied on this abstraction to create the final representation repr^{short} (of length 8) defined as:

$$\boldsymbol{repr}^{short} = \left\{ \begin{array}{l} max_1 = max. \text{ from run lengths of ones,} \\ sum_1 = sum \text{ of run lengths of ones,} \\ max_0 = max. \text{ from run lengths of zeros,} \\ crossings = \text{length of RLE encoding } - 1, \\ f_0 = \text{number of first zeros,} \\ l_0 = \text{number of last zeros,} \\ f_1 = \text{number of first ones,} \\ l_0 = \text{number of last ones,} \end{array} \right. \tag{10}$$

Finally, the ultimate $\boldsymbol{repr}^{long}$ abstraction is an union of d short representations $\boldsymbol{repr}^{short}_d$ which has length $d * 8$. Whenever a new window b^{short}_{d+1} is arrived, first 8 features from $\boldsymbol{repr}^{long}$ are removed and new $\boldsymbol{repr}^{short}_{d+1}$ is attached to the end of $\boldsymbol{repr}^{long}$.

Based on the calculated FeaClip abstractions of all available time series data streams, outlying values can be easily and automatically detected by using domain knowledge. To automatize this, mean values of $crossings$ and sum_1 are calculated for each stream and corresponding $\boldsymbol{repr}^{long}$. Bead on these statistics, lower and upper quartiles and IQR (interquartile range) are calculated to create box-and-whisker diagrams, with threshold value λ set at 1.5. Time series with the characteristics that meet the following conditions: $Q_1^{sum_1} - \lambda * IQR^{sum_1} \leq x \geq Q_3^{sum_1} + \lambda * IQR^{sum_1}$ and $x \geq Q_3^{crossings} + \lambda * IQR^{crossings}$, are considered as non-outliers. Outlying values are not deleted from the whole clustering, they are simply stored in memory, and after the clusters are determined, those objects are assigned to the nearest ones.

Once the data representation phase is completed second offline stage follows to create the final grouping. Only filtered (without outliers) $\boldsymbol{repr}^{long}$ representations are subject to clustering using K-medoids method with Partition Around Medoids (PAM) algorithm [50] with Euclidean distance. To capture the dynamic and evolving nature of time series data streams, the number of clusters should also be determined dynamically. Therefore, the optimal number of clusters is determined on the basis of the internal measure of Davies–Bouldin index [51]. During the first iteration of clustering the number of possible clusters is determined in the range $p_{min} - p_{max}$, where p that minimizes the Davies–Bouldin index is chosen. To speed up further iterations of clustering the optimal number of clusters is selected from $\langle p - 2, p + 2 \rangle$, where p is the number of clusters from the previous iteration.

In order to carry out the process of grouping time series data streams only when it is necessary, i.e., only when data streams evolve and change of distributions occur, a stage for detecting concept drift is conducted. It detects changes of the Empirical Distribution Function (EDF) of the normalized aggregated data stream within each cluster, using K-sample Anderson–Darling test, defined as:

$$A_{kw}^2 = \frac{1}{w} \sum_{k=1}^{K} \frac{1}{z} \sum_{t=1}^{w-1} \frac{(wN_{kt} - tz)^2}{t(w-t)}, \tag{11}$$

where (according to the Section 3.1 and notation introduced at the beginning of this section) $s_{j,t}$ is the t-th recording in the k-th sample, N_{kt} denotes the number of observations in the k-th sample that are not greater than x_t, where $x_t < \cdots < x_w$ is the pooled ordered sample (long window). Concept drift is detected if p-value is less than the significance level α set at 0.05, however clustering is updated only if one of these conditions are meet: (1) The number of detected changes is more than half of the grouped p time series (number of clusters); (2) the number of detected changes is higher than in the previous step of the sliding window.

According to the authors [35] the representation phase has the linear time complexity $O(w)$ with respect to the length of the time window. Outlier detection phase is linear $O(m)$. The offline phase consists of the PAM clustering algorithm that for each iteration has the quadratic complexity of $O(p((m - m_o) - p)^2)$, where m_o denotes number of outliers.

3.4. Extended TS-Stream Algorithm

The algorithm presented by [45] is an extended (improved) version of the algorithm presented in [37]. In general, it evokes a model with a structure similar to the decision tree, but built in an unsupervised manner. The top-down strategy is employed to build the tree, starting from all times series data streams in the same main cluster (root) and gradually creating partition or aggregations. Each indirect node executes a binary test of a type $feature_{value} \leq x$ for a specific time series descriptive measure. Once a leaf is reached, the time series is stored together with other time series which belong to the same leaf.

During the first step the algorithm calculates descriptive measures (here also called coefficients, characteristics, indices) for each time series data stream. This gives a matrix of characteristics of the dimension $m \times f$, where f is the number of characteristics. To make all features comparable (which is required when variance minimization criterion is used), for each column of the matrix the z-score normalization of the form $x = (x - \mu)\sigma$ is performed. A simple and natural way to model each time series data stream is to use generating functions to depict their behavior in time domain. Unfortunately, many of the existing grouping techniques do not take into account specific characteristics of the generating function, e.g., stochasticity, linearity, and stationarity. So, the algorithm employs many descriptive measures in order to obtain the appropriate characteristics of the generating function to better describe the resemblance between the series.

Authors in [37] claim that after their investigation of several descriptive measures such as Discrete and Continuous Wavelet Transforms, Recurrence Quantification Analysis measures, Empirical Mode Decomposition, Lyapuno, Discrete Cosine Transform, Detrended Fluctuation, Autocorrelation function and Box and Jenkins model parameters, the best ones were Hurst exponent, Auto Mutual Information (AMI) and Discrete Fourier Transform (DFT). Those indices have been chosen because they are efficient to compute and provide high information gain (see Formulas (12)–(14), below).

The Hurst's exponent, is a measure of long-term memory of the time series. It refers to the auto-correlation of the time series and the rate at which it decreases as the delay between value pairs increases. There are different estimating approaches of the exponent; the Scaled Range approach is most often used. The Hurst, H exponent is defined in terms of the asymptotic behavior of the Scaled Range as a function of the time series time interval, as follows [37]:

$$\frac{R_t}{S_t} = ct^H, \tag{12}$$

where t stands for the time span of the observation, c is a constant, R_t is the range of the first t cumulative deviations from the mean, and S_t is their standard deviation.

The second measure, which is Auto Mutual Information (AMI), provides insight of how much one random variable explains the other variable. To calculate this characteristic, a histogram (with intervals) has to be created. Let p_i be the probability that the signal has a value inside the i-th intervals, and let $p_{ij}(\tau)$ be the probability that s_t is in intervals i and $s_{t+\tau}$ is in intervals j. Then, the AMI for time delay, τ, is defined as [37]:

$$AMI(\tau) = \sum p_{ij}(\tau) \log\left(\frac{p_{ij}(\tau)}{p_i p_j}\right). \tag{13}$$

The last one is the Discrete Fourier Transform (DFT) [52] which describes time series in the frequency domain. This transform, after receiving a time series s_t as input, provides a new series X_m of n complex numbers, each one describing a sine function at a given frequency [37]:

$$DFT = X_m = \sum_{t=0}^{n-1} s_t e^{-j2\pi tm/n}, \tag{14}$$

where $j = \sqrt{-1}$. The Fourier transform helps to characterize the generating function of this time series by indicating the most relevant frequencies, i.e., first 20 DFT coefficients of every time series in each window with the highest energy have been retained.

To split the times series into different clusters/nodes, each time a dedicated function is called which is accountable for finding the best coefficient for the binary test of the current node. This function takes as its input normalized matrix of characteristics and aims to minimize the weighted variance criterion of the form:

$$Gain = \sigma^2(V) - \frac{n_{left} * \sigma^2(V_{left}) + n_{right} * \sigma^2(V_{right})}{n}, \tag{15}$$

where V is the current node consisting of n time series data streams, $\sigma^2(\cdot)$ is the variance function, V_{right} and V_{left} are the nodes established after the split, each with n_{right} and n_{left} series, respectively.

In each consecutive iteration after obtaining a new time window the algorithm maintains the current tree model (structure from the previous iteration) and clusters time series based on the new batch of data. After this, the update stage begins, in which the breakdowns and/or aggregations are checked and executed, if necessary and/or possible [37], which is controlled by a set of parameters, i.e., $\alpha \in [0,1]$, $\lambda \in [0,1]$, and *minSeries*. Two sibling leaves (denotes as *LeftChild* and *RightChild*) must be aggregated if their weighted variance (denoted as *WVC*) is greater than or equal to λ of the parent node variance (*VP*) computed from its test feature. This makes the structure of the tree simpler and more resistant to noise/outliers. If aggregation did not occur the algorithm checks for possible leaf splits, which is done if the weighted variance of its potential children decreases by at least α times its variance. Finally, to prevent a split when two possible children have less than a certain percentage of all observations, *minSeries* parameter controlling the complexity/depth of the tree is set by default at 5%.

The overall time complexity is $O(m^2 w)$. It is important to note that the quadratic term in the algorithm refers to the number of time series, which is typically low (order of tenths) [45].

4. Research Framework and Settings

4.1. Numerical Implementation

As presented below, numerical experiments were prepared using R programming language working on Ubuntu 18.04 operating system on a personal computer equipped with Intel Core i7-9750H 2.6 GHz processor (12 threads) and 32 GB of RAM.

The first algorithm, which is histogram-based clustering, was implemented using several libraries. To represent each time series as a histogram and to compute the L_2 Wasserstein distance the *HistDAWass* package was used [47], which implements a framework of Symbolic Data Analysis, a relatively new approach for the statistical analysis of multi-valued data. Next, to get a local data partition based on a set of histograms a modification of *BR_BIRCH* package was used [53]. Finally, a *symbolicDA* [54] package was utilized to obtain a global clustering using DCLUST algorithm. The second algorithm, which is ClipStream, was entirely implemented using *ClipStream* library which is a software strictly connected to the article [55]. Finally, the extended TS-Stream algorithm was implemented in line with the following work [45].

4.2. Algorithms Parameters Setting

In order to have robust and consistent results all algorithms parameters settings are in line with the source articles and libraries. Since for the extended TS-Stream algorithm the parameters α and λ have a similar influence, it is not recommended to set one value as a function of the other. During the research preparation stage, it was observed that setting these two parameters to values smaller than 0.6 resulted in almost no splits. On the other hand, values greater than 0.6 could result in a too wide and too deep tree. Next, *minSeries* parameter which is responsible for controlling the size of a tree, is set at 5% (50 time series). Due to the fact that there are 1000 time series in the investigated data set (see Section 5) the final tree structure might have up to 20 leaves, i.e., clusters.

For ClipStream algorithm, long (b^{long}) and short windows (b^{short}) length were set to 1008 or 48 for overlapping windows and to 1440 or 48 for non-overlapping windows (see Section 4.4), while threshold value λ determining outliers was set at 1.5. The optimal number of cluster derived by the Davies–Bouldin measure was determined in the range 5 and 11. The latter number was determined as an average number of clusters obtained for each batch (for both overlapping and non-overlapping windows) for extended TS-Stream algorithm. Finally, concept drift is detected if p-value is less than the significance level α set at 0.05.

Histogram-based clustering algorithm has the following changeable parameters: P, which determines number of bins for each histograms, was set at 10 (average number of clusters obtained for both aforementioned algorithms), u, which is a threshold on the micro-cluster size, was set at 0.01, and because other two remaining algorithms usually provided maximal number of clusters, o parameter, which defines number of clusters, was set at 11.

4.3. Tested Changeable Components

One of the main goals of the article is to find the best clustering algorithm and, if possible, to propose some improvements with regards to different components adopted from other algorithms. To do so, firstly, a comparative study between overlapping windows and non-overlapping windows was be conducted, i.e., research was conducted in two different variants (see also Figure 1):

- Using non-overlapping window: This approach is in line with our previous study where the window length w of each block B_l, has been set to 30 days. As the electricity consumption data were recorded at 30-min intervals, each window has length of 1,440 (2×24 h $\times 30$ days);
- Using overlapping window: This approach is in line with the article [35] implementing ClipStream algorithm where window is of length 21 days (3 weeks). In this case, each time there are two overlapping weeks led by the new arriving week (2×24 h $\times 21$ days = 1008).

Secondly, a new Fast Fourier Transformation based features (calculated in liner time) is proposed, allowing to compress and represent time series using the business context. In our previous paper a set of 20 dominating Fourier coefficients have been taken as descriptive measures (see also Section 3.4). To make the usage of Fourier coefficients more intuitive, in this paper, the frequency domain have been divided into four intervals/ranges. Each of them represents electricity consumption behavior changes, respectively, monthly, weekly, daily, and all more frequent (see Table 1). The frequency is calculated with respect to the following equation:

$$f_c(m) = (f_s * m)/w, \tag{16}$$

where $f_c(m)$ is the frequency of m-th coefficient f_s is the frequency of sampling, w number of samples (i.e., window length) used in Fourier transform. A period is calculated as $1/f_c(m)$. As it can be noted, end of an interval is not a beginning of the another one. One should remember about discrete nature of values of DFT coefficients. Moreover $f_c(0)$ represents the mean value.

Table 1. Matching between Fourier coefficient and electricity consumption behaviors.

Fourier Coefficients No.	Non Overlapping Windows	Overlapping Windows
1–6	20 days–120 days	14 days–84 days
7–30	4 days–17 days 3 h	2 days 19 h–12 days
31–240	12 h–3 days 21 h	8.4 h–2 days 17 h
>240	<12 h	<8.4 h

Those aforementioned features were used in the extended TS-Stream algorithm (in this case a node partition is performed based on only one feature) and in the ClipStream algorithm. In the latter case instead of FeaClip representation each time series is represented base on those 4 features.

Thirdly, to conduct process of time series data streams clustering only when it is necessary, a stage for detecting concept drift using K-sample Anderson–Darling test (idea taken from ClipStream algorithm) was also implemented in the extended TS-Stream algorithm.

Finally, it is necessary to mention that all above improvements/components were not implemented in the histogram-based clustering algorithm, because it would entirely change the logic and the behavior of this algorithm.

4.4. Framework and Measures for Clustering Comparison

The main problem existing in the investigated area is the fact that there are no explicit frameworks, measures, criteria allowing to assess the performance, effectiveness and to compare algorithms to each other. To overcome this issue, we have proposed the following framework.

To compare the results of the grouping against external criteria, a measure of consensus is needed. Since it is assumed that each time series is assigned to only one cluster a natural way is to utilize the Adjusted Rand Index which is a measure of the similarity between two data clusterings. However, the practical aim in this article is to propose an optimal tariff for each time series. In this context we would like to know which clustering algorithm provides stable results, i.e., clusterings that are similar to each other. To do so we reformulated standard ARI measure as follows:

$$ARI = \frac{\sum_{uo}\binom{n_{ou}}{2} - \left[\sum_{o}\binom{n_{o*}}{2}\sum_{u}\binom{n_{*u}}{2}\right]\Big/\binom{n}{2}}{\frac{1}{2}\left[\sum_{o}\binom{n_{o*}}{2} + \sum_{u}\binom{n_{*u}}{2}\right] - \left[\sum_{o}\binom{n_{o*}}{2}\sum_{u}\binom{n_{*u}}{2}\right]\Big/\binom{n}{2}},\tag{17}$$

where n_{ou} denotes the number of objects that are in both, cluster l_o form l-th time window and cluster l_u from the $l+1$ time window (l_u is simply the same cluster as l_o but from consecutive window), with the marginal distributions denoted as n_{o*} and n_{*u}. After comparing each batch to each other an upper triangle matrix is created [45] (for an example please see Table 4).

The second measure is closely related to the selection of an optimal tariff for each customer. Let us assume that a particular customer has a base tariff G11 (single time zone with flat price rate per kWh) over an entire year. From the customer perspective it might be better to change a tariff to G12 for an entire year. Furthermore, one may analyze more frequent changes of the tariff e.g., after each month or even after each week. To answer that question we propose the following approach:

(1)   For a particular time window l apply a given clustering algorithm;
(2)   Assign a particular customer to his cluster;
(3)   Determine an optimal tariff for the entire cluster, i.e., the lowest price for an aggregate consumption of all customers in cluster by calculating the total electricity cost if they would belong to G11, G12, G12r or G12w tariff plan;
(4)   Select an optimal tariff from the previous step as an optimal tariff for a given customer;
(5)   Deploy an optimal tariff for each customer as a tariff for the next time window $l+1$;
(6)   Return to the first step.

According to the above procedure it might happen that for a given customer an optimal tariff for an entire year is G12. However, on the other hand it might happen that an optimal tariff will change after each time a new batch of data arrives. Next, to assess whether application of a particular clustering algorithm and aforementioned procedure make sense, we propose to derive, as previously, a similar upper tringle matrix having the following values:

$$\text{Tariff improvement} = \frac{\text{dynamic optimal tariff}}{\text{static optimal tariff}}.\tag{18}$$

To clarify that, let us consider first data batch l in a given year (for non-overlapping windows there would be 12 batches). This case is represented as the first top row in the upper triangle table (Table 4). Based on that particular window it was decided that an optimal tariff for the entire year is G12w (an optimal tariff for a cluster where a particular customer belongs), therefore, for this investigated row, denominator in the above equation takes always the same value, i.e., price of this fixed tariff for a particular customer calculated for each month separately. On the other hand, nominator is determined dynamically. For the first column it takes the same value as the denominator. For the remaining eleven columns (batches from $l + 1, \ldots, l + 11$) it takes dynamically changeable price of the tariff determined in the 5th step of the mentioned earlier procedure. Such table is prepared for each customer, therefore to have only one global table, as in case of the ARI, each field in the final table was calculated as the mean value of the 1000 customer-wise matrices.

The last measure is the weighted volatility of time series for a given block B_l. After the division, the time series are spread over several groups. It is assumed that the variation (standard deviation) of electricity consumption in each group is to be less than the variation of time series in only one group (root) [45]. Furthermore, because of the difference in the size of each group, the measure takes into account this fact by assigning smaller weights to a smaller leaf—as in the right-hand side of the Equation (19):

$$\text{Weighted volatility}_{B_l} = \sum_{L_{l_0} \in L_l} \frac{\#L_{l_0}}{m} * \sigma\left(L_{l_0}\right), \tag{19}$$

where $\#L_{l_0}$ denotes the number of time series for a given cluster, m denotes the number of time series in a block B_l and $\sigma(\cdot)$ is the standard deviation of all times series assigned to a given cluster L_{l_0}.

5. Empirical Analysis

5.1. Data and Tariffs Characteristics

The dataset used in this research is originated from the Irish Commission for Energy Regulation (CER) project where the measurements of the electricity load where recorded for 4182 households between July 2009 and December 2010. In total, time span covers 75 weeks where each reading was recorded with 30 min data granularity [5]. Due to the missing recordings in the time series and computational complexity of the investigated algorithms, the research was conducted using data from 1000 households selected randomly.

Unfortunately, CER dataset does not provide any information regarding tariff plan of each customer. After investigation of several tariffs plans provided by electricity suppliers in the European countries, it can be stated that there are many similarities. Therefore, to conduct simulation of the optimal tariff, all the information and the tariff prices were taken from one of the biggest energy holding company in Poland.

Depending on the tariff plan, the customers can benefit from lower prices per kWh if the usage falls between certain time zones. In Figure 2 the prices for G11, G12, G12w, and G12r tariff are presented. G11 tariff (blue straight line) has the fixed price of 0.35 PLN/kWh. G12r tariff (purple dotted-dashed line) plan has lower rate of 0.21 PLN/kWh between 10 p.m. and 7 a.m. and between 1 p.m. and 4 p.m., while the higher rate of 0.48 PLN/kWh is applicable outside these windows. G12w tariff (green double dotted–dashed line) has lower rate of 0.28 PLN/kWh during the weekends and Monday–Friday between 10 p.m. and 6 a.m. and between 1 p.m. and 3 p.m., while the higher price of 0.43 PLN/kWh is applicable outside these windows.

Let us now simulate what is the relation between the best and the worst tariff for each customer. Table 2 shows various statistics of the simulation (aggregated over 1000 customers) for non-overlapping windows case. When dynamically changing an optimal tariff for each customer a minimal improvement between the best and the worst individual tariff is 2.39%, while the biggest improvement reaches 19.27%. Second row of the table shows what is the improvement between dynamically changing an optimal tariff and one fixed optimal tariff derived based on the entire period. It was observed that dynamic change resulted in average improvement of 0.28%. Finally, it can be concluded that, on average, an optimal tariff would change almost 5 times, out of 17 data batches, each 30 days long, in the analyzed timeframe.

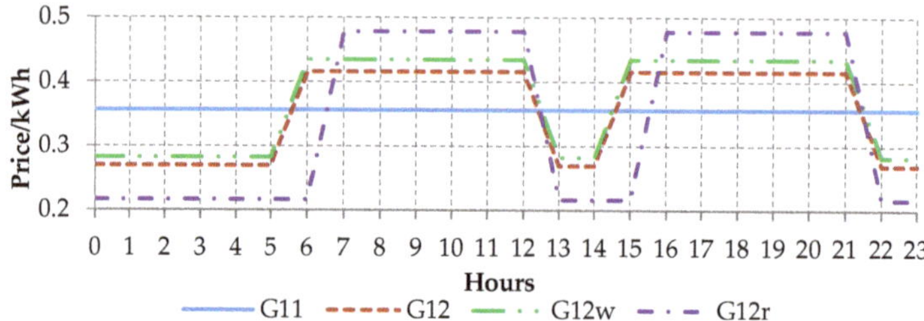

Figure 2. Prices in G11, G12, G12r, and G12w tariff plans (1 Polish PLN~0.22 EUR).

Table 2. Simulation of households' electricity consumption characteristics based on different tariff group rates for non-overlapping windows.

	Min	1st Quartile	Median	Mean	3rd Quartile	Max
Best vs. worst individual tariff for each batch	2.39%	5.76%	7.67%	8.08%	9.88%	19.27%
Best individual tariff for each batch vs. best individual tariff for the entire period	0.00%	0.06%	0.21%	0.28%	0.40%	2.39%
Number of dynamic individual tariff change	0.00	2.00	5.00	4.81	7.00	12.00

When speaking of overlapping windows case (Table 3), results are slightly higher. Average improvement between the best and the worst individual tariff for each batch increases to 8.47%, while the best individual tariff for each batch vs best individual tariff for the entire period increases to 0.51%, on average. Due to the fact that there are 73 batches in this scenario, each batch of 21 days long, the median of dynamic individual tariff changes is 25.

Table 3. Simulation of households' electricity consumption characteristics based on different tariff group rates for overlapping windows.

	Min	1st Quartile	Median	Mean	3rd Quartile	Max
Best vs. worst individual tariff for each batch	2.68%	6.27%	8.13%	8.47%	10.32%	19.28%
Best individual tariff for each batch vs. best individual tariff for the entire period	0.00%	0.23%	0.43%	0.51%	0.69%	3.52%
Number of dynamic individual tariff change	0.00	18.00	25.00	24.40	32.00	50.00

Those results present the best and the worst case scenarios, when an optimal tariff is derived for each customer separately without any clustering algorithm. Therefore, those results provide benchmarking ranges between which the clustering results presented in the following subsections will be included.

5.2. Clustering Results

Let us now investigate which algorithm provide relatively robust results, i.e., overall groupings that are similar to each other (in other words, maintaining time series belonging to the same clusters). For the non-overlapping case, the extended TS-Stream algorithm provides on average 11 clusters, all having more than 5% of all time series. For the 17 investigated batches on average each time series should change his optimal tariff 7.98 times (median is 8; this is determined as the optimal tariff for the cluster to be monitored). The ClipStream algorithm changes the tariff 5.38 times on average (median is 6), while not using the concept drift results in increasing these values to 6.52 and 7. On average, histogram-based algorithm changes the tariff 7.04 times (median is 7). All aforementioned numbers are higher than those reported in Table 2, where the best tariff is chosen separately for each customer without any clustering algorithm, which means that a time series changes its tariff more frequent than it should. For better understanding of the idea, in this article we present only sample matrix of the ARI index obtained for the ClipStream algorithm (in Table 4). Tables 5 and A1 (in the Appendix A) provide various statistics of the ARI and tariffs improvement derived based on the upper-triangular matrixes (described also in Section 4.4) for both non-overlapping and overlapping windows (see Appendix A).

Table 4. Sample of the upper-triangular matrix of the ARI indexes obtained based on the ClipStream algorithm for non-overlapping windows.

	B_2	B_3	B_4	B_5	B_6	B_7	B_8	B_9	B_{10}
B_1	0.100	0.088	0.062	0.062	0.062	0.040	0.040	0.040	0.049
B_2		0.098	0.080	0.067	0.067	0.038	0.038	0.038	0.060
B_3			0.097	0.084	0.084	0.068	0.068	0.068	0.075
B_4				0.166	0.166	0.115	0.115	0.115	0.059
B_5					1	0.199	0.199	0.199	0.064
B_6						0.199	0.199	0.199	0.064
B_7							1	1	0.063
B_8								1	0.063
B_9									0.063

Table 5. Statistics of the ARI indexes for non-overlapping windows.

Clustering Algorithm	Min	1st Quartile	Median	Mean	3rd Quartile	Max
Extended TS-Stream (Fourier coeff.)	0.014	0.024	0.033	0.035	0.043	0.070
Extended TS-Stream (Fourier coeff., concept drift)	0.014	0.024	0.033	0.035	0.043	0.070
ClipStream (concept drift)	0.026	0.057	0.067	0.119	0.097	**1.000**
ClipStream (without concept drift)	0.021	0.054	0.070	0.079	0.091	0.232
ClipStream (Fourier coeff., concept drift)	0.029	0.053	0.066	0.120	0.082	**1.000**
ClipStream (Fourier coeff., without concept drift)	0.025	0.049	0.065	0.065	0.077	0.113
Histogram-based	**0.149**	**0.230**	**0.309**	**0.335**	**0.419**	0.740

In this example, similarity (measured using ARI) between the first batch B_1 and second the batch B_2 is 0.100. Clustering from the first batch is the least similar to batches from seven to nine (0.040). Because algorithm detected no concept drift between batches B_7–B_9, the change of clusters membership did not occur which results in ARI equals 1.

According to the results presented in Table 5 (the best results for each statistic are bolded), it can be seen that, on average, the highest ARI provides histogram-based algorithm. This is impacted by two things, first—it always generates the same number of clusters. Secondly, it divides customers into the clusters based on the iteratively updated (after each batch) global proximity matrix which uses partition from the BIIRCH algorithm (second step of this algorithm). This step provides only a minor modification of the global matrix and once in the last step the DCLUST is incorporated, it provides very similar groupings (customers rarely change their cluster). It should be noted that whenever ClipStream algorithm decides not to make any changes ARI is equal to 1. The worst results are connected with the extended TS-Stream algorithm (median is 0.033).

For the overlapping widows case (Table A1 in the Appendix A) the dependencies are similar. One more time the histogram-based algorithm produces the most stable partitions. In previous case, for the extended TS-Stream algorithm concept drift module was not used. This time for couple of batches the tree preserved the same structure which increased the highest value at 0.326. What is interesting, for ClipStream algorithm the new data representation (Fourier coefficients) increases lower (up to median) statistics.

In the similar manner as for the ARI index the upper triangle matrix has been derived for the tariffs improvement (Equation (18)).

From practical point of view it is better for the electricity provider to have customer groups with relatively similar size [2]. The extended TS-Stream algorithm guaranties that each cluster has no less than 5% of all customers, and after investigation of the group size it can be stated that this algorithm produces clusters with the similar size. On the other hand, both ClipStream and Histogram-based algorithms do not have such restriction. On average, ClipStream algorithm generates one (rarely two) cluster having only couple of customers (1–5 time series). Histogram-based algorithm usually produces three up to four clusters whose are very small. This observation has high influence on the values of the investigated metrics (they are rewarded), since in small groups memberships change rarely and the volatility is small (see Tables 6 and A3).

Table 6. Statistics of the weighted volatility for non-overlapping windows.

Clustering Algorithm	Min	1st Quartile	Median	Mean	3rd Quartile	Max
Extended TS-Stream (Fourier coeff.)	**15.10**	**18.93**	**21.06**	**23.22**	**26.15**	37.18
Extended TS-Stream (Fourier coeff., concept drift)	**15.10**	**18.93**	**21.06**	**23.22**	**26.15**	37.18
ClipStream (concept drift)	20.50	27.40	32.69	37.08	47.93	61.93
ClipStream (without concept drift)	16.87	24.48	32.28	36.15	49.66	69.65
ClipStream (Fourier coeff., concept drift)	39.95	42.87	52.24	58.49	72.21	95.98
ClipStream (Fourier coeff., without concept drift)	39.95	42.87	52.24	55.83	64.88	90.06
Histogram-based	17.01	20.45	24.46	25.71	31.47	**36.05**

According to the results presented in Table 6, the least volatile partitions provides the extended TS-Stream algorithm, median is 21.06 while mean is 23.22 (since there were no batches when the concept drift module was used both versions produce the same results). Seconds place in this ranking takes the Histogram-based algorithm whose maximal volatility is even smaller than for the extended TS-Stream. For the overlapping windows case, the least volatile groups produces the histogram-based algorithm. Slightly worse results are connected with the Extended TS-Stream (with the concept drift module) whose the minimal statistic is even smaller than for the histogram-based algorithm. Finally, in both windows (overlapping and non-overlapping), new data representation and not use the concept drift procedure in ClipStream worsen the results.

5.3. Tariff Evaluation

In this section tariff improvements are discussed. When it comes to the various statistics for non-overlapping windows, it is observed that the all investigated algorithms provide, on average, an improvement of 0.3%–0.4%, please refer to Table 7. The highest improvement is observed for the Extended TS-Stream and the histogram-based algorithms, and for the ClipStream algorithm with the newly proposed data representation (up to 1.8%). Moreover, the first two algorithms mentioned do not produce worse results (please refer to the first column with Min values).

Table 7. Statistics of the tariffs improvement for non-overlapping windows.

Clustering Algorithm	Min	1st Quartile	Median	Mean	3rd Quartile	Max
Extended TS-Stream (Fourier coeff.)	**0.00%**	**0.10%**	**0.20%**	**0.40%**	0.50%	**1.80%**
Extended TS-Stream (Fourier coeff., concept drift)	**0.00%**	**0.10%**	**0.20%**	**0.40%**	0.50%	**1.80%**
ClipStream (concept drift)	−0.20%	**0.10%**	**0.20%**	0.30%	0.40%	1.50%
ClipStream (without concept drift)	−0.10%	**0.10%**	**0.20%**	0.30%	0.50%	1.50%
ClipStream (Fourier coeff., concept drift)	−0.10%	0.00%	0.10%	**0.40%**	**0.90%**	**1.80%**
ClipStream (Fourier coeff., without concept drift)	−0.10%	0.00%	0.10%	**0.40%**	0.80%	**1.80%**
Histogram-based	**0.00%**	**0.10%**	**0.20%**	**0.40%**	0.50%	**1.80%**

For the overlapping windows case, please refer to Table A2, one more time, all algorithms usually provide the improvement, with the mean value between 0.1% and 0.2%. Unfortunately, in the worst-case-scenario each algorithm chose worse tariff, the smallest worsening (−0.1%) is for the extended TS-Stream algorithm without concept drift module.

The last results presented below are to answer the question, whether it is possible and justified to use clustering (and associated optimal tariffs for each group) obtained for a particular batch B_l and the deploy those optimal tariffs as the applicable tariffs in the following period B_{l+1}. Tables 8 and A4, provide statistics of the tariffs improvement compared to the basic (flat) tariff G11 in case when the future optimal tariff for each customer (for the next data batch) is derived as the current optimal tariff for the cluster to which a particular customer belongs. The advantage of this approach is that it does not require training nor the use of any predictive models.

Table 8. Statistics of the predicted tariffs improvement comparing to the G11 for non-overlapping windows.

Clustering Algorithm	Min	1st Quartile	Median	Mean	3rd Quartile	Max
Extended TS-Stream (Fourier coeff.)	−8.89%	−0.56%	−0.09%	0.00%	0.45%	5.73%
Extended TS-Stream (Fourier coeff., concept drift)	−8.89%	-0.56%	−0.09%	0.00%	0.45%	5.73%
ClipStream (concept drift)	−3.17%	−0.34%	**−0.03%**	**0.31%**	**0.61%**	**8.46%**
ClipStream (without concept drift)	**−2.90%**	**−0.32%**	−0.04%	0.28%	0.53%	8.19%
ClipStream (Fourier coeff., concept drift)	−5.76%	−0.49%	−0.10%	−0.05%	0.40%	2.90%
ClipStream (Fourier coeff., without concept drift)	−5.76%	−0.49%	−0.10%	−0.05%	0.40%	3.25%
Histogram-based	−6.44%	−0.51%	−0.11%	−0.01%	0.41%	4.25%

As shown in Table 8, for the non-overlapping windows case, on average, it is possible to achieve some improvement. The ClipStream algorithm provides better results of 0.31% comparing to the base tariff (removing concept drift module gives improvements as well). The mean improvement for both versions of the extended TS-Stream produces no improvement; however, median value equals −0.09%. Unfortunately, the histogram-based algorithm usually provides worse tariff than costs related with the G11. It should be noted that when comparing the optimal predicted tariff to the random tariff (rather than to the G11), on average, the results are always better (see Table A5). For the extended TS-Stream algorithm it is 1.66%, for the ClipStream algorithm (base version) it is 2.17%, and for the histogram-based algorithm it is 1.50%.

For the overlapping windows case (batch size equals 3 weeks while each time new data cover one week), please refer to Table A4, the improvements are more common and clear for all algorithms, i.e., according to the median and to the mean value the improvement is positive. Only for the statistics such as 3rd quartile and above the worsening can be noted. The biggest improvement is noted for the base version of the ClipStream algorithm (7.9%). Second place in terms of the mean value belongs to both versions of the extended TS-Stream algorithm (0.21%; 0.20%).

Finally, when it comes to the comparison to the random assignment of tariff (as an optimal for the future), the extended TS-Stream algorithm (base version) achieves improvement of 2.69%, for the ClipStream algorithm (base version) it is 2.91%, and for the histogram-based algorithm it equals 2.65% (see Table A6).

Based on the results we could summarize the comparative study between overlapping windows and non-overlapping windows and their impact on the choice of an optimal tariff as outlined in Table 9. For the purpose of results discussion the average improvements were considered. It was observed that the implementation of the current best tariff is feasible and could deliver the benefits for both, overlapping and non-overlapping windows. Specifically, for non-overlapping windows the general tariff improvement was up to 0.40%, on average, depending on the algorithm. In case of tariffs improvement comparing to the G11 tariff plan the highest improvement was for overlapping windows, where two ClipStream algorithms (with and without concept drift) were able to deliver up to 0.43% of the improvement, on average.

Importantly, the results, in terms of the tariff improvement, are only the highlight for possible knowledge utilization based on the algorithms that were used for profiling the customers. Nevertheless, the results are promising although the improvements might appear negligible. Please note that the improvement rates of 0.40–0.43%, as provided in Table 9, directly influence the elasticity of electricity demand. In case of Poland, the whole installed capacity of the system is approx. 45,000 MW so the improvement of 0.43% is representing 193.5 MW which is an equivalent of one power block in the power plant. Therefore, if some of the usage can be shifted outside peak hours then the benefit is not only for the customers but also for the electricity operators who can purchase the electricity cheaper.

Table 9. Summary results in terms of the average improvements on non-overlapping and overlapping windows.

Clustering Algorithm	Tariff Improvement		Tariffs Improvement Comparing to the G11	
	Non-Overlapping	Overlapping	Non-Overlapping	Overlapping
Extended TS-Stream (Fourier coeff.)	**0.40%**	**0.20%**	0.00%	0.21%
Extended TS-Stream (Fourier coeff., concept drift)	**0.40%**	0.10%	0.00%	0.20%
ClipStream (concept drift)	0.30%	0.10%	**0.31%**	**0.43%**
ClipStream (without concept drift)	0.30%	0.10%	0.28%	**0.43%**
ClipStream (Fourier coeff., concept drift)	**0.40%**	**0.20%**	−0.05%	0.14%
ClipStream (Fourier coeff., without concept drift)	**0.40%**	**0.20%**	−0.05%	0.15%
Histogram-based	**0.40%**	**0.20%**	−0.01%	0.16%

5.4. Other Applications—Australian Case Study

To proof the applicability of the dynamic profiling approach further analysis was conducted based on the data from the customer trial conducted as part of the Smart Grid Smart City (SGSC) project [56]. It provides sets of customer time of use (half hour increments) and demographic data for Australia between 2010 and 2014. For the purpose of the case study 998 households were randomly extracted covering 1 September 2012–28 February 2014 time frame. The reason to select that time frame was availability of complete data, i.e., without missing values. In total, 25,399 data points were analyzed, each representing half hour readings.

For the purpose of results discussion the average improvements were considered as presented in Table 10. It was observed that the implementation of the current best tariff is feasible and could deliver the benefits for both, overlapping and non-overlapping windows. Specifically, for non-overlapping windows the general tariff improvement was up to 0.96%, on average, depending on the algorithm. In case of tariffs improvement comparing to the G11 tariff plan the highest improvement was for overlapping windows, where two ClipStream algorithms with and without concept drift, were able to deliver up to 1.08% and 1.06% of the improvement, on average, respectively. The results are consistent with the results on Irish data set. However, this time an improvement is considerably higher what can influence directly the elasticity of electricity demand.

Table 10. Summary results in terms of the average improvements on non-overlapping and overlapping windows for Australian data.

Clustering Algorithm	Tariff Improvement		Tariffs Improvement Comparing to the G11	
	Non-Overlapping	Overlapping	Non-Overlapping	Overlapping
Extended TS-Stream (Fourier coeff.)	**0.96%**	0.14%	−0.18%	0.84%
Extended TS-Stream (Fourier coeff., concept drift)	**0.96%**	0.76%	−0.18%	0.84%
ClipStream (concept drift)	0.92%	**0.77%**	**0.19%**	**1.08%**
ClipStream (without concept drift)	0.91%	0.74%	0.18%	1.06%
ClipStream (Fourier coeff., concept drift)	0.93%	0.76%	−0.15%	0.82%
ClipStream (Fourier coeff., without concept drift)	0.93%	0.76%	−0.15%	0.83%
Histogram-based	0.91%	**0.77%**	−0.01%	0.16%

More detailed analysis are presented in Appendix B, please refer to Tables A7–A16.

5.5. Other Applications—London Case Study

Another verification of dynamic profiling approach was conducted based on the data from UK Power Networks led Low Carbon London project [57]. The dataset contains energy consumption in kWh (per half hour) for the sample of 5567 London households observed between November 2011 and February 2014. The customers in the trial were recruited as a balanced sample representative of the Greater London population.

For the purpose of the case study 1000 households were randomly extracted covering 1 September 2012–28 February 2014 time frame. The reason to select that time frame was availability of complete data, i.e., without missing values. In total, 25,440 data points were analyzed, each representing half hour readings.

To enable comparison of the results with previous applications (case studies) the average improvements were considered, as presented in Table 11. It was observed that the implementation of the current best tariff is feasible and could deliver the benefits for both, overlapping and non-overlapping windows. Specifically, for non-overlapping windows the general tariff improvement was equal, on average, to 0.93% for Extended TS-Stream without concept drift. The lower improvements, between 0.15% and 0.39%, were observed for other algorithms. In case of tariffs improvement comparing to the G11 tariff plan the highest improvement was for overlapping windows, where histogram-based approach resulted in the improvement of 0.68%, on average. Other methods were able to deliver improvements between 0.49% and 0.65% which could be considered significant, too. The improvement for non-overlapping windows was slightly lower, i.e., 0.55% and similarly, it was observed for histogram-based clustering approach. The results are consistent with the results on Irish data and Australian data.

Table 11. Summary results in terms of the average improvements on non-overlapping and overlapping windows for London data.

Clustering Algorithm	Tariff Improvement		Tariffs Improvement Comparing to the G11	
	Non-Overlapping	Overlapping	Non-Overlapping	Overlapping
Extended TS-Stream (Fourier coeff.)	**0.93%**	0.10%	0.39%	0.59%
Extended TS-Stream (Fourier coeff., concept drift)	0.19%	0.10%	0.39%	0.57%
ClipStream (concept drift)	0.35%	**0.22%**	0.36%	0.62%
ClipStream (without concept drift)	0.39%	0.21%	0.39%	0.65%
ClipStream (Fourier coeff., concept drift)	0.26%	0.10%	0.23%	0.49%
ClipStream (Fourier coeff., without concept drift)	0.27%	0.10%	0.23%	0.50%
Histogram-based	0.15%	0.07%	**0.55%**	**0.68%**

More detailed results are presented in Appendix C, please refer to Tables A17–A26.

6. Conclusions

Data streams clustering is one of the most common ways of analyzing data that is potentially infinite and evolves over time. Although the literature provides some methods of the data streams clustering, unfortunately, majority of them are not appropriate for the whole time series data streams clustering. Even though electricity consumer objectives are usually based on monetary benefits, electricity providers benefit from the knowledge of consumer' profiles, to create individualized means aimed at consumers with compatible use profiles and socio-economic behavior. The analysis has shown that there are prominent distinction between consumers' behaviors, which allows us to distinguish homogeneous groups.

Through the CER Irish data analysis and two other case studies, i.e., Australian and London data sets, an attempt was made to evaluate different ways of time series data streams clustering by comparative study of the state-of-the-art algorithms, as well as new combinations employing elements from different algorithms. From the technical point of view the results introduce a general guidance on when and where to apply a particular clustering algorithm (along with its improvements).

It was revealed that the extension to the way of ARI index calculation (and its statistics) based on the upper triangle matrix, which compares blocks to each other, provides good evaluation framework, and it also allows to visualize the dependencies. This part of the research has shown that the best results, in terms of the similarity of the clusters, are provided by the histogram-based clustering algorithm. That is due to the fact that the algorithm always performs a partitioning using the same number of clusters and the underlying procedure is less fragile to any distribution changes than other two algorithms. Therefore, if the electricity providers need stable partitions this algorithm would be their first choice.

Furthermore, to obtain a partition which provides clusters with the least weighted volatility the extended TS-Stream algorithm should be applied. It is mainly caused by the fact that this algorithm is able to expand or to shrink the tree structure very quickly according to the distribution changes of the particular phenomenon. On the other side of the pole is the ClipStream algorithm.

As it was presented in our previous work [45], standard TS-Stream algorithm outperforms benchmarking clustering methods and, in addition, this research indicates that these results can be further improved. The new Fast Fourier Transformation based features allow to improve the operation of the base for this algorithm. The new data representation slightly deteriorates the performance of the ClipStream algorithm; however it should be noted that this time a business interpretation is prevailing. Moreover, a much smaller dimension is needed to represent a given time series, i.e., only 5 features instead of 8 multiplied number of weeks (3 weeks for overlapping and 4 weeks for non-overlapping windows).

In terms of the implementation/software requirements all the algorithms are able to work in linear time, however the histogram-based algorithm requires $O(m^2)$ memory space. It also produces fixed number of clusters. For the ClipStream algorithm it is necessary to set up minimal and maximal number of cluster in advance (which sometimes might be impracticable or unfounded). The extended TS-Stream algorithm is the most flexible in its nature what allows to incorporate new descriptive measures, data representation and concept drift detection module.

When it comes to the comparison between overlapping and non-overlapping windows, as it might expect, statistics of the ARI and weighted volatility for the overlapping windows are usually better (base version of each algorithm). This is due to the fact that each time we analyze almost the same time series that differ only with one new added week.

Based on the comparative study between the state-of-the-art time series data streams clustering algorithms and their modifications we could perform the dynamic consumer segmentation and prediction of an optimal tariff. Finally, comparative study between overlapping and non-overlapping windows and their impact on the choice of an optimal tariff was undertaken what revealed that significant improvements could be reported due to tariff changes. Specifically, the percentage improvements, on average, were as follows: Irish data—0.40–0.43%; Australian data—0.96–1.08%; and London data—0.68–0.93%. Assuming that the overall capacity of the system is approx. 45,000 MW in Poland, thus the improvements may deliver elasticity of electricity demand which is between 193.5 MW (0.43%) and 486 MW(1.08%). Those values are considered a significant from market balancing perspective.

The direction for the future work will be to develop a fully scalable system (along with the results which are interpretable) for a large number of time series in the data stream, in the presence of:

- Concept drift of different kinds, such as incremental, recurring, sudden, or gradual;
- unstable number of sources (some sensors are newly created while other removed);
- heterogeneous and missing recordings;
- irregularly spaced data; and
- assuming application of other approaches for classifying incoming continuous data in dynamic systems e.g., stochastic learning weak estimators.

Due to that, we will investigate different incrementally computable time series similarity measures. In the future, we will investigate the influence (sensitivity of the algorithm) of the input parameters on the final results.

Author Contributions: K.G. prepared the simulation and analysis and wrote the Sections 1–6 of the manuscript; M.B. wrote Section 1, Section 3, and Section 4; T.Z. wrote Section 1, Section 2, and Section 6 of the manuscript. All authors have read and agreed to the published version of the manuscript.

Funding: This research received no external funding.

Conflicts of Interest: The authors declare no conflict of interest.

Appendix A. Results Based on Irish Data Set

Table A1. Statistics of the ARI indexes for overlapping windows.

Clustering Algorithm	Min	1st Quartile	Median	Mean	3rd Quartile	Max
Extended TS-Stream (Fourier coeff.)	0.003	0.019	0.025	0.031	0.033	0.264
Extended TS-Stream (Fourier coeff., concept drift)	0.003	0.024	0.034	0.044	0.050	0.326
ClipStream (concept drift)	0.011	0.046	0.059	0.091	0.082	**1.000**
ClipStream (without concept drift)	0.010	0.051	0.067	0.083	0.095	0.547
ClipStream (Fourier coeff., concept drift)	0.019	0.047	0.059	0.079	0.072	**1.000**
ClipStream (Fourier coeff., without concept drift)	0.020	0.049	0.062	0.070	0.076	0.405
Histogram-based	**0.222**	**0.348**	**0.467**	**0.486**	**0.597**	0.991

Table A2. Statistics of the tariffs improvement for overlapping windows.

Clustering Algorithm	Min	1st Quartile	Median	Mean	3rd Quartile	Max
Extended TS-Stream (Fourier coeff.)	−0.20%	**0.00%**	**0.10%**	**0.20%**	0.20%	0.80%
Extended TS-Stream (Fourier coeff., concept drift)	−0.20%	**0.00%**	**0.10%**	0.10%	0.20%	0.80%
ClipStream (concept drift)	−0.30%	**0.00%**	**0.10%**	0.10%	0.20%	0.80%
ClipStream (without concept drift)	−0.20%	**0.00%**	**0.10%**	0.10%	0.20%	0.90%
ClipStream (Fourier coeff., concept drift)	−0.20%	**0.00%**	**0.10%**	**0.20%**	0.30%	**1.00%**
ClipStream (Fourier coeff., without concept drift)	−0.20%	**0.00%**	**0.10%**	**0.20%**	0.30%	**1.00%**
Histogram-based	−0.20%	**0.00%**	**0.10%**	**0.20%**	0.30%	0.90%

Table A3. Statistics of the weighted volatility for overlapping windows.

Clustering Algorithm	Min	1st Quartile	Median	Mean	3rd Quartile	Max
Extended TS-Stream (Fourier coeff.)	15.94	22.27	25.83	26.28	29.97	38.21
Extended TS-Stream (Fourier coeff., concept drift)	**14.09**	20.76	25.22	25.09	28.67	39.20
ClipStream (concept drift)	19.15	24.61	28.41	29.57	32.50	59.14
ClipStream (without concept drift)	21.73	29.45	37.42	39.64	45.89	82.29
ClipStream (Fourier coeff., concept drift)	24.31	42.71	53.23	55.94	68.91	99.47
ClipStream (Fourier coeff., without concept drift)	32.69	45.22	53.85	58.30	69.60	106.51
Histogram-based	15.73	**19.59**	**23.45**	**24.29**	**27.19**	**36.41**

Table A4. Statistics of the predicted tariffs improvement comparing to the G11 for overlapping windows.

Clustering Algorithm	Min	1st Quartile	Median	Mean	3rd Quartile	Max
Extended TS-Stream (Fourier coeff.)	−9.28%	−0.40%	0.10%	0.21%	0.74%	5.45%
Extended TS-Stream (Fourier coeff., concept drift)	−8.52%	−0.40%	0.08%	0.20%	0.73%	6.22%
ClipStream (concept drift)	**−2.77%**	**−0.17%**	0.07%	**0.43%**	0.74%	**7.90%**
ClipStream (without concept drift)	−3.04%	**−0.17%**	0.08%	**0.43%**	0.75%	7.51%
ClipStream (Fourier coeff., concept drift)	−7.71%	−0.42%	0.11%	0.14%	0.68%	4.53%
ClipStream (Fourier coeff., without concept drift)	−7.58%	−0.42%	**0.12%**	0.15%	0.66%	4.69%
Histogram-based	−7.40%	−0.38%	0.08%	0.16%	0.65%	5.14%

Table A5. Statistics of the predicted tariffs improvement comparing to the random tariff for non-overlapping windows.

Clustering Algorithm	Min	1st Quartile	Median	Mean	3rd Quartile	Max
Extended TS-Stream (Fourier coeff.)	−6.28%	−0.10%	1.66%	2.50%	4.63%	13.45%
Extended TS-Stream (Fourier coeff., concept drift)	−6.28%	−0.10%	1.66%	2.50%	4.63%	13.45%
ClipStream (concept drift)	**−2.61%**	**0.00%**	**2.17%**	**2.80%**	**4.81%**	**14.35%**
ClipStream (without concept drift)	−3.69%	−0.01%	2.11%	2.77%	4.76%	**14.35%**
ClipStream (Fourier coeff., concept drift)	−6.53%	−0.12%	1.43%	2.45%	4.60%	13.27%
ClipStream (Fourier coeff., without concept drift)	−6.53%	−0.12%	1.43%	2.45%	4.60%	13.27%
Histogram-based	−6.47%	−0.10%	1.50%	2.49%	4.64%	13.45%

Table A6. Statistics of the predicted tariffs improvement comparing to the random tariff for overlapping windows.

Clustering Algorithm	Min	1st Quartile	Median	Mean	3rd Quartile	Max
Extended TS-Stream (Fourier coeff.)	−6.55%	0.13%	1.89%	2.69%	4.79%	13.48%
Extended TS-Stream (Fourier coeff., concept drift)	−6.63%	0.09%	1.82%	2.68%	4.80%	13.38%
ClipStream (concept drift)	−2.76%	0.12%	2.32%	**2.91%**	4.92%	**14.45%**
ClipStream (without concept drift)	**−2.51%**	0.13%	**2.34%**	**2.91%**	**4.96%**	14.40%
ClipStream (Fourier coeff., concept drift)	−6.53%	0.15%	1.66%	2.63%	4.76%	13.31%
ClipStream (Fourier coeff., without concept drift)	−6.16%	0.15%	1.70%	2.63%	4.74%	13.46%
Histogram-based	−6.64%	**0.17%**	1.81%	2.65%	4.75%	14.02%

Appendix B. Results Based on Australian Data Set

Table A7. Simulation of households' electricity consumption characteristics based on different tariff group rates for non-overlapping windows.

	Min	1st Quartile	Median	Mean	3rd Quartile	Max
Best vs worst individual tariff for each batch	2.86%	5.53%	6.99%	7.60%	8.96%	48.80%
Best individual tariff for each batch vs best individual tariff for the entire period	0.00%	0.66%	1.08%	1.04%	1.363%	4.26%
Number of dynamic individual tariff change	0.00	4.00	6.00	6.22	8.00	13.00

Table A8. Simulation of households' electricity consumption characteristics based on different tariff group rates for overlapping windows.

	Min	1st Quartile	Median	Mean	3rd Quartile	Max
Best vs worst individual tariff for each batch	3.16%	6.60%	8.21%	8.79%	10.27%	49.45%
Best individual tariff for each batch vs best individual tariff for the entire period	0.00%	1.32%	1.70%	1.67%	1.95%	6.63%
Number of dynamic individual tariff change	0.00	16.00	24.00	22.97	30.00	50.00

Table A9. Statistics of the ARI indexes for non-overlapping windows.

Clustering Algorithm	Min	1st Quartile	Median	Mean	3rd Quartile	Max
Extended TS-Stream (Fourier coeff.)	0.024	0.047	0.063	0.066	0.082	0.165
Extended TS-Stream (Fourier coeff., concept drift)	0.024	0.047	0.063	0.066	0.082	0.165
ClipStream (concept drift)	0.001	0.040	0.058	0.078	0.101	**1.000**
ClipStream (without concept drift)	0.001	0.040	0.057	0.071	0.094	0.245
ClipStream (Fourier coeff., concept drift)	0.085	0.134	0.149	0.188	0.169	**1.000**
ClipStream (Fourier coeff., without concept drift)	0.085	0.127	0.149	0.154	0.171	0.300
Histogram-based	**0.286**	**0.400**	**0.457**	**0.491**	**0.524**	0.935

Table A10. Statistics of the tariffs improvement for non-overlapping windows.

Clustering Algorithm	Min	1st Quartile	Median	Mean	3rd Quartile	Max
Extended TS-Stream (Fourier coeff.)	**−0.06%**	0.03%	0.21%	**0.96%**	0.75%	**13.36%**
Extended TS-Stream (Fourier coeff., concept drift)	**−0.06%**	0.03%	0.21%	**0.96%**	0.75%	**13.36%**
ClipStream (concept drift)	−0.22%	0.05%	**0.27%**	0.92%	0.65%	12.13%
ClipStream (without concept drift)	−0.22%	0.04%	0.25%	0.91%	0.65%	12.13%
ClipStream (Fourier coeff., concept drift)	−0.13%	0.05%	0.20%	0.93%	0.74%	**13.36%**
ClipStream (Fourier coeff., without concept drift)	−0.13%	**0.06%**	0.23%	0.93%	0.74%	**13.36%**
Histogram-based	−0.11%	0.03%	0.21%	0.91%	**0.76%**	11.74%

Table A11. Statistics of the ARI indexes for overlapping windows.

Clustering Algorithm	Min	1st Quartile	Median	Mean	3rd Quartile	Max
Extended TS-Stream (Fourier coeff.)	0.015	0.037	0.049	0.067	0.079	0.438
Extended TS-Stream (Fourier coeff., concept drift)	−0.008	0.037	0.052	0.066	0.080	0.433
ClipStream (concept drift)	0.000	0.037	0.055	0.086	0.094	**1.000**
ClipStream (without concept drift)	−0.004	0.038	0.059	0.080	0.099	0.654
ClipStream (Fourier coeff., concept drift)	0.040	0.120	0.142	0.165	0.164	**1.000**
ClipStream (Fourier coeff., without concept drift)	0.040	0.120	0.142	0.152	0.167	0.515
Histogram-based	**0.218**	**0.405**	**0.537**	**0.540**	**0.648**	0.996

Table A12. Statistics of the tariffs improvement for overlapping windows.

Clustering Algorithm	Min	1st Quartile	Median	Mean	3rd Quartile	Max
Extended TS-Stream (Fourier coeff.)	−0.17%	**0.02%**	0.10%	0.14%	**0.23%**	**10.23%**
Extended TS-Stream (Fourier coeff., concept drift)	−0.19%	0.00%	0.04%	0.76%	0.15%	10.15%
ClipStream (concept drift)	−0.32%	0.00%	**0.11%**	**0.77%**	**0.23%**	**10.23%**
ClipStream (without concept drift)	−0.34%	−0.03%	0.07%	0.74%	0.21%	10.21%
ClipStream (Fourier coeff., concept drift)	−0.18%	0.00%	0.04%	0.76%	0.15%	10.15%
ClipStream (Fourier coeff., without concept drift)	−0.15%	0.00%	0.03%	0.76%	0.14%	10.14%
Histogram-based	**−0.13%**	0.01%	0.05%	**0.77%**	0.15%	10.15%

Table A13. Statistics of the weighted volatility for non-overlapping windows.

Clustering Algorithm	Min	1st Quartile	Median	Mean	3rd Quartile	Max
Extended TS-Stream (Fourier coeff.)	**4.33**	**5.79**	**7.70**	**7.65**	**9.13**	**10.89**
Extended TS-Stream (Fourier coeff., concept drift)	**4.33**	**5.79**	**7.70**	**7.65**	**9.13**	**10.89**
ClipStream (concept drift)	5.18	8.06	11.45	12.86	14.33	28.73
ClipStream (without concept drift)	5.18	8.06	12.03	12.9	14.33	28.73
ClipStream (Fourier coeff., concept drift)	12.29	19.83	24.04	23.94	27.91	42.01
ClipStream (Fourier coeff., without concept drift)	12.29	19.83	23.15	24.1	27.91	42.01
Histogram-based	5.19	6.73	8.72	8.64	10.29	11.87

Table A14. Statistics of the weighted volatility for overlapping windows.

Clustering Algorithm	Min	1st Quartile	Median	Mean	3rd Quartile	Max
Extended TS-Stream (Fourier coeff.)	**4.46**	**5.10**	**6.92**	**6.705**	8.05	**9.09**
Extended TS-Stream (Fourier coeff., concept drift)	4.66	5.27	6.94	13.21	**7.61**	70.76
ClipStream (concept drift)	7.88	12.95	16.15	19.38	23.38	37.46
ClipStream (without concept drift)	7.88	12.27	17.82	19.62	25.12	37.46
ClipStream (Fourier coeff., concept drift)	12.85	22.7	24.62	25.02	30.10	37.61
ClipStream (Fourier coeff., without concept drift)	11.94	19.18	24.62	23.63	28.04	33.60
Histogram-based	4.83	6.22	8.251	7.98	9.554	10.32

Table A15. Statistics of the predicted tariffs improvement comparing to the G11 for non-overlapping windows.

Clustering Algorithm	Min	1st Quartile	Median	Mean	3rd Quartile	Max
Extended TS-Stream (Fourier coeff.)	−6.29%	−1.12%	−0.27%	−0.18%	0.62%	14.57%
Extended TS-Stream (Fourier coeff., concept drift)	−6.29%	−1.12%	−0.27%	−0.18%	0.62%	14.57%
ClipStream (concept drift)	**−4.69%**	−0.79%	**−0.09%**	**0.19%**	**0.75%**	23.41%
ClipStream (without concept drift)	**−4.69%**	−0.76%	−0.10%	0.18%	0.73%	**24.80%**
ClipStream (Fourier coeff., concept drift)	−4.97%	−1.12%	−0.21%	−0.15%	0.64%	12.68%
ClipStream (Fourier coeff., without concept drift)	−5.39%	−1.07%	−0.23%	−0.15%	0.62%	13.92%
Histogram-based	−6.44%	**−0.51%**	−0.11%	−0.01%	0.41%	4.25%

Table A16. Statistics of the predicted tariffs improvement comparing to the G11 for overlapping windows.

Clustering Algorithm	Min	1st Quartile	Median	Mean	3rd Quartile	Max
Extended TS-Stream (Fourier coeff.)	−5.09%	−0.33%	**0.74%**	0.84%	1.77%	15.94%
Extended TS-Stream (Fourier coeff., concept drift)	−5.21%	−0.38%	0.73%	0.84%	1.81%	16.34%
ClipStream (concept drift)	−5.03%	0.08%	0.72%	**1.08%**	1.73%	23.12%
ClipStream (without concept drift)	**−5.00%**	**0.03%**	0.71%	1.06%	1.66%	**23.65%**
ClipStream (Fourier coeff., concept drift)	−6.33%	−0.44%	0.70%	0.82%	**1.82%**	16.83%
ClipStream (Fourier coeff., without concept drift)	−6.28%	−0.43%	0.71%	0.83%	**1.82%**	16.98%
Histogram-based	−7.40%	−0.38%	0.08%	0.16%	0.65%	5.14%

Appendix C. Results Based on London Data Set

Table A17. Simulation of households' electricity consumption characteristics based on different tariff group rates for non-overlapping windows.

	Min	1st Quartile	Median	Mean	3rd Quartile	Max
Best vs worst individual tariff for each batch	2.39%	4.79%	6.61%	7.21%	8.82%	33.37%
Best individual tariff for each batch vs best individual tariff for the entire period	0.00%	0.08%	0.24%	0.35%	0.51%	2.87%
Number of dynamic individual tariff change	0.00	3.00	5.00	4.90	7.00	14.00

Table A18. Simulation of households' electricity consumption characteristics based on different tariff group rates for overlapping windows.

	Min	1st Quartile	Median	Mean	3rd Quartile	Max
Best vs worst individual tariff for each batch	2.72%	5.33%	7.22%	7.71%	9.43%	36.14%
Best individual tariff for each batch vs best individual tariff for the entire period	0.00%	0.26%	0.52%	0.62%	0.87%	3.36%
Number of dynamic individual tariff change	0.00	18.00	25.00	23.95	31.00	47.00

Table A19. Statistics of the ARI indexes for non-overlapping windows.

Clustering algorithm	Min	1st Quartile	Median	Mean	3rd Quartile	Max
Extended TS-Stream (Fourier coeff.)	0.000	0.066	0.081	0.080	0.100	0.199
Extended TS-Stream (Fourier coeff., concept drift)	0.000	0.066	0.081	0.080	0.100	0.199
ClipStream (concept drift)	0.080	0.140	0.173	0.213	0.215	**1.000**
ClipStream (without concept drift)	0.078	0.130	0.164	0.173	0.204	0.344
ClipStream (Fourier coeff., concept drift)	0.074	0.112	0.143	0.162	0.174	**1.000**
ClipStream (Fourier coeff., without concept drift)	−0.001	0.110	0.136	0.136	0.168	0.368
Histogram-based	**0.224**	**0.333**	**0.368**	**0.417**	**0.488**	0.889

Table A20. Statistics of the tariffs improvement for non-overlapping windows.

Clustering Algorithm	Min	1st Quartile	Median	Mean	3rd Quartile	Max
Extended TS-Stream (Fourier coeff.)	−0.13%	0.06%	**0.23%**	**0.93%**	**0.74%**	1.74%
Extended TS-Stream (Fourier coeff., concept drift)	−0.25%	0.01%	0.07%	0.19%	0.20%	1.72%
ClipStream (concept drift)	**−0.05%**	0.08%	0.20%	0.35%	0.41%	1.58%
ClipStream (without concept drift)	−0.05%	**0.11%**	0.22%	0.39%	0.47%	1.71%
ClipStream (Fourier coeff., concept drift)	−0.20%	0.01%	0.13%	0.26%	0.22%	**1.92%**
ClipStream (Fourier coeff., without concept drift)	−0.16%	0.02%	0.13%	0.27%	0.22%	**1.92%**
Histogram-based	−0.09%	0.03%	0.09%	0.15%	0.27%	0.72%

Table A21. Statistics of the ARI indexes for overlapping windows.

Clustering Algorithm	Min	1st Quartile	Median	Mean	3rd Quartile	Max
Extended TS-Stream (Fourier coeff.)	0.036	0.064	0.077	0.090	0.099	0.442
Extended TS-Stream (Fourier coeff., concept drift)	0.031	0.064	0.078	0.090	0.098	0.447
ClipStream (concept drift)	0.060	0.125	0.150	0.176	0.190	**1.000**
ClipStream (without concept drift)	0.060	0.121	0.149	0.168	0.191	0.744
ClipStream (Fourier coeff., concept drift)	0.053	0.111	0.143	0.164	0.183	**1.000**
ClipStream (Fourier coeff., without concept drift)	0.055	0.110	0.139	0.155	0.177	0.658
Histogram-based	**0.209**	**0.321**	**0.395**	**0.426**	**0.504**	0.984

Table A22. Statistics of the tariffs improvement for overlapping windows.

Clustering Algorithm	Min	1st Quartile	Median	Mean	3rd Quartile	Max
Extended TS-Stream (Fourier coeff.)	−0.21%	**0.02%**	0.08%	0.10%	0.16%	1.39%
Extended TS-Stream (Fourier coeff., concept drift)	−0.21%	0.00%	0.06%	0.10%	0.16%	1.39%
ClipStream (concept drift)	−0.37%	**0.02%**	0.17%	**0.22%**	0.34%	2.13%
ClipStream (without concept drift)	−0.41%	0.01%	**0.17%**	0.21%	**0.35%**	2.13%
ClipStream (Fourier coeff., concept drift)	−0.55%	−0.03%	0.04%	0.10%	0.15%	**2.43%**
ClipStream (Fourier coeff., without concept drift)	−0.38%	−0.03%	0.04%	0.10%	0.15%	**2.43%**
Histogram-based	**−0.17%**	0.00%	0.04%	0.07%	0.11%	1.14%

Table A23. Statistics of the weighted volatility for non-overlapping windows.

Clustering Algorithm	Min	1st Quartile	Median	Mean	3rd Quartile	Max
Extended TS-Stream (Fourier coeff.)	**6.75**	**8.47**	**9.84**	**10.11**	**12.16**	164.98
Extended TS-Stream (Fourier coeff., concept drift)	**6.75**	**8.47**	**9.84**	**10.11**	**12.16**	164.98
ClipStream (concept drift)	19.36	23.72	26.63	29.58	33.91	48.24
ClipStream (without concept drift)	19.36	24.10	26.48	29.81	37.19	48.24
ClipStream (Fourier coeff., concept drift)	27.91	43.52	48.81	51.61	63.26	85.77
ClipStream (Fourier coeff., without concept drift)	27.91	43.52	48.81	57.82	70.38	164.32
Histogram-based	9.36	10.37	11.50	12.46	14.73	**18.77**

Table A24. Statistics of the weighted volatility for overlapping windows.

Clustering Algorithm	Min	1st Quartile	Median	Mean	3rd Quartile	Max
Extended TS-Stream (Fourier coeff.)	7.03	8.94	9.46	10.54	11.14	17.08
Extended TS-Stream (Fourier coeff., concept drift)	7.03	8.94	9.46	10.54	11.14	17.08
ClipStream (concept drift)	16.79	20.30	23.10	25.41	29.59	43.31
ClipStream (without concept drift)	16.79	20.30	23.10	25.15	29.59	39.00
ClipStream (Fourier coeff., concept drift)	24.99	28.92	41.69	42.91	53.07	62.66
ClipStream (Fourier coeff., without concept drift)	24.99	28.92	41.69	42.33	51.61	62.66
Histogram-based	8.86	10.91	11.77	12.76	13.55	21.39

Table A25. Statistics of the predicted tariffs improvement comparing to the G11 for non-overlapping windows.

Clustering Algorithm	Min	1st Quartile	Median	Mean	3rd Quartile	Max
Extended TS-Stream (Fourier coeff.)	−5.37%	−0.67%	0.27%	0.39%	1.27%	7.14%
Extended TS-Stream (Fourier coeff., concept drift)	−5.37%	−0.67%	0.27%	0.39%	1.27%	7.14%
ClipStream (concept drift)	**−3.96%**	**−0.46%**	0.11%	0.36%	0.78%	12.72%
ClipStream (without concept drift)	**−3.96%**	−0.48%	0.10%	0.39%	0.79%	**13.81%**
ClipStream (Fourier coeff., concept drift)	−6.62%	−0.76%	0.18%	0.23%	1.08%	7.89%
ClipStream (Fourier coeff., without concept drift)	−6.62%	−0.75%	0.18%	0.23%	1.10%	7.89%
Histogram-based	−6.43%	−0.68%	**0.31%**	**0.55%**	**1.52%**	10.97%

Table A26. Statistics of the predicted tariffs improvement comparing to the G11 for overlapping windows.

Clustering Algorithm	Min	1st Quartile	Median	Mean	3rd Quartile	Max
Extended TS-Stream (Fourier coeff.)	−7.41%	−0.53%	0.41%	0.59%	1.53%	8.66%
Extended TS-Stream (Fourier coeff., concept drift)	−5.83%	−0.50%	**0.42%**	0.57%	1.51%	7.66%
ClipStream (concept drift)	−3.95%	**−0.32%**	0.28%	0.62%	1.10%	13.26%
ClipStream (without concept drift)	**−5.14%**	**−0.32%**	0.31%	0.65%	1.22%	13.27%
ClipStream (Fourier coeff., concept drift)	−6.08%	−0.51%	0.37%	0.49%	1.33%	7.86%
ClipStream (Fourier coeff., without concept drift)	−6.71%	−0.48%	0.37%	0.50%	1.37%	8.43%
Histogram-based	−7.61%	−0.48%	0.37%	**0.68%**	**1.64%**	**10.90%**

References

1. Zabkowski, T.; Gajowniczek, K.; Szupiluk, R. Grade analysis for energy usage patterns segmentation based on smart meter data. In Proceedings of the 2015 IEEE 2nd International Conference on Cybernetics (CYBCONF), Gdynia, Poland, 24–26 June 2015. [CrossRef]

2. Nafkha, R.; Gajowniczek, K.; Ząbkowski, T. Do Customers Choose Proper Tariff? Empirical Analysis Based on Polish Data Using Unsupervised Techniques. *Energies* **2018**, *11*, 514. [CrossRef]

3. Silva, J.A.; Faria, E.R.; Barros, R.C.; Hruschka, E.R.; Carvalho, A.C.P.L.F.; de Gama, J. Data stream clustering. *ACM Comput. Surv.* **2013**, *46*, 1–31. [CrossRef]

4. Bhaduri, M.; Zhan, J.; Chiu, C.; Zhan, F. A Novel Online and Non-Parametric Approach for Drift Detection in Big Data. *IEEE Access* **2017**, *5*, 15883–15892. [CrossRef]

5. Gajowniczek, K.; Ząbkowski, T.; Sodenkamp, M. Revealing Household Characteristics from Electricity Meter Data with Grade Analysis and Machine Learning Algorithms. *Appl. Sci.* **2018**, *8*, 1654. [CrossRef]

6. Bhaduri, M.; Zhan, J.; Chiu, C. A Novel Weak Estimator for Dynamic Systems. *IEEE Access* **2017**, *5*, 27354–27365. [CrossRef]

7. Bhaduri, M.; Zhan, J. Using Empirical Recurrence Rates Ratio for Time Series Data Similarity. *IEEE Access.* **2018**, *6*, 30855–30864. [CrossRef]

8. Balzanella, A.; Verde, R. Histogram-based clustering of multiple data streams. *Knowl. Inf. Syst.* **2019**, *62*, 203–238. [CrossRef]

9. Macedo, M.N.; Galo, J.J.; Almeida, L.A.; Lima, A.C. Typification of load curves for DSM in Brazil for a smart grid environment. *Int. J. Electr. Power Energy Syst.* **2015**, *67*, 216–221. [CrossRef]

10. Gajowniczek, K.; Ząbkowski, T. Simulation Study on Clustering Approaches for Short-Term Electricity Forecasting. *Complexity* **2018**, *2018*, 3683969. [CrossRef]

11. Jain, A.K.; Murty, M.N.; Flynn, P.J. Data clustering: A review. *ACM Comput. Surv.* **1999**, *31*, 264–323. [CrossRef]

12. Pitt, B.D.; Kitschen, D.S. Application of data mining techniques to load profiling. In Proceedings of the 21st 1999 IEEE International Conference on Power Industry Computer Applications–PICA'99, Santa Clara, CA, USA, 21 May 1999; pp. 131–136.

13. Gerbec, D.; Gasperic, S.; Simon, I.; Gubina, F. Hierarchic clustering methods for consumers load profile determination. In Proceedings of the 2nd Balkan Power Conference, Belgrade, SR Yugoslavia, 19 June 2002; pp. 9–15.

14. Nazarko, J.; Styczynski, Z.A. Application of statistical and neural approaches to the daily load profiles modelling in power distribution systems. In Proceedings of the 1999 IEEE Transmission and Distribution Conference, New Orleans, LA, USA, 11 April 1999; Volume 1, pp. 320–325.

15. Espinoza, M.; Joye, C.; Belmans, R.; De Moor, B. Short-term load forecasting, profile identification, and customer segmentation: A methodology based on periodic time series. *IEEE Transact. Power Syst.* **2005**, *20*, 1622–1630. [CrossRef]

16. Suganthi, L.; Samuel, A.A. Energy models for demand forecasting—A review. *Renew. Sustain. Energy Rev.* **2012**, *16*, 1223–1240. [CrossRef]

17. McLoughlin, F.; Duffy, A.; Conlon, M. A clustering approach to domestic electricity load profile characterisation using smart metering data. *Appl. Energy* **2015**, *141*, 190–199. [CrossRef]

18. Lamedica, R.; Santolamazza, L.; Fracassi, G.; Martinelli, G.; Prudenzi, A. A novel methodology based on clustering techniques for automatic processing of MV feeder daily load patterns. In Proceedings of the IEEE Power Engineering Society Summer Meeting, Seattle, WA, USA, 16–20 July 2000; Volume 1, pp. 96–101.

19. Chicco, G.; Napoli, R.; Postolache, P.; Scutariu, M.; Toader, C. Customer characterization options for improving the tariff offer. *IEEE Transact. Power Syst.* **2003**, *18*, 381–387. [CrossRef]

20. Benítez, I.; Quijano, A.; Díez, J.L.; Delgado, I. Dynamic clustering segmentation applied to load profiles of energy consumption from Spanish customers. *Int. J. Electr. Power Energy Syst.* **2014**, *55*, 437–448. [CrossRef]

21. Rhodes, J.D.; Cole, W.J.; Upshaw, C.R.; Edgar, T.F.; Webber, M.E. Clustering analysis of residential electricity demand profiles. *Appl. Energy* **2014**, *135*, 461–471. [CrossRef]

22. Tsekouras, G.J.; Hatziargyriou, N.D.; Dialynas, E.N. Two-stage pattern recognition of load curves for classification of electricity customers. *IEEE Transact. Power Syst.* **2007**, *22*, 1120–1128. [CrossRef]

23. Chicco, G.; Napoli, R.; Piglione, F. Comparisons among clustering techniques for electricity customer classification. *IEEE Transact. Power Syst.* **2006**, *21*, 933–940. [CrossRef]

24. Chen, J.-Y.; He, H.-H. A fast density-based data stream clustering algorithm with cluster centers self-determined for mixed data. *Inf. Sci.* **2016**, *345*, 271–293. [CrossRef]

25. Amini, A.; Saboohi, H.; Herawan, T.; Wah, T.Y. MuDi-Stream: A multi density clustering algorithm for evolving data stream. *J. Netw. Comput. Appl.* **2016**, *59*, 370–385. [CrossRef]

26. Chen, Y.; Tu, L. Density-based clustering for real-time stream data. In Proceedings of the 13th ACM SIGKDD International Conference on Knowledge Discovery and Data Mining—KDD '07, San Jose, CA, USA, 12–15 August 2007. [CrossRef]

27. Aggarwal, C.C.; Yu, P.S.; Han, J.; Wang, J. A Framework for Clustering Evolving Data Streams. In Proceedings of the 2003 VLDB Conference, Berlin, Germany, 9–12 September 2003; pp. 81–92. [CrossRef]

28. Hahsler, M.; Bolaos, M. Clustering Data Streams Based on Shared Density between Micro-Clusters. *IEEE Trans. Knowl. Data Eng.* **2016**, *28*, 1449–1461. [CrossRef]

29. Zhang, T.; Ramakrishnan, R.; Livny, M. BIRCH: An efficient data clustering method for very large databases. *ACM SIGMOD Rec.* **1996**, *25*, 103–114. [CrossRef]

30. Udommanetanakit, K.; Rakthanmanon, T.; Waiyamai, K. E-Stream: Evolution-Based Technique for Stream Clustering. *Lect. Notes Comput. Sci.* **2007**, 605–615. [CrossRef]

31. Ackermann, M.R.; Märtens, M.; Raupach, C.; Swierkot, K.; Lammersen, C.; Sohler, C. StreamKM++. *J. Exp. Algorithmics* **2012**, *17*, 173–187. [CrossRef]

32. Beringer, J.; Hllermeier, E. Fuzzy Clustering of Parallel Data Streams. *Adv. Fuzzy Clust. Appl.* **2007**, 333–352. [CrossRef]

33. Chen, Y. Clustering Parallel Data Streams. *Data Min. Knowl. Discov. Real Life Appl.* **2009**. [CrossRef]

34. Dai, B.R.; Huang, J.W.; Yeh, M.Y.; Chen, M.S. Adaptive Clustering for Multiple Evolving Streams. *IEEE Trans. Knowl. Data Eng.* **2006**, *18*, 1166–1180. [CrossRef]

35. Laurinec, P.; Lucká, M. Interpretable multiple data streams clustering with clipped streams representation for the improvement of electricity consumption forecasting. *Data Min. Knowl. Discov.* **2018**, *33*, 413–445. [CrossRef]

36. Khan, I.; Huang, J.Z.; Ivanov, K. Incremental density-based ensemble clustering over evolving data streams. *Neurocomputing* **2016**, *191*, 34–43. [CrossRef]

37. Pereira, C.M.M.; de Mello, R.F. TS-stream: Clustering time series on data streams. *J. Intell. Inf. Syst.* **2014**, *42*, 531–566. [CrossRef]

38. Rodrigues, P.P.; Gama, J.; Pedroso, J.P. Hierarchical Clustering of Time-Series Data Streams. *IEEE Trans. Knowl. Data Eng.* **2008**, *20*, 615–627. [CrossRef]

39. Chen, L.; Zou, L.-J.; Tu, L. A clustering algorithm for multiple data streams based on spectral component similarity. *Inf. Sci.* **2012**, *183*, 35–47. [CrossRef]

40. Alseghayer, R.; Petrov, D.; Chrysanthis, P.K.; Sharaf, M.; Labrinidis, A. Detection of Highly Correlated Live Data Streams. In Proceedings of the International Workshop on Real-Time Business Intelligence and Analytics, Munich, Germany, 28 August 2017; pp. 1–8. [CrossRef]

41. Sakurai, Y.; Papadimitriou, S.; Faloutsos, C. BRAID: Stream mining through group lag correlations. In Proceedings of the 2005 ACM SIGMOD International Conference on Management of Data, Baltimore, MD, USA, 14–16 June 2005; pp. 599–610. [CrossRef]

42. Shafer, I.; Ren, K.; Boddeti, V.N.; Abe, Y.; Ganger, G.R.; Faloutsos, C. RainMon: An integrated approach to mining bursty timeseries monitoring data. In Proceedings of the 18th ACM SIGKDD International Conference on Knowledge Discovery and Data Mining-KDD 2012, Beijing, China, 12–16 August 2012; pp. 1158–1166. [CrossRef]

43. Zhu, Y.; Shasha, D. Statstream: Statistical monitoring of thousands of data streams in real time. In Proceedings of the 28th International Conference on Very Large Databases 2002–VLDB'02, Hong Kong, China, 20–23 August 2002; pp. 358–369.

44. Wu, Y.; Liu, Y.; Ahmed, S.H.; Peng, J.; Abd El-Latif, A.A. Dominant Data Set Selection Algorithms for Electricity Consumption Time-Series Data Analysis Based on Affine Transformation. *IEEE Internet Things J.* **2020**, *7*, 4347–4360. [CrossRef]

45. Gajowniczek, K.; Bator, M.; Ząbkowski, T.; Orłowski, A.; Loo, C.K. Simulation Study on the Electricity Data Streams Time Series Clustering. *Energies* **2020**, *13*, 924. [CrossRef]

46. Irpino, A.; Verde, R. Basic statistics for distributional symbolic variables: A new metric-based approach. *Adv. Data Anal. Classif.* **2014**, *9*, 143–175. [CrossRef]

47. Verde, R.; Irpino, A. Dynamic Clustering of Histogram Data: Using the Right Metric. Studies in Classification. *Data Anal. Knowl. Organ.* **2007**, 123–134. [CrossRef]

48. Diday, E.; Noirhomme-Fraiture, M. *Symbolic Data Analysis and the SODAS Software*; John Wiley & Sons: Chichester, UK, 2007; pp. 191–204. [CrossRef]

49. Robinson, A.H.; Cherry, C. Results of a prototype television bandwidth compression scheme. *Proc. IEEE* **1967**, *55*, 356–364. [CrossRef]

50. Kaufman, L.; Rousseeuw, P.J. *Finding Groups in Data: An Introduction to Cluster Analysis*; John Wiley & Sons: Chichester, UK, 2009; Volume 344.

51. Davies, D.L.; Bouldin, D.W. A Cluster Separation Measure. *IEEE Transact. Pattern Anal. Mach. Intell.* **1979**, *1*, 224–227. [CrossRef]

52. Lyons, R.G. *Understanding Digital Signal Processing, 2/E*; Prentice Hall PTR Upper: Saddle River, NJ, USA, 2004.

53. BIRCH-Clustering-R-Package. Available online: https://github.com/rohitkata/BIRCH-Clustering-R-package (accessed on 10 March 2020).

54. SymbolicDA: Analysis of Symbolic Data. Available online: https://rdrr.io/cran/symbolicDA/ (accessed on 10 March 2020).

55. ClipStream. Available online: https://github.com/PetoLau/ClipStream (accessed on 10 March 2020).

56. Langham, E.; Downes, J.; Brennan, T.; Fyfe, J.; Mohr, S.; Rickwood, P.; White, S. *Smart Grid, Smart City, Customer Research Report*; Institute for Sustainable Futures: Ultimo, NSW, Australia, June 2014.

57. UK Power Networks Led Low Carbon London. Available online: https://data.london.gov.uk/dataset/smartmeter-energy-use-data-in-london-households (accessed on 1 December 2020).

Publisher's Note: MDPI stays neutral with regard to jurisdictional claims in published maps and institutional affiliations.

Article

Deep Learning for Walking Behaviour Detection in Elderly People Using Smart Footwear

Rocío Aznar-Gimeno *, Gorka Labata-Lezaun, Ana Adell-Lamora, David Abadía-Gallego, Rafael del-Hoyo-Alonso and Carlos González-Muñoz

Department of BigData and Cognitive Systems, Instituto Tecnológico de Aragón, ITAINNOVA, María de Luna 7-8, 50018 Zaragoza, Spain; glabata@itainnova.es (G.L.-L.); adell@itainnova.es (A.A.-L.); dabadia@itainnova.es (D.A.-G.); rdelhoyo@itainnova.es (R.d.-H.-A.); cgonzalez@itainnova.es (C.G.-M.)
* Correspondence: raznar@itainnova.es

Abstract: The increase in the proportion of elderly in Europe brings with it certain challenges that society needs to address, such as custodial care. We propose a scalable, easily modulated and live assistive technology system, based on a comfortable smart footwear capable of detecting walking behaviour, in order to prevent possible health problems in the elderly, facilitating their urban life as independently and safety as possible. This brings with it the challenge of handling the large amounts of data generated, transmitting and pre-processing that information and analysing it with the aim of obtaining useful information in real/near-real time. This is the basis of information theory. This work presents a complete system aiming at elderly people that can detect different user behaviours/events (sitting, standing without imbalance, standing with imbalance, walking, running, tripping) through information acquired from 20 types of sensor measurements (16 piezoelectric pressure sensors, one accelerometer returning reading for the 3 axis and one temperature sensor) and warn the relatives about possible risks in near-real time. For the detection of these events, a hierarchical structure of cascading binary models is designed and applied using artificial neural network (ANN) algorithms and deep learning techniques. The best models are achieved with convolutional layered ANN and multilayer perceptrons. The overall event detection performance achieves an average accuracy and area under the ROC curve of 0.84 and 0.96, respectively.

Keywords: assistive technology; elderly people; wearable devices; smart footwear; deep learning; artificial neural networks

Citation: Aznar-Gimeno, R.; Labata-Lezaun, G.; Adell-Lamora, A.; Abadía-Gallego, D.; del-Hoyo-Alonso, R.; González-Muñoz, C. Deep Learning for Walking Behaviour Detection in Elderly People Using Smart Footwear. *Entropy* **2021**, *23*, 777. https://doi.org/10.3390/e23060777

Academic Editors: Felipe Ortega and Emilio López Cano

Received: 10 May 2021
Accepted: 15 June 2021
Published: 19 June 2021

Publisher's Note: MDPI stays neutral with regard to jurisdictional claims in published maps and institutional affiliations.

1. Introduction

1.1. Context

The proportion of elderly people in Europe has been increasing in recent years and is expected to follow a clear upward trend in the coming years, reaching 29.4% of the total population in 2050 [1]. This ageing population is due to falling fertility rates and increasing life expectancy, the latter due to numerous advances in science, technology, medicine and public health, combined with increased awareness of nutrition and personal hygiene [2,3]. Although the increase in demographic longevity can be seen as one of history's great success stories, it has social consequences and challenges that need to be addressed, such as custodial care. One third of people over 75 have physical, mental or sensory impairments [4] and therefore need long-term custodial care. This care can be provided in institutional care or at home. Studies have shown that older people living in institutional care experience a higher level of dependency, loneliness and decreased life satisfaction and that they prefer to live in their own homes [5–7]. Living in their own home provides them with greater independence, reduces social isolation with a positive effect on the elderly [8]. However, ageing at home implies addressing certain aspects of home care.

The use of approaches and techniques for the care of the elderly has become an emerging challenge that needs to be addressed in a way that supports, facilitates and enables them to age with a better quality of life and as independently as possible. To this end, smart home systems [2,9–12] and assistive technologies (AT) [13,14] have been developed. They are based on the implementation of different sensors and devices (Internet of Things, IoT). However, their adoption may raise certain barriers [15] and concerns on the part of older adults related to how they perceive these technologies [16], such as privacy, ease of use, lack of training or suitability for everyday use [17]. This leads to high rates of dissatisfaction and abandonment of assistive technology and its use. Another aspect of abandonment is related to the design, aesthetics or unobtrusiveness of the device [18].

The integration of sensors and devices generates large amounts of real data and information, which brings with it the challenge of handling this data, pre-processing and analysing it in order to obtain useful information in real time. This is the basis of the fundamentals of information theory, which was conceived by Claude Shannon in 1948 [19]. Information theory is a subfield of mathematics that deals with the quantification of information, the representation of information and the ability of communication systems to transmit and process information. The need for this theoretical basis arose in the face of the increase in complexity and the massification of communication channels in the mid-20th century. Extrapolating it to the 21st century, with the development of concepts such as IoT, Artificial Intelligence, Big Data, Machine Learning, Deep Learning, the fundamentals of information theory remain basic foundations today. One of the important concepts of information theory is the quantification of the amount of information through the use of probabilities ("entropy"). This concept of information theory has had great contributions in areas such as machine learning and neural networks. In particular, the computation of information and entropy is a useful tool in machine learning and is used as a basis for techniques such as feature selection, decision tree construction, imbalance calculations in the target class distribution and, in general, when optimising classification models (e.g., artificial neural networks) considering cross-entropy as a loss function. The application of models based on artificial neural networks and particularly Deep Learning has become widespread in recent years due to its ability to automatically detect the most particular features of data, which has led to promising performance in many areas such as, in particular, sensor-based activity recognition [20–23] and in the application of smart homes and wearable devices [24–26].

Thanks to the fundamentals related to information theory, the miniaturisation of sensors and the improvement of data storage and transmission systems have been possible and is one of the reasons for the success of monitoring and pattern detection through IoT devices and sensors, particularly in the integration of fabrics and textiles ("smart fabrics/wearable") [27–30]. Particularly, the data retrieved by sensors can be used to monitor the elderly in real time and predict their behaviour, preventing potential health problems, while providing them with independence and facilitating them urban living. Furthermore, ensuring that the electronics are fully integrated into the fabric ensures truly wearable products without discomfort.

The research problem of this article is based on the detection of walking behaviour of elderly people using wearable AT prototype for everyday use by using deep learning algorithms. This work is result of the European project MATUROLIFE (Metallisation of Textiles to make Urban living for Older people more Independent Fashionable) [31], which has been carried out within the framework of the Horizon 2020 (The EU-Horizon2020 (H2020-EU.2.1.3. Leadership in enabling and Leadership in enabling and industrial technologies—Advanced materials)). The project aim was to research, innovate and develop a more integrated assistive technology in textiles and fabrics through the use of advanced materials, allowing sensors and electronic devices to be fully integrated into intelligent fabrics in a discreet, fashionable and comfortable way [32–36]. The project studied the incorporation of sensors in three prototypes of AT for everyday use: clothing, furniture and footwear, that will make urban living for older adults easier, more independent, fashionable and comfortable. In the article we focus on the smart footwear prototype.

1.2. Related Work

Research and study of the walking behaviour detection through the use of footwear incorporating sensors (smart footwear) has been widely explored in last years [37]. Abnormal walking behaviour can indicate danger and detecting it can prevent potential health problems, such as injuries that can be caused by falls [38–42]. This is of great interest to the elderly population, as it allows them to lead a comfortable, more independent and safer urban life by monitoring their activity.

De Pinho et al. [43] presented the results of an experiment aimed at detecting 6 types of activities (walking straight, walking slope up, walking slope down, ascending stairs, descending stairs and sitting) from information retrieved from smart shoes. The experiment involved 11 participants, 2 of whom were elders. The classifier used a Random Forest algorithm with leave-one-out cross validation, achieving good average accuracy. A set of 12 features were considered as model inputs: 2 axis of the gyroscope, 2 axis of the magnetometer, 1 axis of the accelerometer, 4 force-sensitive resistors (FSRs), 2 Euler angles and the cumulative difference between samples of the barometer. el Achkar et al. [44] studied also the recognition of daily activities (level walking, sitting, standing, up/down stairs, up/down hill, elevator use) of older people. For this purpose, ten elderly people wearing the instrumented footwear system carried out the activities in a semi-structured protocol. The smart footwear included inertial and barometric pressure sensors, a sensorised insole to measure foot pressure and a box with electronics that strapped to the ankle. A decision tree incorporating rules inspired by movement biomechanics was applied as activity classification algorithm, achiving a high overall accuracy.

Zitouni et al. [45] designed a discreet, comfortable and highly effective device that is housed in the insole and a fall detection algorithm based on acceleration and time thresholds. Six subjects between 25 and 30 years of age were tested for possible falls that an elderly person may have while performing daily activities of daily living. They validated the proposed prototype and algorithm in real time (in a real public demonstration) confirming satisfactory performance. Montanini et al. [46] presented a low complexity and threshold-based methodology capable of detecting a fall and notifying a monitoring system. The smart shoes were equipped with 3 FSRs and a tri-axial accelerometer and tied to a belt an external processing unit box. These devices enabled the analysis of the subject's motion and foot orientation, recognizing abnormal configurations. Laboratory tests involved 17 healthy subjects (aged between 21 and 55 years) and provided satisfactory performances in falls detection. The proposed method was also validated with two elderly users in a real-life scenario. Light et al. [47] mentioned the need for the use of monitoring systems for older people because of their high risk of falls and other mobility problems. They developed an optimized layout of pressure sensors for a smart- shoe fall monitoring application by analysing various machine learning algorithms with 10-fold cross validation that classify fall types. Subjects between the ages of 20 and 45 years participated in the data collection. The activities carried out in this experiment were falling-left, falling right, falling-forward, falling-backward, standing, walking, and kneeling down. Sim et al. [48] attached an accelerometer on the shoes (tongue) to detect fall in the elderly. 3 axis- acceleration signals were measured in three young subjects (2 young males, 1 young female, aged between 24 and 28). The fall types used in this study were the most common fall types in elderly people. The results of the fall detection algorithm showed that this shoe-based fall detection system had relatively high sensitivity.

1.3. Limitations of Existing Practices

The related studies have some limitations related to different aspects such as the comfort and usability of the smart footwear, the set of events capable of detection, the provision of a real-time detection and notification system, and the modularity and scalability of the system.

For greater comfort and ease of use of the footwear, it is desirable that sensors and electronic devices are fully integrated into the shoe without direct contact with the rest of the person's body. This aspect is not fully addressed by some of the related works. The prototype of el Achkar et al. [44] housed an electronic box strapped directly to the ankle. Montanini et al. [46] did not fully integrate everything into the shoe either. Light et al. [47] mentioned that the insole was tied to the user's leg using paper tapes, which could cause some discomfort, especially when removing these tapes from the subject's leg. The design of the prototype of Sim et al. [48] also had some limitations. The accelerometer module of the prototype could easily detached because it was attached to the outside of shoes, as well as encumbering some activities, and the battery was slightly heavy. The authors suggested reducing the size of the module and embedding it under the insole, as well as incorporating piezoelectric elements to solve the problem with energy harvesting. All these factors represent a clear limitation, as they are less comfortable and intrusive devices that prevent certain activities from being carried out with complete normality.

Regarding the considered events, some of the studies focus on detecting more common events related to the user daily activity (sitting, walking, going up/down stairs, etc.) [43,44], while other studies focus only on immediate risk events such as falls [45–48]. However, there is a lack of studies proposing a system capable of detecting a broader spectrum of events, both hazard events and daily activity events.

Concerning the detection algorithms used, most studies proposed simple rule-based Machine Learning algorithms (decision tree [44], random forest [43]) or threshold-based methodologies [45,46]. Although in all cases the authors reported achieving good performance, they did not mention guarantees of modularity and scalability for the detection of new events or new functionalities and models. In addition, rule-based algorithms may be not very robust and have a higher risk of overfitting, with the risk of not generalising well to a different population.

There are also limitations in the related studies in terms of the provision of a system with practical real-time applicability. Most papers studied the scope of the system (algorithm and smart footwear) in terms of event detection but either did not validate it in real-time or did not provide a complete real-time detection and notification system. Montanini et al. [46] foreseed as future work the integration of a notification service for caregivers and Sim et al. [48] to develop a better smart-shoes system that shows the fall information on a smartphone and is therefore able to detect falls only with shoes and a smartphone.

1.4. Proposed Solution

We propose a system based on smart footwear capable of detecting different walking behaviours and warning the person responsible for the elderly person of possible risks in near-real time via a Telegram message. Figure 1 shows a general outline of the proposed system.

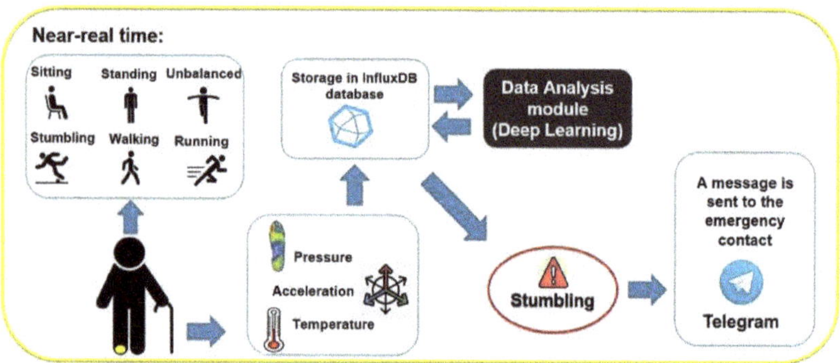

Figure 1. General outline of the proposed system.

The system is able to detect 6 events (sitting, standing still without imbalance, standing with imbalance, walking, running and stumbling) from the information retrieved from the pressure, acceleration and temperature sensors incorporated in the smart footwear. For this purpose, Deep Learning techniques and neural networks algorithms are applied, following an architecture that allows the system to be easily modulated, scalable and robust. In addition, advanced materials were used in the design of the smart footwear, so that the sensors and electronic devices are fully integrated into the shoe in a discreet, elegant and comfortable way. This is important to encourage their use because, as mentioned above, suitability for everyday use, design, aesthetics, discretion and comfort are important aspects to their adoption. Many of the related studies introduced in the previous section do not fully address this aspect.

In conclusion, the system we propose includes a smart footwear, comfortable for daily use, that retrieves real-time information of the elderly person walking and is able to detect a wide set of events and warn the responsible person of possible dangers, allowing to act quickly to avoid potential health problems. Therefore, unlike most related works, in addition to studying the retrieved data and generating classification models based on deep learning, we developed a system useful for real practice that allows act and send notifications to the mobile phone in near real time. Furthermore, our system detects both events of possible immediate risk (imbalance, stumbling) and more common events in daily activity (walking, for example), making it a more complete behaviour detection system with greater practical interest. Another of the differential aspects that our proposal addresses is to ensure a scalable and easily modular system (it allows to be recreated with new functionalities) and alive, external to possible errors or failures in the sensors, for example.

To realise this whole system, the fundamentals of information theory were elementary and data analytics played a key role. Our work considered the methodological framework related to data analytics CRISP-DM (Cross-Industry Process for Data Mining) [49] which is based on an agile and iterative methodology whose approach consists of several interrelated phases: Business Understanding (understanding the context from the business perspective), Data Understanding (acquisition and exploration of the data), Data Preparation (pre-processing of the corresponding data for the subsequent use of models), Modelling (generation of Machine Learning and Artificial Intelligence models), Evaluation (evaluation of the results of the models related to the definition of the business objectives) and Deployment (deployment of the application). Specifically, the article focuses on the analysis and modelling of data from smart footwear sensors using Deep Learning techniques and artificial neural networks algorithms with cross-entropy loss function. The following sections present the proposed system in more detail, explaining the different modules it integrates, such as the data analysis module carried out, as well as the results and scopes obtained in the project.

2. Materials and Methods

2.1. System Architecture

The final objective of the designed prototype is to be able to use the information retrieved by the sensors implemented in the shoe in order to control the walking and movement behaviour of the user so that the relative responsible for the user or professional health carers can be alerted if a possible risk is detected. The system architecture achieving this is presented in Figure 2. The system architecture has been designed taking into account the potential scalability of the system with a possible growth potential. Kubernetes [50] is the platform used to manage the different components of the architecture.

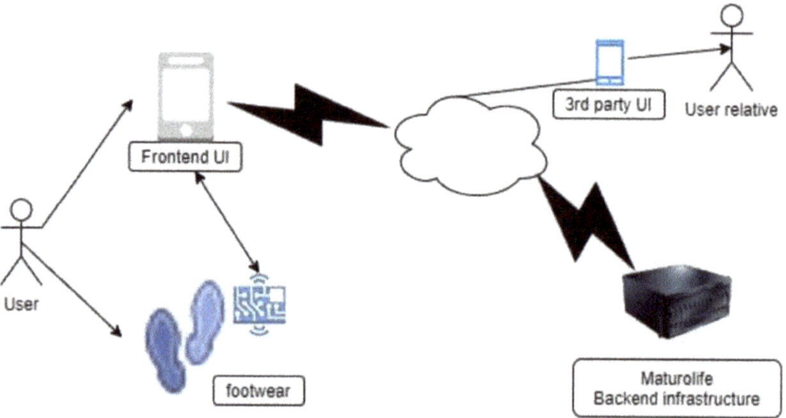

Figure 2. Scheme of system architecture.

The components involved in the architecture and therefore necessary to guarantee the information flow are: smart shoe, an Android smartphone for the user (in particular, having installed the MaturoApp application available on the PlayStore), mobile phone for the responsible person (contact person) with the Telegram application [51], internet connection (4G–5G) and an infrastructure (server) that supports the management, preprocessing and modelling of the data. The system is mainly composed of two information transmission processes: information retrieval flow from the user's smart footwear to the database and information exploitation flow from the database to the end user (user relative).

The information retrieval process involves the following components: the smart footwear, the Android smartphone for the user (MaturoApp), a cloud data manager (MQTT [52]) and a database for storage. The process is as follows. The smart shoe generates information about the user gait through the measurements of the implemented sensors retrieved with an average frequency of 4 Hz. The raw data is coded by the PCB using scientific notation fixed-point coding. It is then transferred via Bluetooth to the user's mobile phone (MaturoApp application), where the values of the pressure sensors can be displayed in real time. The mobile application transforms the received data chunks and transmits them via an MQTT message as follows:

$$\{\text{"timestamp"}:\text{"2020-10-25T17:12:24:6847"},\text{"data"}:[X_1, X_2, X_3, X_4, X_5, X_6, X_7, X_8, X_9, X_{10}, X_{11}, X_{12}, X_{13}, X_{14}, X_{15}, X_{16}]\} \tag{1}$$

where the timestamp and the values of the 16 pressure sensors (X_i) are displayed. These messages are grouped and inserted into an InfluxDB time series database [53].

Simultaneously, the information exploitation process is carried out. It involves the following components: the database with the retrieved information stored, the Data Analysis and Modelling module and the mobile phone of the contact person with the Telegram application. The process is as follows. The latest data stored in the database are retrieved with an appropriate frequency from the data Analysis module and preprocessed and prepared in a suitable way for the subsequent application of the artificial neural networks and Deep Learning models. The model output, as well as its timestamp, are stored in a table in the InfluxDB database. If the outcome indicates a risky event for the user, a Telegram message is sent to the emergency contact via the Telegram bot so an action can be taken. The information retrieval and exploitation system ensures near real-time event detection.

The following sections explain in detail the sensors implemented in the smart shoe, the experimental setup and data collection process necessary for the generation of the models and the data analysis module including sections on data pre-processing, model architecture and trained artificial neural networks.

2.2. Smart Footwear Design

Prototyped shoes were designed in 2 different models (male and female) and include sensorization and electronics in one shoe of each pair (right foot). In particular, the sensorized pairs of shoes that were made available for data collection and analysis were an European size 38 of the women's shoe model and an European size 41 of the men's shoe model.

The sensors implemented in the shoe retrieve information from three physical magnitudes: pressure (16 sensors), temperature (1) and acceleration (one measurement for each axis). The pressure sensors are piezoelectric sensors that are housed along the insole (designed with metallised textile) as presented in Figure 3. In particular, the material used for the produced insoles followed a coating process called electroless copper plating used for the selective metallization of textiles for electro-magnetic interference shielding [33]. The printed materials were used on a multilayer solution in which pressure sensors based on Printed Electronic technology are sandwiched between two layers with printed electrodes. The temperature sensor and the accelerometer are embedded in a printed circuit board (PCB). In addition to retrieving the temperature and acceleration measurements, the PCB is designed to connect and pre-process the data retrieved from the smart insole (pressure sensors), after a signal conditioning to convert pressure sensors signals into required values for the Analog Digital converters.

Figure 3. Location of the piezoelectric sensors on the insole.

Besides the sensors, the shoes have a Bluetooth antenna, a battery and a micro-usb connector for charging. All these components and the PCB are housed in a 3D printed box incorporated in the heel area, for which it was necessary to drill a cavity. As it can be seen in Figure 4, the box is divided into two parts: one to hold the battery and the other to hold the PCB. The box has access to charging port and insole connector from outside.

Being the insole independent of the PCB allows the insoles to be easily removed and substituted with newer ones by simply unplugging them from the connector. Thus, users can easily replace the insoles if damaged or if newer versions arrive to market.

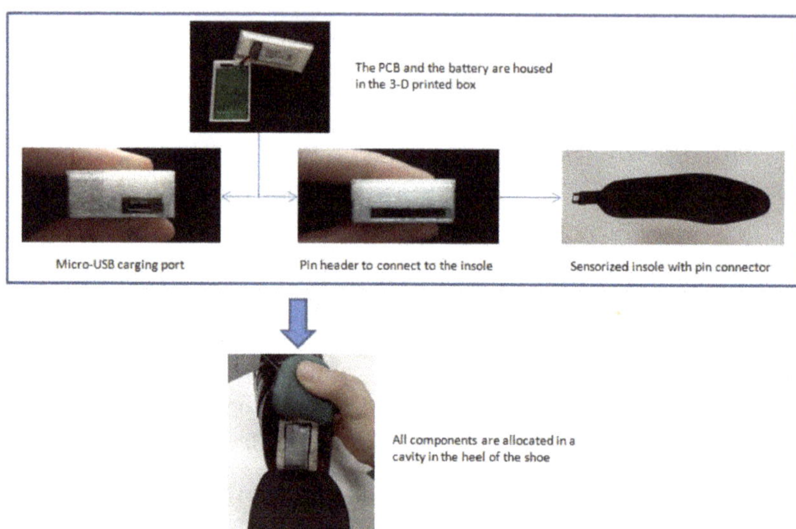

Figure 4. Components of the sensorised shoe.

In order to check that no interactions with other signals occur that could affect the proper functioning of other surrounding devices, laboratory tests of the implemented sensorics were performed. The tests performed compared insoles produced with commercial textile sets (made with a flexible substrate circuit) and the ones produced with the metallized textiles with no variations detected regarding electro-magnetic interferences and resistance variations for the different forces applied. In addition, the communication system was designed and verified in such a way that it transmits information correctly and without error. Specifically, data transmission error rate is near null at the average distance a potential user could keep the mobile while walking (a pocket, on the hand, etc.), besides, the PCB is able to store several data chunks at an internal buffer and retransmit them on error till properly received by the mobile device.

2.3. Experimental Setup. Data Collection

2.3.1. Events

In order to detect possible risks based on the user behaviour, a set of representative events of the gait to be modeled (supervised learning) was defined. The final classification models will allow the user behaviour to be related to one of the defined events. The events considered are the following:

- Sitting: sitting on a chair. May also include movements of the feet or the crossing of the legs.
- Standing still without imbalance: standing without moving forwards, backwards or sideways. May also include small foot taps.
- Standing with imbalance: includes lateral, frontal and random imbalances.
- Walking: includes different walking speeds, from slower to more normal.
- Running: running with a higher gait than walking.
- Stumbling: stumbling with the right foot. Includes both more violent and softer stumbles.

The selection criteria for the events was to consider a wide heterogeneous range of possible user behaviours including both more immediate hazard events and more common events of daily activity, in order to monitor and prevent possible health problems of the user. The stumbling and imbalance detection is important as these events indicate a possible risk of a fall and lack of body control by any user that may lead to a dangerous fall and negatively affect the user health. On the other hand, although the other events may relate to more common and less dangerous events in principle, their detection can also help to inform us of certain abnormal user behaviours in specific time periods. For example, the detection of the event "sitting" in a certain time period where the user usually walks may be an indication that the user has suffered a health problem (e.g., stroke), or the detection of the event "standing" without movement for a long period of time could indicate disorientation. As the data retrieval and exploitation system, discussed above, allows the detection of these events in a small time period (near real time), it enables emergency contacts to act rather quickly in order to avoid these potential health-related problems.

2.3.2. Data Collection

For the application of the final classification models to detect the events presented above, it was necessary to generate and retrieve data (temporal information from the sensors) for each event. At the beginning the project activities were scheduled and addressed to a testing group of elderly participants at the village of Arnedo -La Rioja-. However, due to the COVID19 pandemic in 2020 that brought an initial lockdown across all Europe followed by mobility restrictions, they could only participate in the identification of needs and contribution of ideas to the product design and interaction teams. Therefore, as a consequence of these pandemic limitations, the generation or retrieval of (anonymised) data from the different defined events had to be finally performed by a group of participants from the project technical team. The group consisted of 3 people (2 women and 1 man), aged 26, 27 and 26 years, respectively, and weighting approximately 70 kg each, which remained stable throughout the study period. As they had different foot sizes, two of the subjects wore the same pair of shoes (male model) and one of the women wore the other pair (female model) throughout the study period.

The data collection process was as follows. The subject put the shoe on his right foot, logged into the MaturoApp application on his mobile phone and performed one of the events mentioned for a time. While this action lasted, the sensors implemented in the shoes were capturing acceleration, pressure and temperature values with the frequency mentioned above. This information was sent via Bluetooth to the Maturoapp application on the user's mobile phone where the values of the pressure sensors could be visualised in real time. This data was stored in InfluxDB as described above with a manually defined tag identifying the subject, the event to which the data corresponded and the timestamp, in order to have complete traceability. This process was carried out multiple times by the 3 subjects and for the 6 events.

The database stored a total of 2.5 h of captured data. The first few minutes corresponded to initial recording tests in order to test and become familiar with the data capture system. Also, as explained later in the article, the models use as input the historical information for each time instant. Therefore, the first data captured in each recording were also not used in the generation of the model because they did not have sufficient historical information. Finally, the labelled dataset used for the generation of the models corresponded to a total of approximately 2 h of recording. The number of samples corresponding to each of the events is shown in Table 1.

Table 1. Number of observations for each event.

Events	No of Observations
Sitting	3020
Standing still without imbalance	6920
Standing with imbalance	5230
Walking	9620
Running	3480
Stumbling	820

Figure 5 shows an example of data generation for the event "walking". The figure displays the person with the sensorized shoe and the image of the insole (shown in the

MaturoApp) with the measurement information from the pressure sensors in real time. On the left of the figure is the person with the foot resting on the floor (active sensors with measurements in green) and on the right with the foot lifted (inactive sensors in red).

Figure 5. Recording data from sensorized shoe. **Left**: Shoe resting on the floor. **Right**: Shoe lifted.

2.4. Data Analysis Module

Data Preprocessing

Once the data are captured and stored, a pre-processing of the data is carried out in order to unify the temporal information retrieved from the sensors before applying models.

Firstly, given that the acceleration sensors and the temperature sensor sent the data with a certain delay with respect to the pressure sensors, the times recorded by the accelerometer were assigned to the times of the pressure sensors. Thus all sensor readings ended up having the same time stamp each time a measurement was taken. Secondly, the maximum number of previous values that the model would use to predict the event was defined. After testing for computational speed and after having discussed and verified the time window to detect the event, a maximum of 32 previous values was chosen.

Therefore, the final data structure for each event was as follows (number of samples, 32, 20), the last component being the total number of sensor measurements. That is, each sample corresponded to a matrix of dimensions 32×20 of the form:

$$
\begin{pmatrix}
P^0_{t-1} & P^1_{t-1} & \cdots & P^{15}_{t-1} & T^0_{t-1} & A^0_{t-1} & \cdots & A^2_{t-1} \\
P^0_{t-2} & P^1_{t-2} & \cdots & P^{15}_{t-2} & T^0_{t-2} & A^0_{t-2} & \cdots & A^2_{t-2} \\
\cdots & \cdots & \cdots & \cdots & \cdots & \cdots & \cdots & \cdots \\
P^0_{t-32} & P^1_{t-32} & \cdots & P^{15}_{t-32} & T^0_{t-32} & A^0_{t-32} & \cdots & A^2_{t-32}
\end{pmatrix}
\tag{2}
$$

where "P" refers to the pressure sensor measurements, "A" to the acceleration and "T" to the temperature, the superscript corresponds, for each type, the sensor number and the subscript to the instant in time, being "t" the actual instant. Therefore, for each sample, historical information in the form of the Equation (2) was obtained.

Finally, the data associated with each of the events were divided into three separate data sets: training (60%), validation (20%) and test (20%), ensuring the same proportion of samples of each class (event) in each set. The training data was used for model training/tuning, the validation data was used for selection of the best model configuration (hyperparameter set) and the test data was used to provide unbiased evaluation metrics to give a generalized value of the performance of the chosen fitted model.

2.5. Model Architecture

For the prediction of the user state (defined events), different binary models were generated. The final outcome is the consequence of the application of these binary models in cascade following the hierarchical structure shown in Figure 6. The underlying idea was to start from more general binary models of behaviour that include more particular groups of events and, depending on their outcomes, to continue in the tree with more specific models, following a hierarchy.

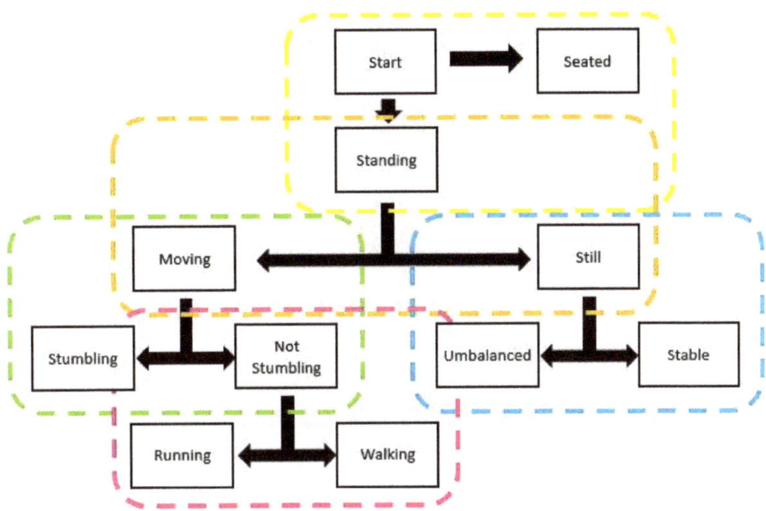

Figure 6. Hierarchical structure of models.

A total of 5 types of binary models were generated: (1) binary model generated from recorded event information that determines whether the user remains seated or standing (standing still without imbalance, standing still with imbalance, stumbling, walking, running); (2) binary model generated from recorded standing event information that determines whether the user remains still (standing still without imbalance, standing still with imbalance) or moving (stumbling, walking, running); (3) binary model generated from the information of the recorded non-moving standing events that determines whether the user is unbalanced or not (standing still stable); (4) binary model generated from the information of the recorded moving events that determines whether the user stumbles or does not stumble (walking, running); (5) binary model generated from the information of the recorded non-stumbling moving events that determines whether the user is walking or running.

The training and validation of the different binary models is explained in the following section.

Artificial Neural Networks

Artificial neural networks architectures with different types of layers including dense layers, time-distributed dense layers, convolutional layers and long and short term memory (LSTM) layers were used to train the models.

The dense layer is the regular layer of the deep-connected neural network and the time-distributed dense layer applies the same dense layer to every temporal slice of an input. The structure of the convolutional layers has a connection between neurons that is not fully complete but parameters are shared between different neurons. This particular structure implies, on the one hand, the ability to learn general and invariant representations of the data and, on the other hand, the training of complex architectures with less computational time. The convolution structure used also allowed the use of pooling layers. The LSTM layers allow for a recurrence and learning of dependencies not only in the short term but also in the long term.

For the fitting of the models, the concept of "entropy" from information theory was used. In particular, cross-entropy was used as a loss function:

$$E = -\frac{1}{N}\sum_{i=1}^{N} y_i \cdot log(p(y_i)) + (1 - y_i) \cdot log(1 - p(y_i)) \tag{3}$$

where y_i is the class (1 or 0), and $p(y_i)$ the predicted probability of belonging class 1 for observation i out of N observations.

In order to avoid overfitting, regularisation techniques such as dropout and EarlyStopping were used so that the model would stop training if it did not improve within a certain number of iterations.

In the training of each binary event detection model, a hyperparameter search was performed, allowing to select the number of dense layers to be introduced, whether an LSTM layer or a convolutional layer was included as well as the value of the hyperparameters defining each of these neural network layers. The search space is presented in Table 2. The possibility to select or not the different sensors as inputs was also allowed. This implies a selection of features that may differ depending on the event to be modelled. Another parameter to choose was the number of previous timestamps for each event, this value being a maximum of 32 and a minimum of 4.

Table 2. Search space for each of the parameters when training the model.

Parameters	Search Space
No. of hidden Dense/Time Distr. Dense layers	[1, 11]
No. of units of each layer	[1, 64]
Activation function of each layer	{tanh,selu}
Use of Conv. layer	{True,False}
No. of filters in Conv.	[1, 50]
Size window in Conv.	[2, 32]
Activation function of Conv layer	{selu,sigmoid}
Use of pooling layers	[True,False]
Use of LSTM layer	[True,False]
No. of units of LSTM	[1, 50]
Learning rate	{0.1, 0.01, 0.001}
Optimizer	{sgd,adam,rmsprop}
Batch size	[1, 50]

This hyperparameter search was carried out automatically using the framework called Optuna [54]. Optuna is a define-by-run API that allows users to construct the parameter search space dynamically and implements both searching and pruning strategies. Particularly, the Tree-structured Parzen Estimator (TPE) algorithm [55] was used. Thus, Optuna allowed training different models considering multiple combinations of hyperparameters. The selection of the best model configuration was carried out by the validation set. The metric considered was the area under the receiver operating characteristics curve (AUC). The discrimination ability of the final chosen models was calculated with the test set.

All data analysis and implementation of the models were performed using the Python programming language v. 3.8. [56].

3. Results

Table 3 shows the neural network architecture configuration of the best models for the 5 classification problems considered. Two of the models (stumble model and unbalanced model) selected a neural network with a convolutional layer and depth of 11 hidden layers as the best network configuration. For the stumbling problem 21 filters were chosen and 6 for the imbalance problem. For the rest of the problems, simpler architectures were chosen, namely multilayer perceptrons with one layer and 3 hidden layers (running vs.

walking model). The model that discerns between running and walking and the one that discerns between standing with movement and standing still consider the 32 previous timestamps of each sensor as input; the model that discerns between sitting and standing and the unbalance model consider the previous 16 and the stumbling model the previous 4. Regarding the inputs of the best models, the Table shows the pressure sensors (P) and acceleration measurements (A) selected. Temperature was not selected in any of the models as an explanatory variable. In the case of the pressure sensors, the superscript indicates the sensor number, whose corresponding distribution along the insole is displayed in the Figure 3. The superscript in the acceleration indicates each of the three axes of the coordinate system. It is observed that the model that discerns between moving and still and the one that discerns between running and walking includes information from two acceleration axes as input to the model, while the rest of the models are fed with information from a single acceleration axis. Regarding the pressure sensors, in general the models considered as inputs a subset of pressure sensors housed along the entire insole. It could be noted that the model that discerns between standing and sitting considers more sensors from the bottom of the foot insole (heel) as inputs than the rest of the models. This result seems reasonable since the heel is the part of the foot that tends to bear a greater difference in load when standing compared to sitting.

Table 3. Best models configuration.

	Model Standing vs. Seated	Model Moving vs. Still	Model Stumbling vs. Not Stumbling	Model Unbalanced vs. Stable	Model Running vs. Walking
No. of hidden Dense/ Time Distr. Dense layers	1	1	11	11	3
No. of units of each layer	32	42	[26,54,16,38, 46,54,18,38, 36,54,1]	[32,4,38,50 34,38,22,6 24,58,58]	[46,6,9]
Activation function of each layer	selu	tanh	[tanh,selu,selu,selu, selu,tanh,selu,selu, selu,tanh,selu]	[tanh,tanh,selu,tanh, tanh,tanh,selu,tanh, selu,tanh,selu]	[selu,tanh,tanh]
Use of Conv. layer	False	False	True	True	False
No. of filters in Conv.	-	-	21	6	-
Size window in Conv.	-	-	10	4	-
Activation function of Conv layer	-	-	selu	sigmoid	-
Use of pooling layers	False	False	False	True	False
Use of LSTM layer	False	False	False	False	False
No. of units of LSTM	-	-	-	-	-
Learning rate	0.01	0.01	0.01	0.001	0.01
Optimizer	adam	sgd	sgd	adam	adam
Batch size	33	48	21	19	47
Previous timestamps	16	32	4	16	32
Selected sensors	$\{P^0, P^1, P^2, P^5, P^6, P^7, P^{15}, A^0\}$	$\{P^3, P^4, P^6, P^8, P^{10}, P^{12}, P^{13}, A^0, A^2\}$	$\{P^0, P^1, P^4, P^7, P^{10}, P^{11}, P^{12}, P^{14}, A^1\}$	$\{P^0, P^6, P^9, P^{11}, P^{12}, P^{13}, P^{15}, A^0\}$	$\{P^1, P^3, P^4, P^6, P^8, P^{11}, P^{12}, A^0, A^2\}$

This difference in the complexity of the neural networks between the problems addressed is sensible, as a stumble or an imbalance are more difficult to detect with the information provided by the wearable sensorised device than the other behaviours (sitting, standing without imbalance, walking, running), possibly due to their greater heterogeneity. The difference in the prior information needed may lie in the type of activity, some of them involving more continuous events over time (e.g., walking and running) and others shorter ones, such as stumbles.

The evaluation metrics of the best binary models are shown in Table 4. The metrics represented are accuracy, AUC, precision or positive predictive value (PPV), recall or sensitivity, f1-score, specificity and negative predictive value (NPV):

$$Accuracy = \frac{tp + tn}{tp + tn + fp + fn} \tag{4}$$

$$Precision = \frac{tp}{tp + fp} \tag{5}$$

$$Recall = \frac{tp}{tp + fn} \tag{6}$$

$$Specificity = \frac{tn}{tn + fp} \tag{7}$$

$$NPV = \frac{tn}{tn + fn} \tag{8}$$

$$F1 - score = 2 \cdot \frac{precision \cdot recall}{precision + recall} \tag{9}$$

where tp, tn, fp and fn are the number of true positives, true negatives, false positives and false negatives, respectively. The metric AUC is the area under ROC curve, being ROC curve a graphic representation of sensibility against 1-specificity depending on the discrimination threshold.

Table 4. Metrics obtained by the best models.

	Model Standing vs. Seated	Model Moving vs. Still	Model Stumbling vs. Not Stumbling	Model Unbalanced vs. Stable	Model Running vs. Walking
Accuracy	0.99	0.98	0.78	0.91	0.96
AUC	0.98	0.98	0.77	0.91	0.95
Precision	1	0.98	0.64	0.91	0.95
Recall	1	0.99	0.75	0.89	0.91
F1-score	1	0.98	0.69	0.9	0.93
Specificity	0.97	0.98	0.79	0.93	0.98
NPV	0.97	0.99	0.86	0.91	0.97

The model that discerns between sitting (class 0) and standing (class 1) and the one that discerns between standing without movement (class 0) and with movement (class 1) obtain a high performance close to 1. Although slightly lower, the model discerning between walking (class 0) or running (class 1) also achieves a high discriminative ability. However, the model discerning between no stumbling (class 0) and stumbling (class 1) and the one discerning between no unbalance (class 0) and unbalance (class 1) perform worse, with an AUC of 0.77 and 0.91 respectively. Consequently, the metrics show that problems dealing with stumble and imbalance detection are more difficult to model. As noted above, both problems were modeled with more complex network architectures.

In general, the metrics show that the binary models achieve a good discriminative ability. However, to calculate an evaluation metric for the general problem (detection of several events) it is necessary to apply the hierarchical structure (Figure 6) of the binary models. The average result achieved is an accuracy of 0.84 and an AUC of 0.96, which also indicates a remarkable overall performance.

4. Discussion and Conclusions

Although the increase in life expectancy could be considered one of history's great success stories, it brings with it certain societal challenges that need to be addressed, such as the care of the elderly. Thanks to the advancement of technology (IoT, Big Data, Artificial Intelligence, etc.), highly innovative and attractive assistive technology (AT) products can be developed to enable a more independent, comfortable and safe life for the elderly.

The work presented is part of the result of the European project MATUROLIFE whose ultimate goal is to enable the elderly to age with the highest possible quality of life and independence through the implementation and development of an assistive technology integrated in wearable devices (clothing, furniture and footwear) in a discreet, fashionable and comfortable way. This sensorisation allows the remote monitoring of elderly people and the analysis of the large amount of data generated in order to prevent certain health problems. This article presents the prototype of the footwear (smart insole) that incorporates a total of 20 sensors that measure physical magnitudes such as temperature, pressure and acceleration. A scalable and easily modular system architecture was designed and implemented. Such architecture manages and updates the data retrieved from the smart shoes through an Android application (MaturoApp) via Bluetooth protocol, stores the information in a database on which the generated models are fed and sends a warning message via Telegram to the user's contact person (responsible person) in the event of an indication of risk or anomalous behaviour. The fundamentals of information theory were essential to enable the system of consistent communication to transmit, process, analyse data and obtain useful information in near real time. The paper focuses especially on the part of detecting different walking behaviours by analysing and modelling the data retrieved from the smart footwear using deep learning techniques.

There are several studies that have explored algorithms that achieved good performance for human activity recognition based on smart footwear and focused on providing greater independence to elderly people. De Pinho Andre et al. [43] showed an average accuracy of 93.34%, el Achkar et al. [44] achieved a total algorithm precision of 97.41%, Montanini et al. [46] achieved an accuracy of 97.1%, Zitouni et al. [45] reached 100% sensibility and more than 93% sensitivity, Light et al. [47] achieved a 88% of accuracy approximately and Sim et al. [48] a 81.5% sensitivity. However, many of the studies focused on a single event such as falling [45–48] and others detected more common events related to the user daily activity but they do not include events of inmediate risk such as unbalancing,stumbling or falling [43,44].

Our proposal allows for the identification of 6 types of representative gait events that include events of immediate interest such as stumbling and imbalance and other more common events such as sitting, standing, walking or running. In the data collection process, some flexibility was allowed for in the conduct of these events. The criterion to consider these events was motivated in order to monitor and prevent possible user health problems by considering a wide range of possible user behaviours, both more immediate hazard events and more common everyday activity events that may also indicate abnormal behaviour depending on the patient and even the time of day. For example, although the event "walking" may be a completely normal event during the day, the detection of such an event at night may indicate abnormal behaviour of the patient at that time of the day when he should be sleeping. This may indicate disorientation and possible danger if prolonged over time.

Artificial neural network techniques and algorithms capable of detecting these events, which may be related to health problems such as disorientation, loss of control, among others, were explored and applied. In particular, 5 binary models were generated for the detection of such events through a hierarchical cascade structure. This cascade structure was designed so that the system could be easily modulated, allowing, for example, to be re-created with new binary models for the detection of more particular events, keeping the rest of the models or substituting only some of them. Optimisation in the training of the models allowed a choice of built-in sensors as inputs. The best models, which included different sensors as inputs, were stored sorted by performance. The aim of this was to ensure that the event detection system was always kept alive even in circumstances where a certain sensor stopped working, which may be possible in practice. Thus, if a sensor fails, the system uses the best model that does not use information from that sensor as input.

The results showed a high overall discrimination ability, reaching an average accuracy and AUC of 0.84 and 0.96, respectively. The worst performing binary models were those detecting stumble and imbalance with an AUC value of 0.77 and 0.91, respectively. This may be due to the fact that they are more difficult events to model than the others, as they are less constant and possibly more heterogeneous walking behaviours. This complexity is also reflected in the selected network architecture, where these models follow a network architecture with convolutional layers and considerable depth (11 hidden layers), while the rest of the models are multilayer perceptrons with 1 or 3 hidden layers.

Thanks to the fundamentals of information theory and the combination of different technologies such as sensing techniques, data acquisition and analytics, machine learning and deep learning, it is possible to improve the state of the art and develop new sensors and smarter systems. This is achieved by integrating intelligence techniques and deployment in wearables and related edge computing, where all related phases take place inside the sensorised device. This is the case of the work we present, which focuses on a wearable smart footwear comprising a scalable, easily modulated and live system that allows, through artificial neural network modelling, to detect with high accuracy a wide heterogeneity of walking behaviours and to warn the relatives or healthcare professionals about anomalous user behaviours so that they can act quickly. This system was designed in such a way that it can be implemented on any current embedded system with a lower CPU.

However, this work has some limitations and future work to consider. Due to the COVID19 pandemic in which we are immersed and the timelines set in the project, data collection by the end users (elderly population) was not possible. Consequently, data collection had to be carried out by the technical team with only 3 subjects of approximately the same age and weight. Other related studies included more heterogeneity in this regard and some involved elderly population in their experiments. De Pinho Andre et al. [43] used data from 11 subjects, two of whom were elders, el Alchkar et al. [44] involved ten elderly subjects (8 men, 2 women, age 65–75 years, weight 62–114 kg, height 162–184 cm), and Montanini et al. [46] conducted laboratory tests with 17 healthy subjects and demonstrated the effectiveness of their method with two older users in a real-life setting. Zitouni et al. [45] involved six subjects between 25 and 30 years of age, Light et al. [47] collected data from subjects aged between 20 and 45 years and Sim et al. [48] three young subjects (2 young males, 1 young female, aged between 24 and 28 years).

As future work, we propose to validate our system by including greater heterogeneity in the data, incorporating information from elderly population of different ages, weights and physical shape and in different environments where humidity, external temperature or the relief of the terrain may have an effect on the measurements. Thanks to the scalability and modularity that the designed system allows, another of the future lines of work to be explored could be to include clustering modules with the aim of grouping behaviours and applying specific models to each group. The application of modules for detecting changes in user behaviour (trend models) or the inclusion of more specific event models such as fall detection could also be studied.

Author Contributions: Conceptualization: R.A.-G., G.L.-L., A.A.-L. and D.A.-G.; Data Curation: R.A.-G., G.L.-L. , A.A.-L. and D.A.-G.; Formal analysis: R.A.-G., G.L.-L. and A.A.-L.; Investigation: R.A.-G., G.L.-L. and A.A.-L.; Methodology: R.A.-G., G.L.-L. and A.A.-L.; Software: R.A.-G., G.L.-L., A.A.-L. and C.G.-M.; Supervision: C.G.-M., D.A.-G. and R.d.-H.-A.; Validation: R.A.-G., G.L.-L. and A.A.-L.; Writing—original draft: R.A.-G., G.L.-L., A.A.-L., D.A.-G. and C.G.-M.; Writing—review & editing: R.A.-G., G.L.-L., A.A.-L., D.A.-G., R.d.-H.-A. and C.G.-M. All authors have read and agreed to the published version of the manuscript.

Funding: This publication is based on work performed in MATUROLIFE (Metallisation of Textiles to make Urban Living for Older People more Independent and Fashionable), which has received funding from the European Union's Horizon 2020 research and innovation programme under Grant Agreement No. 760789. The work has also been partially funded by the European Social Fund.

Institutional Review Board Statement: Not applicable.

Informed Consent Statement: Informed consent was obtained from all subjects involved in the study.

Acknowledgments: The authors wish to sincerely thank all the consortium members of european MATUROLIFE project who provided the technology and the support to enable the hereby described experiments, specially the ones directly involved in the development of the shoe prototypes:

- Coventry University -https://www.coventry.ac.uk/- (accessed on 10 May 2021) for the Maturolife project leadership and the aplication of the state of the art technologies for Selective Catalysation and Metallisation of Fabrics and Textiles.
- CTCR -https://www.ctcr.es/en- (accessed on 10 May 2021) for their work in the integration of the hardware and software components needed to seamlessly transfer information from the sensorized devices and specially for their support during the experiments.
- Sensing Tex -http://sensingtex.com- (accessed on 10 May 2021) for the integration of their innovative Pressure Sensing Mat solutions in the field of smart textiles based on printed electronics.
- Printed Electronics -https://www.printedelectronics.com/- (accessed on 10 May 2021) for their work in the analysis and implementation of printing methods and the selection of compatible material to interconnect with the SensingTex sensing mat solutions.
- Pitillos(r) -https://www.calzadospitillos.com/en/- (accessed on 10 May 2021) for the manufacturing of the shoes and their knowledge and experience in the design of usable models to house the above described technologies.

Any dissemination reflects the authors' view only and the European Commission is not responsible for any use that may be made of the information it contains. The views and opinions expressed in this paper are those of the authors and are not intended to represent the position or opinions of the MATUROLIFE consortium or any of the individual partner organisations.

Conflicts of Interest: The authors declare no conflict of interest.

References

1. Eurostat. Ageing Europe. Looking at the Lives of Older People in the EU. 2020. Available online: https://ec.europa.eu/eurostat/web/products-statistical-books/-/ks-02-20-655 (accessed on 3 May 2021).
2. Majumder, S.; Aghayi, E.; Noferesti, M.; Memarzadeh-Tehran, H.; Mondal, T.; Pang, Z.; Deen, M.J. Smart homes for elderly healthcare—Recent advances and research challenges. *Sensors* **2017**, *17*, 2496. [CrossRef]
3. Deen, M.J. Information and communications technologies for elderly ubiquitous healthcare in a smart home. *Pers. Ubiquitous Comput.* **2015**, *19*, 573–599. [CrossRef]
4. Communication from the Commission to the European Parliament, the Council, the European Economic and Social Committee and the Committee of the Regions, European Disability Strategy 2010–2020: A Renewed Commitment to a Barrier-Free Europe. Available online: https://eur-lex.europa.eu/legal-content/EN/TXT/?uri=celex%3A52010DC0636 (accessed on 3 May 2021).
5. Canjuga, I.; Železnik, D.; Neuberg, M.; Božicevic, M.; Cikac, T. Does an impaired capacity for self-care impact the prevalence of social and emotional loneliness among elderly people? *Work. Older People* **2018**, *22*, 211–223. [CrossRef]
6. Borg, C.; Hallberg, I.R.; Blomqvist, K. Life satisfaction among older people (65+) with reduced self-care capacity: The relationship to social, health and financial aspects. *J. Clin. Nurs.* **2006**, *15*, 607–618. [CrossRef] [PubMed]
7. Kahya, N.C.; Zorlu, T.; Ozgen, S.; Sari, R.M.; Sen, D.E.; Sagsoz, A. Psychological effects of physical deficiencies in the residences on elderly persons: A case study in Trabzon Old Person's Home in Turkey. *Appl. Ergon.* **2009**, *40*, 840–851. [CrossRef] [PubMed]
8. Brookes, N.; Palmer, S.; Callaghan, L. I live with other people and not alone: A survey of the views and experiences of older people using Shared Lives (adult placement). *Work. Older People* **2016**, *20*, 179–186. [CrossRef]
9. Balta-Ozkan, N.; Davidson, R.; Bicket, M.; Whitmarsh, L. Social barriers to the adoption of smart homes. *Energy Policy* **2013**, *63*, 363–374. [CrossRef]
10. De Silva, L.C.; Morikawa, C.; Petra, I.M. State of the art of smart homes. *Eng. Appl. Artif. Intell.* **2012**, *25*, 1313–1321. [CrossRef]
11. Lutolf, R. Smart Home Concept and the Integration of Energy Meters into a Home Based System. In Proceedings of the Seventh International Conference on Metering Apparatus and Tariffs for Electricity Supply, Glasgow, UK, 17–19 November 1992; pp. 277–278.
12. Aldrich, F.K. Smart Homes: Past, Present and Future. In *Inside the Smart Home*; Springer: London, UK, 2006; pp. 17–39.
13. Shi, W.V. A survey on assistive technologies for elderly and disabled people. *J. Mechatron.* **2015**, *3*, 121–125. [CrossRef]
14. Troncone, A.; Saturno, R.; Buonanno, M.; Pugliese, L.; Cordasco, G.; Vogel, C.; Esposito, A. Advanced Assistive Technologies for Elderly People: A Psychological Perspective on Older Users' Needs and Preferences (Part B). *Acta Polytech. Hung.* **2021**, *18*, 29–44. [CrossRef]
15. Peek, S.T.M.; Wouters, E.J.M.; van Hoof, J.; Luijkx, K.G.; Boeije, H.R.; Vrijhoef, H.J.M. Factors influencing acceptance of technology for aging in place: A systematic review. *Int. J. Med. Inform.* **2014**, *83*, 235–248. [CrossRef]
16. Jo, T.H.; Ma, J.H.; Cha, S.H. Elderly Perception on the Internet of Things-Based Integrated Smart-Home System. *Sensors* **2021**, *21*, 1284. [CrossRef] [PubMed]
17. Yusif, S.; Soar, J.; Hafeez-Baig, A. Older people, assistive technologies, and the barriers to adoption: A systematic review. *Int. J. Med. Inform.* **2016**, *94*, 112–116. [CrossRef]
18. Davies, K.N.; Mulley, G.P. The views of elderly people on emergency alarm use. *Clin. Rehabil.* **1993**, *7*, 278–282. [CrossRef]
19. Shannon, C.E. A mathematical theory of communication. *Bell Syst. Tech. J.* **1948**, *27*, 379–423. [CrossRef]

20. Bevilacqua, A.; MacDonald, K.; Rangarej, A.; Widjaya, V.; Caulfield, B.; Kechadi, T. Human Activity Recognition with Convolutional Neural Networks. In *Joint European Conference on Machine Learning and Knowledge Discovery in Databases* ; Lecture Notes in Computer Science; Springer: Cham, Switzerland, 2018; Volume 11053.

21. Domínguez-Morales, M.J.; Luna-Perejón, F.; Miró-Amarante, L.; Hernández-Velázquez, M.; Sevillano-Ramos, J.L. Smart footwear insole for recognition of foot pronation and supination using neural networks. *Appl. Sci.* **2019**, *9*, 3970. [CrossRef]

22. Wang, J.; Chen, Y.; Hao, S.; Peng, X.; Hu, L. Deep learning for sensor-based activity recognition: A survey. *Pattern Recognit. Lett.* **2019**, *119*, 3–11. [CrossRef]

23. Papagiannaki, A.; Zacharaki, E.I.; Kalouris, G.; Kalogiannis, S.; Deltouzos, K.; Ellul, J.; Megalooikonomou, V. Recognizing physical activity of older people from wearable sensors and inconsistent data. *Sensors* **2019**, *19*, 880. [CrossRef]

24. Shi, Q.; Zhang, Z.; He, T.; Sun, Z.; Wang, B.; Feng, Y.; Shan, X.; Salam, B.; Lee, C. Deep learning enabled smart mats as a scalable floor monitoring system. *Nat. Commun.* **2020**, *11*, 1–11. [CrossRef]

25. Lee, S.S.; Choi, S.T.; Choi, S.I. Classification of gait type based on deep learning using various sensors with smart insole. *Sensors* **2019**, *19*, 1757. [CrossRef] [PubMed]

26. Maitre, J.; Bouchard, K.; Bertuglia, C.; Gaboury, S. Recognizing activities of daily living from UWB radars and deep learning. *Expert Syst. Appl.* **2021**, *164*, 113994. [CrossRef]

27. Scataglini, S.; Moorhead, A.P.; Feletti, F. A Systematic Review of Smart Clothing in Sports: Possible Applications to Extreme Sports. *Muscles Ligaments Tendons J. MLTJ* **2020**, *10*, 333–342. [CrossRef]

28. Shiang, T.Y.; Hsieh, T.Y.; Lee, Y.S.; Wu, C.C.; Yu, M.C.; Mei, C.H.; Tai, I.H. Determine the foot strike pattern using inertial sensors. *J. Sens.* **2016**. [CrossRef]

29. Moore, S.R.; Kranzinger, C.; Fritz, J.; Stöggl, T.; Kröll, J.; Schwameder, H. Foot strike angle prediction and pattern classification using loadsoltm wearable sensors: A comparison of machine learning techniques. *Sensors* **2020**, *20*, 6737. [CrossRef]

30. Sazonov, E.S.; Fulk, G.; Hill, J.; Schutz, Y.; Browning, R. Monitoring of posture allocations and activities by a shoe-based wearable sensor. *IEEE Trans. Biomed. Eng.* **2010**, *58*, 983–990. [CrossRef] [PubMed]

31. European Maturolife Project Website. Available online: http://maturolife.eu (accessed on 3 May 2021).

32. Moody, L.; York, N.; Ozkan, G.; Cobley, A. Bringing assistive technology innovation and material science together through design. In *Innovation in Medicine and Healthcare Systems, and Multimedia*; Springer: Singapore, 2019; pp. 305–315.

33. Moody, L.; Cobley, A.J. MATUROLIFE: Using Advanced Material Science to Develop the Future of Assistive Technologies. In *Design of Assistive Technology for Ageing Populations*; Springer: Berlin/Heidelberg, Germany, 2019; Volume 167, p. 189.

34. Yang, D.; Moody, L.; Cobley, A. Integrating Cooperative Design and Innovative Technology to Create Assistive Products for Older Adults. In Proceedings of the International Association of Societies of Design Research Conference 2019: DESIGN REVOLUTIONS, Manchester, UK, 2–5 September 2019.

35. Callari, T.C.; Moody, L.; Magee, P.; Yang, D.; Ozkan, G.; Martinez, D. MATUROLIFE. Combining Design Innovation and Material Science to Support Independent Ageing. In *Design Journal*; Taylor & Francis: Dundee, UK, 10–13 April 2019; pp. 2161–2162.

36. Callari, T.C.; Moody, L.; Magee, P.; Yang, D. 'Smart—not only intelligent' Co-creating priorities and design direction for 'smart' footwear to support independent ageing. *Int. J. Fash. Des. Technol. Educ.* **2019**, *12*, 313–324. [CrossRef]

37. Hegde, N.; Bries, M.; Sazonov, E. A comparative review of footwear-based wearable systems. *Electronics* **2016**, *5*, 48. [CrossRef]

38. Ma, X.; Wang, H.; Xue, B.; Zhou, M.; Ji, B.; Li, Y. Depth-based human fall detection via shape features and improved extreme learning machine. *IEEE J. Biomed. Health Inform.* **2014**, *18*, 1915–1922. [CrossRef]

39. Kangas, M.; Konttila, A.; Lindgren, P.; Winblad, I.; Jämsä, T. Comparison of low-complexity fall detection algorithms for body attached accelerometers. *Gait Posture* **2008**, *28*, 285–291. [CrossRef] [PubMed]

40. Li, Q.; Stankovic, J.A.; Hanson, M.A.; Barth, A.T.; Lach, J.; Zhou, G. Accurate, fast fall detection using gyroscopes and accelerometer-derived posture information. In *Sixth International Workshop on Wearable and Implantable Body Sensor Networks*; IEEE: Washington, DC, USA, June 2009; pp. 138–143.

41. Tao, Y.; Qian, H.; Chen, M.; Shi, X.; Xu, Y. A Real-time intelligent shoe system for fall detection. In Proceedings of the 2011 IEEE International Conference on Robotics and Biomimetics, Karon Beach, Thailand, 7–11 December 2011; pp. 2253–2258.

42. Santos, G.L.; Endo, P.T.; Monteiro, K.H.D.C.; Rocha, E.D.S.; Silva, I.; Lynn, T. Accelerometer-based human fall detection using convolutional neural networks. *Sensors* **2019**, *19*, 1644. [CrossRef] [PubMed]

43. De Pinho André, R.; Diniz, P.; Fuks, H. Bottom-up Investigation: Human Activity Recognition Based on Feet Movement and Posture Information. In Proceedings of the 4th International Workshop on Sensor-Based Activity Recognition and Interaction, Rostock, Germany, 21–22 September 2017; pp. 1–6.

44. el Achkar, C.M.; Lenoble-Hoskovec, C.; Paraschiv-Ionescu, A.; Major, K.; Büla, C.; Aminian, K. Instrumented shoes for activity classification in the elderly. *Gait Posture* **2016**, *44*, 12–17. [CrossRef] [PubMed]

45. Zitouni, M.; Pan, Q.; Brulin, D.; Campo, E. Design of a smart sole with advanced fall detection algorithm. *J. Sens. Technol.* **2019**, *9*, 71. [CrossRef]

46. Montanini, L.; Del Campo, A.; Perla, D.; Spinsante, S.; Gambi, E. A footwear-based methodology for fall detection. *IEEE Sens. J.* **2017**, *18*, 1233–1242. [CrossRef]

47. Light, J.; Cha, S.; Chowdhury, M. Optimizing pressure sensor array data for a smart-shoe fall monitoring system. In Proceedings of the 2015 IEEE SENSORS, Busan, Korea, 1–4 November 2015; pp. 1–4.

48. Sim, S.Y.; Jeon, H.S.; Chung, G.S.; Kim, S.K.; Kwon, S.J.; Lee, W.K.; Park, K.S. Fall detection algorithm for the elderly using acceleration sensors on the shoes. In Proceedings of the 2011 Annual International Conference of the IEEE Engineering in Medicine and Biology Society, Boston, MA, USA, 30 August–3 September 2011; IEEE: Boston, MA, USA, August 2011; pp. 4935–4938.
49. Wirth, R.; Hipp, J. CRISP-DM: Towards a standard process model for data mining. In Proceedings of the 4th International Conference on the Practical Applications of Knowledge Discovery and Data Mining (Vol. 1), London, UK, 11 April 2000; Springer: Berlin/Heidelberg, Germany, 2000.
50. Kubernetes. Available online: https://kubernetes.io/ (accessed on 4 May 2021).
51. Telegram Messenger. Available online: https://telegram.org (accessed on 3 May 2021).
52. MQTT—The Standard for IoT Messaging. Available online: https://mqtt.org/ (accessed on 3 May 2021).
53. InfluxDB Time Series Platform | InfluxData. Available online: https://www.influxdata.com/products/influxdb/ (accessed on 3 May 2021).
54. Akiba, T.; Sano, S.; Yanase, T.; Ohta, T.; Koyama, M. Optuna: A next-generation hyperparameter optimization framework. In Proceedings of the 25th ACM SIGKDD International Conference on Knowledge Discovery & Data Mining, Anchorage, AK, USA, 4–8 August 2019; Association for Computing Machinery: New York, NY, USA, July 2019; pp. 2623–2631.
55. Bergstra, J.; Bardenet, R.; Bengio, Y.; Kégl, B. Algorithms for hyper-parameter optimization. In Proceedings of the 25th Annual Conference on Neural Information Processing Systems; Granada, Spain, 12–17 December 2011; Neural Information Processing Systems Foundation, Inc. (NIPS); Volume 24.
56. The Python Tutorial. Available online: https://docs.python.org/3/tutorial/ (accessed on 4 May 2021).

Article

Driving Risk Assessment Using Near-Miss Events Based on Panel Poisson Regression and Panel Negative Binomial Regression

Shuai Sun [1], Jun Bi [1,*], Montserrat Guillen [2,*] and Ana M. Pérez-Marín [2]

[1] Key Laboratory of Transport Industry of Big Data Application Technologies for Comprehensive Transport, School of Traffic and Transportation, Beijing Jiaotong University, Beijing 100044, China; sunshuai@bjtu.edu.cn

[2] Department of Econometrics, Riskcenter-IREA, Universitat de Barcelona, 08034 Barcelona, Spain; amperez@ub.edu

* Correspondence: bilinghc@163.com (J.B.); mguillen@ub.edu (M.G.); Tel.: +86-13488812321 (J.B.); +34-934037039 (M.G.)

Abstract: This study proposes a method for identifying and evaluating driving risk as a first step towards calculating premiums in the newly emerging context of usage-based insurance. Telematics data gathered by the Internet of Vehicles (IoV) contain a large number of near-miss events which can be regarded as an alternative for modeling claims or accidents for estimating a driving risk score for a particular vehicle and its driver. Poisson regression and negative binomial regression are applied to a summary data set of 182 vehicles with one record per vehicle and to a panel data set of daily vehicle data containing four near-miss events, i.e., counts of excess speed, high speed brake, harsh acceleration or deceleration and additional driving behavior parameters that do not result in accidents. Negative binomial regression ($AIC_{overspeed} = 997.0$, $BIC_{overspeed} = 1022.7$) is seen to perform better than Poisson regression ($AIC_{overspeed} = 7051.8$, $BIC_{overspeed} = 7074.3$). Vehicles are separately classified to five driving risk levels with a driving risk score computed from individual effects of the corresponding panel model. This study provides a research basis for actuarial insurance premium calculations, even if no accident information is available, and enables a precise supervision of dangerous driving behaviors based on driving risk scores.

Keywords: driving risk assessment; usage-based insurance; driving risk score; telematics; near-miss event; driving behavior; panel data analysis; count data model; econometrics; generalized linear model

Citation: Sun, S.; Bi, J.; Guillen, M.; Pérez-Marín, A.M. Driving Risk Assessment Using Near-Miss Events Based on Panel Poisson Regression and Panel Negative Binomial Regression. *Entropy* **2021**, *23*, 829. https://doi.org/10.3390/e23070829

Academic Editors: Emilio López Cano and Felipe Ortega

Received: 10 May 2021
Accepted: 24 June 2021
Published: 29 June 2021

Publisher's Note: MDPI stays neutral with regard to jurisdictional claims in published maps and institutional affiliations.

1. Introduction

Near-miss events are incidents that denote the existence of danger, even if no accident occurs. Reporting of near-miss events is an established error reduction technique that has been used by many industries to manage risk and reduce accidents. In the auto insurance industry, insurers traditionally calculate premiums by analyzing past claims reported by the insured policy holders, and reward those drivers that do not report accidents with a no-claims bonus. However, this may be a rather incorrect approach to the assessment of accident risk, especially when the insured has suffered accidents but chooses not to make a claim so as not to lose the no-claims bonus. Fortunately, the advent of the Internet of Vehicles (IoV) offers a better solution to this problem, using near-miss events to identify driving risk. Near-miss events ultimately provide information that can lead to actuarial premium calculations in the auto insurance industry [1,2].

This study explores how to evaluate driving risks, in the short term, and to score drivers without claims and accidents based on information on near-miss counts over a short period of time. One of the main novelties of this approach, in the absence of claims, is to use telematics sensors for observation of drivers over a given period. The model obtained in this study offers an important alternative for driving risk identification. Not only can the model reflect risk factors that influence each near-miss event but it can also help to evaluate drivers' risks, and fixed-effects panel count data models can be used to rank drivers according to their individual effects. The modeling method and results are invaluable for insurance companies for developing usage-based insurance (UBI) to personalize premiums. They are also of interest to traffic regulatory authorities for promoting safe driving and the prevention of accidents.

Near-miss events are incidents that need to be defined and extracted from the original raw data files for further processing and analysis. By dealing only with near-miss events, and excluding claims or accidents, this study aims to specifically identify driving patterns. This study is carried out both on a per driver summary data set and on a panel data set where a daily summary is shown for each driver. Our data contain counts of the four types of near-miss events in our study. Speeding, high speed braking, harsh acceleration and harsh deceleration have been defined based on actual driving conditions and local laws and regulations. Other high-risk events, e.g., sharp turning, dangerous lane changing and unexpected maneuvers, proved by previous studies to be related to driving risk, are not included in this study due to the dimension and precision limitations of the original data set.

Our interest is to model the frequency of near-miss events given the drivers' characteristics. The simplest statistical model that links a count data dependent variable with explanatory factors is the Poisson model. Essentially, the Poisson model is similar to linear regression, where a response depends on some others inputs. Here we think that distance driven or mean speed among others, influence the expected frequency of near-miss events. A Poisson model, which is also known as a Poisson regression model, is easily interpretable and provides a way to elucidate the significant effects on the conditional expected frequency. Poisson models are constrained by the fact that conditional expectation and conditional variance are equal. Negative binomial regression models are a natural extension that overcomes this restriction. More details on the models are provided in the Methods section below.

Since the extracted frequency of near-miss events is an unbounded non-negative integer, Poisson regression and negative binomial regression are both suitable for modelization. Poisson regression, negative binomial regression, zero-inflated Poisson regression and zero-inflated negative binomial regression are respectively applied to the summary data set. Average speed, brake times, accelerator pedal position, engine fuel rate etc., are selected as independent variables. Either mileage or fuel consumption can be chosen as the exposure variable to offset the model. In order to reach a clear understanding of risk factors of different near-miss events, each near-miss event is individually used as a dependent variable. However, regardless of which one is selected as the dependent variable, negative binomial regression is shown to provide the best fit in the summary data in this study.

Negative binomial regression also performs better than Poisson regression on the panel data sets. Individual effects and time effects are estimated using panel Poisson regression and panel negative binomial regression on a short panel data set of six days in length. The regression results confirm the existence of individual effects and time effects, and also enable the driving risk of each vehicle to be ranked. The driving risk level of vehicles can then be classified by converting the individual effects into scores, thus providing an important reference for further accurate calculation of premiums.

The rest of this article is organized as follows. The development of UBI and previous efforts on driving risk assessment are summarized in Section 2. Section 3 describes the data and introduces the key parameters used in modeling. Section 4 presents the model expression of Poisson regression and negative binomial regression used in the study. The results of negative binomial regression using the summary data set and the panel data set are reported and analyzed in Section 5. The results are discussed and the conclusions are presented in Section 6.

2. Literature Review

The auto insurance industry is continuously pursuing new ways to calculate more accurate actuarial premiums. However, traditional auto insurance calculations are limited by the difficulty of obtaining information on policy holders, so classical ratemaking uses simple information on drivers (age gender,), vehicles (type of car, model and brand) and driving sections [3]. With current advances in information technology, a new type of insurance business, UBI, based on multi-source data and personalized premium calculation is becoming the mainstream. The Pay-as-you-drive (PAYD) mode of charging premiums is based on mileage or fuel consumption, on the premise that mileage or fuel consumption correlates with the probability of suffering an accident [4]. PAYD has evolved into a newer scheme, called the pay-how-you-drive (PHYD) ratemaking mode, which is based on multiple sources of data, including driving behavior data [5]. Following the development of 5G communication technology, it may now be possible to implement an even more sophisticated monitoring and pricing strategy, known as the manage-how-you-drive (MHYD) principle, i.e., real-time calculation of premiums based on multi-source data and providing real-time information to drivers to restrain from bad driving behavior [3,6]. However, due to technological, regulatory and other issues regarding privacy [7], there is still no mature PHYD product on the market at present [8,9] and, in terms of MHYD, further research is necessary on driving risk to produce products that better reflect the driver profile [10].

Traffic accidents all over the world result in a large number of casualties every year, and high-risk driving is one of the main factors behind these incidents [3]. Consequently, research on driving risk has been a topic of interest over recent decades. Simulation experiments to evaluate driving risk have been designed in the laboratory setting to identify driving risk factors [11–14] as well as experiments using actual vehicles on the road [15–19]. Questionnaire surveys for driving risk assessment have also been studied [20,21]. In fact, the naturalistic type of driving data collected by the IoV or smart phones, known as telematics data, can effectively reduce the influence of subjective factors and unreasonable assumptions in producing effective risk-mitigating actions [22–26].

In research related to driving risk assessment in the auto insurance industry, machine learning and generalized linear models feature equally. Machine learning, with its strong ability to process big data efficiently, is increasingly gaining ground in its application in the auto insurance business. Logistic regression [27], cluster analysis [28], decision tree [5], support vector machine [29], neural network [30] and other machine learning models [31–33] have been widely studied in the field of driving risk assessment, and the results have shown machine learning to be a powerful tool [34]. However, since most machine learning procedures, being black box algorithms, do not offer a high degree of interpretability, they cannot completely replace the conventional generalized linear models implemented for decades in the auto insurance industry [8].

Conventional generalized linear models discern the correlation between influencing factors and claims or accidents in frequency and severity models [9,24,25,35]. However, the study of near-miss events even when there is a lack of information on claims and accidents should not be ignored [2,15]; on the contrary, since near-misses are more frequent than accidents and are positively associated with them, they can be considered a good alternative for risk modeling for driving risk assessment [1]. Compared with previous studies, this study not only conducts regression on the summary data set to model and analyze the factors causing near-miss events, but also conducts panel data regression on the panel data set to consider individual effects and time effects. The regression results can not only make more accurate causal inference, but also carry out risk scoring.

3. Data Description

The telematics data used in this study are collected from an IoV information service provider in China. While we cannot obtain more data due to the commercial privacy of the data, the limited data also contains valuable driving risk information, which is worth studying. The original data set contains 182 data files, representing sensor data for 182 vehicles observed from 3–8 July 2018 [10]. Each data file contains 62 different measurements but, after data processing [36], less than one-third of them can be used due to recording errors and inconsistencies. The original data are transformed for modeling into a summary data set with information on each driver (see details in Table 1).

Table 1. Descriptive statistics of the summary data set for 182 drivers observed from 3–8 July 2018.

Variable	Mean	Standard Deviation	Minimum	Median	Maximum	Defination
overspeed	19.19	45.37	0	0	330	Frequency of driving speed greater than 100 km/h
highspeedbrake	44.23	108.3	0	4	942	Frequency of braking when the driving speed is greater than 90 km/h
harshacceleration	139.0	134.7	0	101	899	Frequency of cases when acceleration is greater than $6~\mathrm{m/s^2}$
harshdeceleration	141.9	137.8	1	105	913	Frequency of cases when acceleration is less than $6~\mathrm{m/s^2}$
kilo	2223	1674	3.73	1832.175	7164	Total driving distance (km)
fuel	621.7	470.9	10.25	487.295	2018	Total fuel consumption (L)
brakes	1588	1426	6	1138.5	9243	Total number of brakes
range	5.201	5.021	0.027	3.399	26.78	Range of driving (geographical units)
speed	36.88	16.37	0.297	36.657	67.84	Mean of speed (km/h)
rpm	1028	188.3	233.1	1009.301	1620	Mean of revolutions per minute (r/min)
acceleratorpedalposition	21.05	7.110	0.187	21.26	39.29	Mean of acceleration pedal position (%)
enginefuelrate	11.52	4.464	1.868	11.203	22.01	Mean of engine fuel rate (%)

The number of each parameter is 182.

The variables overspeed, highspeedbrake, harshacceleration and harshdeceleration are individually filtered by combining the rules of traffic law and driving code. Previous studies have confirmed that speeding is a dangerous driving behavior which is likely to cause traffic accidents [3]. In China, traffic safety regulations stipulate a maximum speed for each type of vehicle on all types of roads. The maximum speed limit for the vehicles in this study is 90 km/h; exceeding this by 10% is not deemed to be a traffic offense. Therefore, 100 km/h is taken as the threshold value of the overspeed near-miss event. Another high risk near-miss event that deserves attention is that of emergency braking; at high speed (>90 km/h), if the brake is not used correctly or is subjected to lateral force, the car is prone to side-slip or even cartwheel. Lastly, both harsh acceleration and harsh deceleration are near-miss events that compromise driving safety and fuel economy. Based on previous research experience [1,2,37] and the filter analysis of the extreme values of this data set by box graph method, $6~\mathrm{m/s^2}$ is determined as the filtering threshold value of harsh acceleration and harsh deceleration. Figure 1 shows that near-miss events are all non-negative integers. Combined with the relationship between expectation and variance shown in Table 1, the four near-miss events are shown to be suitable as dependent variables of a Poisson regression or a negative binomial regression.

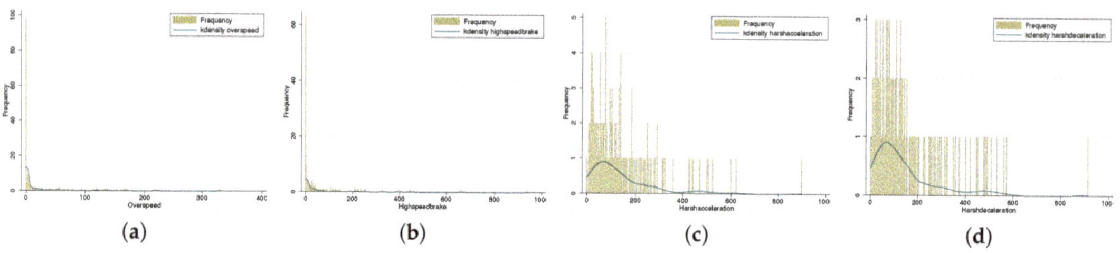

Figure 1. Histogram of frequency distribution of four near-miss events: (**a**) Over speed; (**b**) High speed brake; (**c**) Harsh acceleration; (**d**) Harsh deceleration.

The panel data set has one summary per day for each driver. The statistics of the panel data set are shown in Table 2.

Table 2. Descriptive statistics of a panel data set for 182 drivers observed over six days (total cases 1092).

Variable	N	Mean	Standard Deviation	Minimum	Median	Maximum
overspeed	1092	3.199	14.37	0	0	315
highspeedbrake	1092	7.435	21.74	0	0	215
harshacceleration	1092	23.37	29.78	0	14	223
harshdeceleration	1092	23.86	30.16	0	13.5	233
kilo	1092	372.6	373.2	0	263.24	1739
fuel	1092	104.1	105.7	0	72.15	565.8
brakes	1092	264.7	291.0	0	178	1940
range	1092	2.406	2.963	0	1.243	14.07
speed	1092	31.96	21.58	0	31.514	77.74
rpm	1092	894.3	346.9	0	973.714	1731
acceleratorpedalposition	1092	17.51	10.19	0	18.613	45.74
enginefuelrate	1092	9.794	5.835	0	10.018	26.18

4. Methods

Poisson regression is a generalized linear model. Negative binomial regression can be considered as a generalization of Poisson regression with overdispersion of the dependent variable Y_i where subindex i refers to the i-th observation in the data set. The probability density function of the Poisson distribution is:

$$P(Y_i = y_i \mid x_i) = \frac{e^{-\lambda_i} \lambda_i^{y_i}}{y_i!} \tag{1}$$

where λ_i is the Poisson arrival rate and is determined by explanatory variable x_i in Poisson regression to represent the average number of events, which is equal to the expectation and variance of the explained variable $E(Y_i \mid x_i) = V(Y_i \mid x_i) = \lambda_i$.

The negative binomial distribution is a mixture of a Poisson (λ) and a Gamma (a,b) distribution. The probability density function of the negative binomial distribution is:

$$f(y \mid a, b) = \int_0^\infty f(y \mid \lambda) g(\lambda \mid a, b) d\lambda = \frac{\Gamma(y+a)}{\Gamma(y+1)\Gamma(a)} \left(\frac{b}{1+b}\right)^a \left(\frac{1}{1+b}\right)^y \tag{2}$$

where λ is the mean and variance of the Poisson distribution, a is the shape parameter of the Gamma distribution, b is the inverse scale parameter of the Gamma distribution, $E(y) = \frac{a}{b} = \bar{\lambda}$ and $V(y) = \frac{a}{b}\left(1 + \frac{1}{b}\right) = \bar{\lambda}\left(1 + \frac{\bar{\lambda}}{a}\right)$.

The zero-inflated model is applicable when the counting data contains a large number of zero values. Theoretically, it is a two-stage decision. First, it decides whether to choose zero or a positive integer, and then it determines which positive integer to choose. Therefore, the probability distribution of Y_i is a mixed distribution:

$$Pr(Y_i = y_i \mid x_i) = \begin{cases} \theta + (1 - \theta) P(K_i = y_i \mid x_i) & y_i = 0 \\ (1 - \theta) P(K_i = y_i \mid x_i) & y_i > 0 \end{cases} \tag{3}$$

where θ is the probability of an extra zero value, K_i can follow a Poisson distribution or a negative binomial distribution depending on the characteristics of the dependent variable.

The conditional expectation function of a negative binomial regression model depends on a vector of explanatory variables x_i and, similar to Poisson, is usually defined by a log-link as:

$$E(y_i \mid x_i) = \lambda_i = T_i \times \exp(\alpha + \beta_1 x_{1i} + \cdots + \beta_k x_{ki}) \tag{4}$$

where i is the number of the observation, k depends on the number of independent variables, T_i denotes the offset variables (so, in our application, $kilo_i$ or $fuel_i$ is the exposure variable), $x_{1i} \ldots x_{ki}$ represent the independent variables such as $brakes_i$, $range_i$, $speed_i$, rpm_i,

$acceleratorpedalposition_i$ and $enginefuelrate_i$, α and $\beta_1 \ldots \beta_k$ are unknown parameters that need to be estimated.

The two-way fixed effect model of panel Poisson regression and panel negative binomial regression is specified as:

$$E(y_{it} \mid x_{it}) = \lambda_{it} = T_{it} \times \exp(\alpha + \beta_1 x_{1it} + \cdots + \beta_k x_{kit} + d_i + p_t) \tag{5}$$

where i is the number of the observation, t is of time reference, k depends on the number of independent variables, T_{it} is the offset and equals $kilo_{it}$ or $fuel_{it}$ as the exposure variable of the ith observation at time t, $x_{1it} \ldots x_{kit}$ represent the independent variables of the ith observation at time t such as $brakes_{it}$, $range_{it}$, $speed_{it}$, rpm_{it}, $acceleratorpedalposition_{it}$ and $enginefuelrate_{it}$, α and $\beta_1 \ldots \beta_k$ are unknown parameters that need to be estimated, d_i represents the individual effect and p_t represents the time effect. To avoid identification problems in the model specification, $d_1 = p_1 = 0$.

The methodology of this study involves data preparation, modeling, risk scoring of driving risk, etc. The whole technical process is shown in Figure 2.

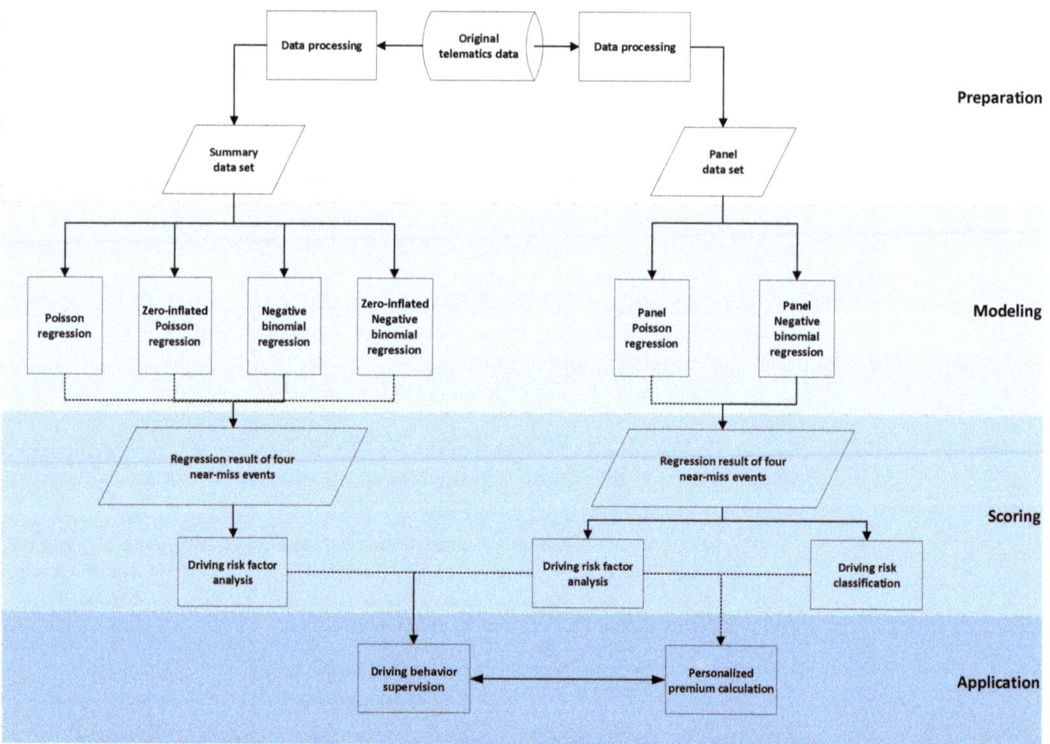

Figure 2. Technical flow chart.

In the data preparation stage, the original data need to be preprocessed, including multi-source data fusion, data cleaning, missing processing, etc. Then the summary data set and panel data set required in this study are obtained through statistical calculation. In the modeling phase, multiple count data models are used on two data sets for regression analysis, which follows certain premises. Our observed drivers can be considered independent of each other. Even if they drive in a similar area, they do not have any apparent relationship between each other. When we observe one driver over time, we have taken care of temporal correlation using the panel model that considers that one individual is observed repeatedly, here each day. In the scoring stage, the regression results obtained from the regression model most suitable for the data in this study can be used for causal analysis of near-miss events and driving risk scoring and rating. In the research field of telematics data application, the results of this study show this application has potential in, for example, driving behavior supervision and personalized premium calculation. The work at this stage has yet to be completed. Data processing in the preparation and Poisson regression and negative binomial regression on different data sets in the modeling can be implemented with data tools such as Stata, Python, R, etc.

5. Results

Before regression, multicollinearity tests are carried out on all explanatory variables to eliminate the influence of multicollinearity on the model. As shown in Table 3, the variance inflation factors (VIF) of all selected independent variables are less than 5, while the correlation coefficients are generally less than 0.7. This indicates that the multicollinearity among variables is weak, so all of them can be included in the regression equation and robust estimates can be made.

Table 3. Variance inflation factor and correlation of explanatory variables.

Variable	VIF	Brakes	Range	Speed	rpm	Accelerator Pedal Positon
brakes	3.07					
range	2.65	0.1213				
speed	2.30	0.0536	0.6262			
rpm	2.13	−0.0254	−0.0203	0.1804		
accelerator pedal positon	2.03	0.0154	0.1174	0.3458	0.7695	
engine fuel rate	1.04	0.1687	0.6313	0.6490	0.1075	0.3529

Both Poisson regression and negative binomial regression are applicable to this study, and the zero-inflated model is taken as a consideration for the large number of zero values of dependent variables. In order to determine the regression model which is most suitable for this study, the performance of the two models on different dependent variables is compared. All the estimated results are obtained by regression after standardization of the original values.

5.1. Results of the Summary Data Set

In the summary data set, four near-miss events are respectively treated as dependent variables while the independent variables are brakes, speed, rpm, accelerator pedal position and engine fuel rate, where kilo is chosen as the exposure variable or offset. Poisson regression, zero-inflated Poisson regression, negative binomial regression and zero-inflated negative binomial regression are estimated (see Table 4). Regardless of which near-miss event is the dependent variable, negative binomial regression has maximum log-likelihood value, and minimum AIC value and BIC value. That is, negative binomial regression has the best performance in this data set.

According to the results of negative binomial regression in different dependent variables (see Table 5 and Figure 3a), different near-miss events are affected by different driving risk factors with different influences. Overall, the average speed has the most obvious influence on near-miss events, with a significant negative effect on harsh acceleration (−0.776) and harsh deceleration (−0.658). The impact of braking event number on near-miss events is also positive significant. The higher the number of braking, the more high speed braking (0.272), harsh acceleration (0.189) and harsh deceleration (0.180) occur. In addition, average RPM is positively correlated with harsh acceleration (0.178), and average accelerator pedal position is positively correlated with harsh acceleration (0.152) and harsh deceleration (0.235). Interestingly, some influencing factors have opposite effects on different dependent variables. Range of driving has a positive effect on high speed brake (0.272) but a negative effect on harsh deceleration (-0.153) while average engine fuel rate has a significant positive effect on high speed braking (0.705) but a negative effect on sharp deceleration (-0.157). Furthermore, the significance of the constant term indicates that, in addition to the factors considered in this study, there are other factors that also influence near-miss events. The results of the other three regression models on the summary data set are shown in Tables A1–A3, and discussed in the Discussion section.

Table 4. Model performances of Poisson, zero-inflated Poisson, negative binomial and zero-inflated negative binomial in summary data set.

Variable	Model	N	Log-Likelihood	df	AIC	BIC
	POS	182	−3518.92	7	7051.846	7074.274
overspeed	ZIP	182	−2369.82	8	4755.64	4781.272
	NB	182	−490.517	8	997.0338	1022.666
	ZINB	182	−490.516	9	999.0315	1027.868
	POS	182	−2830.75	7	5675.498	5697.926
highspeedbrake	ZIP	182	−2667.02	8	5350.034	5375.666
	NB	182	−627.422	8	1270.843	1296.476
	ZINB	182	−627.422	9	1272.843	1301.68
	POS	182	−5857.26	7	11,728.51	11,750.94
harshacceleration	ZIP	182	−5857.26	8	11,730.51	11,756.14
	NB	182	−1032.81	8	2081.623	2107.255
	ZINB	182	−1032.81	9	2083.623	2112.459
	POS	182	−6269.47	7	12,552.93	12,575.36
harshdeceleration	ZIP	182	−6269.47	8	12,554.93	12,580.56
	NB	182	−1037.14	8	2090.285	2115.917
	ZINB	182	−1037.14	9	2092.285	2121.121

Table 5. The results of negative binomial regression for four near-miss events in the summary data set of drivers.

Variable	Overspeed		Highspeedbrake		Harshacceleration		Harshdeceleration	
	Coefficient	z	Coefficient	z	Coefficient	z	Coefficient	z
constant	−5.175 ***	−15.87	−5.114 ***	−35.36	−2.548 ***	−43.39	−2.525 ***	−42.99
brakes	0.264	1.29	0.272 **	2.60	0.189 ***	3.38	0.180 ***	3.45
range	0.185	0.79	0.272 *	2.05	−0.100	−1.29	−0.153	−1.94
speed	−0.113	−0.20	0.249	1.28	−0.776 ***	−8.81	−0.658 ***	−7.20
rpm	0.125	0.43	−0.0241	−0.11	0.178	1.90	0.0969	1.07
acceleratorpedalposition	0.290	1.13	0.171	1.03	0.152	1.87	0.235 **	2.82
enginefuelrate	0.227	0.99	0.705 ***	4.49	−0.0883	−1.12	−0.157 *	−2.07
log-likelihood	−490.5		−627.4		−1032.8		−1037.1	
AIC	997.0		1270.8		2081.6		2090.3	
BIC	1022.7		1296.5		2107.3		2115.9	
Observation	182		182		182		182	

*** $p < 0.01$, ** $p < 0.05$, * $p < 0.1$.

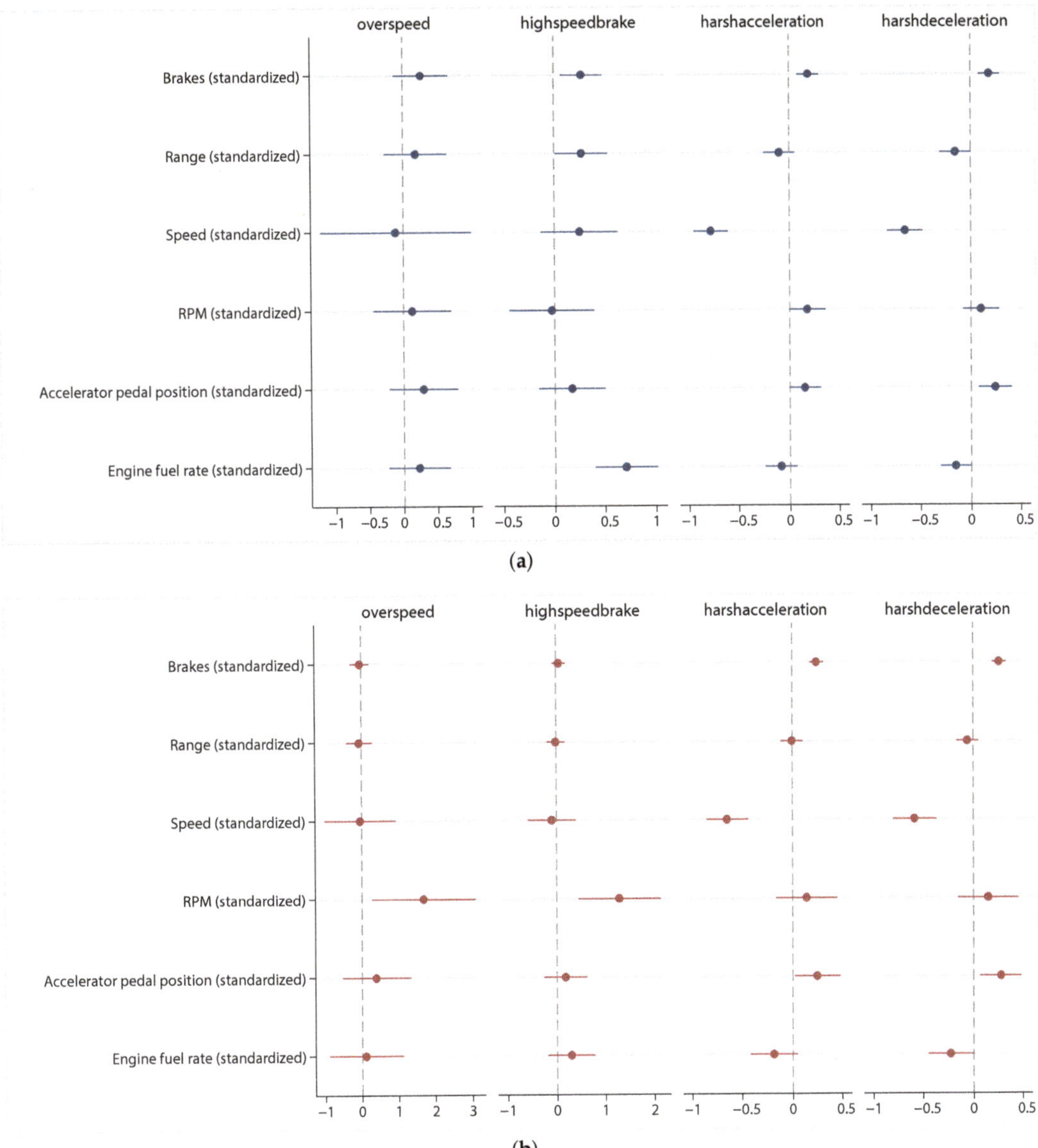

Figure 3. Partial coefficient estimation results of (**a**) negative binomial regression; (**b**) Panel negative binomial regression.

5.2. Results of the Panel Data Set

As shown in Table 6, the evaluation index (log-likelihood, AIC and BIC) of negative binomial regression is lower than that of Poisson regression for each dependent variable. Therefore, negative binomial regression is better than Poisson regression on panel data.

The panel negative binomial regression is used to estimate the two-way fixed effect model, considering both individual effect and time effect on four dependent variables. The influencing factors reflected by this (see Table A4 and Figure 3b) differ from those shown in the results of the summary data. For example, harsh acceleration and harsh deceleration are positively affected by the number of brakes (0.246 and 0.253) and average accelerator pedal position (0.249 and 0.270) but negatively affected by the average speed (−0.645 and −0.586) and average engine fuel rate (−0.188 and −0.229). However, RPM, which is not significant in the summary data, is significantly positive for overspeed (1.683) and high speed braking (1.287). The brakes (0.0505) and engine fuel rate (0.295), which had a significant positive effect on the summary data, become insignificant.

Table 6. Model performances of Poisson and negative binomial in the panel data set of drivers with six observations per driver.

Variable	Model	N	Log-Likelihood	df	AIC	BIC
overspeed	XTPOS	1092	−1926.78	188	4229.559	5168.763
	XTNB	1092	−957.497	189	2292.993	3237.193
highspeedbrake	XTPOS	1092	−2594.37	188	5564.733	6503.937
	XTNB	1092	−1527.05	189	3432.105	4376.305
harshacceleration	XTPOS	1092	−6117.44	188	12,610.89	13,550.09
	XTNB	1092	−3526.09	189	7430.186	8374.386
harshdeceleration	XTPOS	1092	−6042.02	188	12,460.03	13,399.24
	XTNB	1092	−3547.66	189	7473.311	8417.51

The advantage of panel data over summary data is that fixed effects can be estimated and thus individual effects and time effects can be interpreted. The time effect is significant in most cases for high speed braking, harsh acceleration and harsh deceleration, which indicates that these three near-miss events are greatly influenced by time. The time effect on the overspeed event is significant for only one day, suggesting that it is less influenced by time. More importantly, the individual effects of the four near-miss events can be used to score each observation. It should be noted that the first observation has been omitted in the regression to avoid complete multicollinearity, and its value is expected to be zero in the subsequent driving risk score.

6. Discussion

The regression results of Poisson regression (see Table A1), zero-inflated Poisson regression (see Table A2), negative binomial regression (see Table 5) and zero-inflated negative binomial regression (see Table A3) on the summary data set show the importance of driving behavior variables in driving risk. The high significance of two variables, braking times and average speed, in the four regression models indicates that these two factors have a very important impact on the generation of near-miss events. Moreover, the significant performance of specific independent variables in the regression model of specific dependent variables indicates that near-miss events are affected by a variety of driving behavior factors and the formation mechanism of each near-miss event is different. For example, the positive effect of RPM on harsh acceleration events, the positive effect of accelerator pedal position on harsh deceleration events and the positive effect of engine fuel rate on high speed braking events are shown in Tables A1–A3, Table 5.

The results obtained by panel regression are more reliable than those obtained by pooled regression. Tables 5 and A4 and Figure 3 show that some coefficients that are not significant in the pooled negative binomial regression become significant in the panel negative binomial regression, while some significant parameters in the pooled negative binomial regression are not significant in the panel negative binomial regression. This means that the dependent variables are affected by individual effects and time effects. In the panel negative binomial regression, most of the individual and time coefficients are significant, which indicates the suitability of this type of regression analysis.

Driving risks can be evaluated by the regression coefficients of negative binomial models on panel data. The value of the individual coefficients within a regression indicates the individual's deviance from the level of the expected occurrence of a particular near-miss event, given the information on all the other explanatory variables. In other words, the individual effect coefficient can be understood as the effect utility of each vehicle on the occurrence of the corresponding near-miss event. Geometrically, the effect coefficient of each individual is a change in the intercept.

Four near-miss events are used as dependent variables to obtain four sets of regression coefficients. Given that the influencing factors and generating mechanisms of different near-miss events are different, combining the four groups of regression coefficients into one group is not recommended. However, harsh acceleration and harsh deceleration show very similar characteristics in terms of data description before regression (Tables 1 and 5), after regression (Figure 3 and Table A4) and in distribution of driving risk score (Table A5). Even so, it is not recommended to combine them into a single near-miss event for study, because the occurrence conditions and coping operations of them are different, and it is the most appropriate choice to study each near-miss event separately.

In order to transform individual effect estimates of near-miss models into a driving risk grading, several steps need to be followed. Firstly, winsorization avoids the influence of possibly spurious outliers (the double tail was winsorized with the threshold 0.01 in this study). Secondly, the regression coefficient can be compressed to the interval of [0,1] through normalization. Each group of coefficients is then mapped into an interval of [0,5] (see Table A5), and each observation then is given a driving risk level from 1 to 5, i.e., excellent, good, medium, bad and terrible (see Figure 4). The values of exactly 0 and 5 are included because the corresponding observations are the minimum and the maximum values in their group and are Min-Max scaled. In overspeed and *highspeedbrake* groups, two types of observations with high risk or low risk can be clearly seen. This indicates that these two near-miss events are more sensitive to driving behavior than *harshacceleration* and *harshdeceleration* and can be considered as a higher priority and weight in subsequent studies. Note that the same observation (id125) has different risk levels for different near-miss events, which also explains why multiple near-miss events cannot be analyzed together. Ultimately, the premium would be charged individually according to the driving risk level of the insured person.

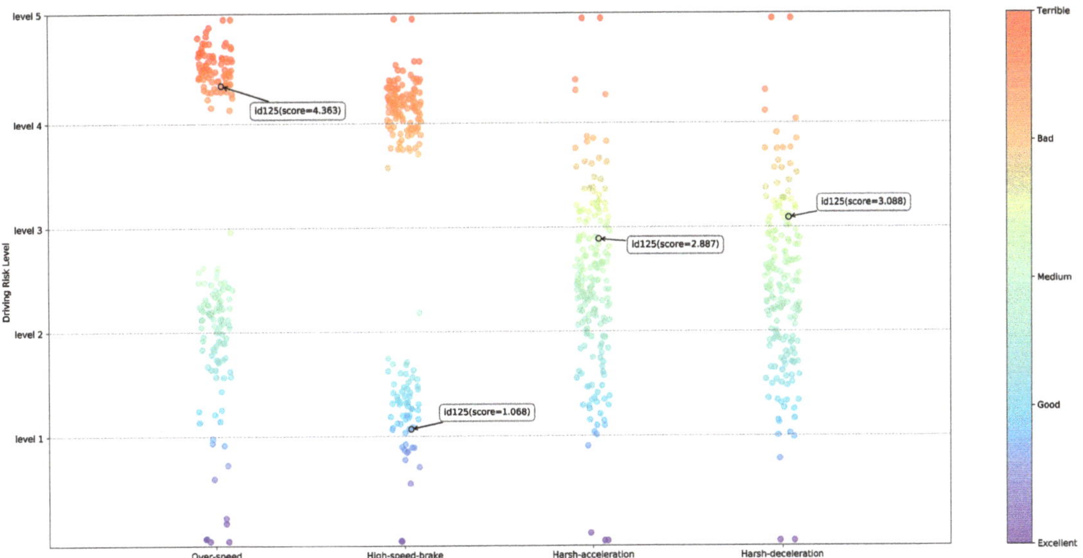

Figure 4. Driving risk ranking of four near-miss events.

7. Conclusions

The number and type of dependent variables and independent variables selected in this study are limited by the size and quality of the original data. With the promotion and innovation of IoV and of new energy vehicles, the amount and dimension of data will be greatly increased. Therefore, application of near-miss events as dependent variables could be easily increased or decreased, according to needs. For example, sharp turn should be included, if possible, as a near-miss event because sharp turn is a highly studied and accident-proven pattern of high driving risk. For the same reason, more driving behavior indicators, such as steering wheel angle speed, brake pedal position, and so on, could be used as independent variables in the regression model. In addition, traditional auto insurance factors, such as driver information, vehicle information, road information, environment information and the health status of batteries (of new energy vehicles) should be considered to provide more optional independent variables for the model.

In practical applications, near-miss events can be combined with claims and accidents to accurately evaluate driving risks. This study proves that near-miss events can be used as driving risk scores when there are no claims or accidents. However, when claims or accidents exist, the driving risk score obtained from claims or accidents can be used as the basis for premium calculation, while the driving risk rating obtained from near-miss events can be used to remind and warn drivers to reduce the corresponding dangerous driving habits.

In this study, the best performing negative binomial regression (see Table 4) was selected as the main method for modeling on our data set. The model is suitable for similar causal analysis of similar data sets. However, in case of risk event prediction or analysis on other data sets, it is necessary to reevaluate the goodness of fit of various models, and even machine learning methods with good prediction performance should be taken into consideration. The optimal method is not fixed, but depends on the data, conditions and purposes.

Econometrics and machine learning complement each other. The generalized linear model established in this study reveals the relationship between driving behavior factors and near-miss events, and gives a driving risk score for each observation. This model has strong explanatory power, but its generalization degree and robustness need to be further tested, especially on larger data volume and data dimension. The successful application of machine learning methods in many fields shows that they are often effective in dealing with big data problems but that their results cannot always be easily interpreted, and this interpretation is exactly what the insurance field values. Therefore, telematics data application offers a new way to help find a balance between econometrics and machine learning so as to have good explainability, good generalization ability, quick response ability, and so on [38,39].

In general, near-miss events can provide insurers with effective risk information in the absence of claims and accident data. In our real case study, negative binomial regression is the most suitable modeling method for near-miss events as dependent variables. This study provides a technical reference for the promotion and development of PHYD ratemaking schemes.

Author Contributions: Conceptualization, S.S. and M.G.; methodology, M.G.; software, S.S.; validation, J.B., M.G. and A.M.P.-M.; formal analysis, S.S.; investigation, S.S.; resources, M.G. and A.M.P.-M.; data curation, J.B., S.S. and M.G.; writing—original draft preparation, S.S.; writing—review and editing, S.S. and M.G.; visualization, S.S.; supervision, J.B. and M.G.; project administration, J.B.; funding acquisition, J.B. and S.S. All authors have read and agreed to the published version of the manuscript.

Funding: This research was granted by the Fundamental Research Funds for the Central Universities 2019YJS091.

Institutional Review Board Statement: Not applicable.

Informed Consent Statement: Not applicable.

Data Availability Statement: This study is based on the original telematics data from China Satellite Navigation and Communications Co., Ltd., which could not be made public due to confidentiality agreements.

Acknowledgments: M.G. thanks Fundación BBVA and ICREA Academia for providing research funds. S.S. thanks the China Scholarship Council for providing visiting research funds in the second institution. All the authors thank China Satellite Navigation and Communications Co., Ltd. for providing the original telematics data set.

Conflicts of Interest: The authors declare no conflict of interest. The funders had no role in the design of the study; in the collection, analyses, or interpretation of data; in the writing of the manuscript, or in the decision to publish the results.

Abbreviations

Abbreviations

The following abbreviations are used in this manuscript:

UBI	usage-based insurance
IoV	Internet of vehicles
PAYD	pay as you drive
PHYD	pay how you drive
MHYD	manage how you drive
VIF	variance inflation factor
POS	Poisson
ZIP	Zero-inflated Poisson
NB	Negative binomial
ZINB	Zero-inflated negative binomial
XTPOS	Panel Poisson
XTNB	Panel negative binomial
AIC	Akaike information criterion
BIC	Bayesian information criterion

Appendix A

Table A1. The results of Poisson regression for four near-miss events in the summary data set of drivers.

Variable	Overspeed		Highspeedbrake		Harshacceleration		Harshdeceleration	
	Coefficient	z	Coefficient	z	Coefficient	z	Coefficient	z
constant	−5.191 ***	−21.45	−5.194 ***	−29.98	−2.612 ***	−40.34	−2.591 ***	−40.05
brakes	0.279	1.82	0.349 ***	5.93	0.191 ***	3.66	0.186 ***	3.78
range	0.0437	0.21	0.0741	0.78	−0.157	−1.65	−0.208 *	−2.04
speed	−0.175	−0.92	0.489 **	3.22	−0.717 ***	−8.57	−0.601 ***	−6.51
rpm	0.514	1.59	0.202	0.87	0.272 **	2.94	0.183 *	1.98
acceleratorpedalposition	0.0467	0.18	−0.0337	−0.23	0.169	1.91	0.238 **	2.67
enginefuelrate	0.540 *	2.24	0.755 ***	4.32	−0.0499	−0.59	−0.119	−1.48
log-likelihood	−3518.9		−2830.7		−5857.3		−6269.5	
AIC	7051.8		5675.5		11,728.5		12,552.9	
BIC	7074.3		5697.9		11,750.9		12,575.4	
Observation	182		182		182		182	

*** $p < 0.01$, ** $p < 0.05$, * $p < 0.1$.

Table A2. The results of zero-inflated Poisson regression for four near-miss events in the summary data set of drivers.

Variable	Overspeed		Highspeedbrake		Harshacceleration		Harshdeceleration	
	Coefficient	z	Coefficient	z	Coefficient	z	Coefficient	z
constant	−4.388 ***	−18.08	−5.006 ***	−24.96	−2.612 ***	−40.34	−2.591 ***	−40.05
brakes	0.167	1.08	0.339 ***	5.34	0.191 ***	3.66	0.186 ***	3.78
range	0.0365	0.20	0.0755	0.81	−0.157	−1.65	−0.208 *	−2.04
speed	−0.391 *	−2.12	0.408 *	2.48	−0.717 ***	−8.57	−0.601 ***	−6.51
rpm	0.607 *	2.18	0.274	1.11	0.272 **	2.94	0.183 *	1.98
acceleratorpedalposition	−0.117	−0.52	−0.0563	−0.38	0.169	1.91	0.238 **	2.67
enginefuelrate	0.346	1.59	0.700 ***	4.01	−0.0499	−0.59	−0.119	−1.48
inflate-constant	0.101	0.66	−1.183 ***	−4.72	−27.29 ***	−295.09	−27.00 ***	−363.25
log-likelihood	−2369.8		−2667.0		−5857.3		−6269.5	
AIC	4755.6		5350.0		11,730.5		12,554.9	
BIC	4781.3		5375.7		11,756.1		12,580.6	
Observation	182		182		182		182	

*** $p < 0.01$, ** $p < 0.05$, * $p < 0.1$.

Table A3. The results of zero-inflated negative binomial regression for four near-miss events in the summary data set of drivers.

Variable	Overspeed		Highspeedbrake		Harshacceleration		Harshdeceleration	
	Coefficient	z	Coefficient	z	Coefficient	z	Coefficient	z
constant	−5.153 ***	−7.37	−5.114 ***	−35.36	−2.548 ***	−43.39	−2.525 ***	−42.99
brakes	0.263	1.25	0.272 **	2.60	0.189 ***	3.38	0.180 ***	3.45
range	0.180	0.73	0.272 *	2.05	−0.100	−1.29	−0.153	−1.94
speed	−0.114	−0.20	0.249	1.28	−0.776 ***	−8.81	−0.658 ***	−7.20
rpm	0.130	0.41	−0.0241	−0.11	0.178	1.90	0.0969	1.07
acceleratorpedalposition	0.284	0.98	0.171	1.03	0.152	1.87	0.235 **	2.82
enginefuelrate	0.227	0.99	0.705 ***	4.49	−0.0883	−1.12	−0.157 *	−2.07
inflate-constant	−3.793	−0.17	−14.62 ***	−6.14	−25.29 ***	−294.22	−23.27 ***	−313.10
log-likelihood	−490.5		−627.4		−1032.8		−1037.1	
AIC	999.0		1272.8		2083.6		2092.3	
BIC	1027.9		1301.7		2112.5		2121.1	
Observation	182		182		182		182	

*** $p < 0.01$, ** $p < 0.05$, * $p < 0.1$.

Table A4. Panel negative binomial regression results for four near-miss events.

Variable	Overspeed		Highspeedbrake		Harshacceleration		Harshdeceleration	
	Coefficient	z	Coefficient	z	Coefficient	z	Coefficient	z
constant	−3.768 ***	(−8.68)	−4.356 ***	(−17.39)	−2.284 ***	(−25.13)	−2.269 ***	(−20.81)
brakes	−0.0400	(−0.31)	0.0505	(0.73)	0.246 ***	(6.93)	0.253 ***	(7.23)
range	−0.0637	(−0.36)	−0.0108	(−0.12)	−0.00410	(−0.07)	−0.0595	(−1.09)
speed	−0.0405	(−0.08)	−0.0965	(−0.39)	−0.645 ***	(−6.09)	−0.586 ***	(−5.24)
rpm	1.683 *	(2.34)	1.287 **	(3.00)	0.143	(0.91)	0.145	(0.93)
acceleratorpedalposition	0.391	(0.83)	0.175	(0.79)	0.249 *	(2.12)	0.270 *	(2.52)
enginefuelrate	0.113	(0.22)	0.295	(1.18)	−0.188	(−1.58)	−0.229 *	(−2.04)
2018-07-04	0.273	(1.23)	0.216	(1.91)	−0.111 *	(−2.12)	−0.216 ***	(−4.33)
2018-07-05	−0.168	(−0.73)	−0.0572	(−0.52)	−0.206 ***	(−4.34)	−0.317 ***	(−6.72)
2018-07-06	−0.00716	(−0.03)	−0.228 *	(−2.08)	−0.257 ***	(−4.84)	−0.370 ***	(−7.19)
2018-07-07	−0.477 *	(−2.11)	−0.200	(−1.68)	−0.485 ***	(−7.41)	−0.600 ***	(−9.27)
2018-07-08	0.206	(0.90)	0.117	(0.95)	−0.694 ***	(−8.63)	−0.784 ***	(−9.58)
id2	−28.81 ***	(−29.17)	−2.001 *	(−2.25)	1.266 ***	(5.24)	1.342 ***	(4.80)
id3	−19.05 ***	(−14.86)	−18.13 ***	(−16.37)	2.004 *	(2.31)	1.740 ***	(4.75)
id4	−18.29 ***	(−15.94)	−17.91 ***	(−20.04)	1.891 ***	(8.66)	1.960 ***	(8.58)
id5	−29.62 ***	(−41.07)	−4.956 ***	(−7.51)	−1.193 ***	(−3.77)	−1.072 ***	(−3.39)
id6	−1.478 *	(−2.40)	−0.554	(−1.79)	1.067 ***	(3.31)	0.935 ***	(4.27)
id7	−3.236 ***	(−4.11)	−0.645	(−1.58)	0.656 **	(3.15)	0.835 **	(3.05)
id8	−20.79 ***	(−24.48)	−2.368 ***	(−4.01)	−0.190	(−0.61)	0.124	(0.35)
id9	−1.156	(−1.10)	−0.0678	(−0.11)	−0.251	(−0.79)	−0.109	(−0.28)
id10	−3.110 ***	(−5.93)	−1.527 ***	(−4.20)	−0.345 *	(−2.17)	−0.256	(−1.63)
id11	−2.026 *	(−2.42)	−1.163 ***	(−3.45)	−0.162	(−0.88)	−0.272	(−1.22)

Table A4. *Cont.*

Variable	Overspeed		Highspeedbrake		Harshacceleration		Harshdeceleration	
	Coefficient	z	Coefficient	z	Coefficient	z	Coefficient	z
id12	−1.342 *	(−2.51)	−0.772 *	(−2.45)	0.0781	(0.34)	0.0981	(0.50)
id13	−2.344 ***	(−5.13)	−0.808 *	(−2.21)	−0.138	(−0.52)	−0.129	(−0.66)
id14	−3.178 ***	(−5.87)	0.442	(1.43)	−0.629 ***	(−3.66)	−0.365 *	(−2.07)
id15	−1.254 *	(−2.31)	0.167	(0.46)	−0.0894	(−0.34)	0.0270	(0.10)
id16	−22.40 ***	(−37.49)	−21.63 ***	(−41.20)	0.271	(1.35)	0.439 *	(2.13)
id17	−21.77 ***	(−25.73)	−2.102 ***	(−4.06)	−0.200	(−0.87)	0.0983	(0.38)
id18	−20.99 ***	(−19.61)	−0.805	(−1.48)	−1.124 ***	(−6.78)	−1.267 ***	(−5.69)
id19	−0.998	(−1.58)	0.380	(1.06)	0.587 **	(2.72)	0.586 ***	(3.89)
id20	−23.99 ***	(−38.51)	−3.749 **	(−3.22)	0.292	(1.19)	0.0926	(0.40)
id21	−21.79 ***	(−35.89)	−2.577 **	(−2.75)	0.322	(1.34)	0.458 **	(2.61)
id22	−2.642 ***	(−3.94)	−0.229	(−0.63)	0.496 **	(3.00)	0.538 **	(2.89)
id23	−0.792	(−1.10)	0.00108	(0.00)	−0.474	(−1.61)	−0.409	(−1.71)
id24	−23.37 ***	(−34.65)	−22.27 ***	(−40.12)	−0.329	(−1.36)	−0.103	(−0.32)
id25	−21.06 ***	(−30.30)	−20.67 ***	(−32.16)	−0.882 ***	(−3.45)	−0.731 *	(−2.56)
id26	−2.739 ***	(−3.82)	−1.000 **	(−2.61)	−0.440	(−1.77)	−0.667 ***	(−3.68)
id27	−23.10 ***	(−41.71)	−22.25 ***	(−45.16)	−0.0464	(−0.15)	0.0656	(0.22)
id28	−17.86 ***	(−17.18)	−17.85 ***	(−18.78)	0.0432	(0.13)	0.309	(1.41)
id29	−1.136	(−1.59)	−0.872 *	(−2.33)	0.591 ***	(3.74)	0.625 ***	(4.09)
id30	−20.56 ***	(−29.51)	−19.84 ***	(−31.89)	−0.223	(−0.53)	−0.102	(−0.25)
id31	−0.407	(−0.91)	−0.633 *	(−2.03)	−1.148 ***	(−3.69)	−0.949 **	(−3.21)
id32	−3.255 **	(−3.24)	−2.923 *	(−2.38)	−0.110	(−0.36)	0.143	(0.49)
id33	−19.05 ***	(−19.59)	−19.50 ***	(−21.06)	−0.177	(−0.37)	−0.153	(−0.32)
id34	−2.431 **	(−3.26)	−1.547 ***	(−3.51)	−0.00573	(−0.02)	−0.0439	(−0.14)
id35	−3.832 ***	(−3.98)	−1.041 *	(−2.18)	−0.607 ***	(−3.58)	−0.552 **	(−3.14)
id36	−4.135 ***	(−4.74)	−2.412 ***	(−4.36)	−0.285	(−1.37)	−0.343	(−1.82)
id37	−38.70 ***	(−32.02)	−1.232	(−1.72)	−0.480	(−1.64)	−0.218	(−0.75)
id38	−20.26 ***	(−29.14)	−1.364 *	(−2.11)	−1.484 ***	(−5.08)	−1.121 ***	(−4.31)
id39	−38.61 ***	(−41.67)	10.89 ***	(14.67)	11.65 ***	(21.85)	11.77 ***	(21.58)
id40	−1.326	(−1.57)	−0.416	(−0.81)	−0.278	(−1.10)	0.0791	(0.30)
id41	−2.443 **	(−3.19)	−1.020 *	(−2.02)	0.180	(0.82)	0.155	(0.61)
id42	−0.467	(−0.57)	0.442	(0.93)	0.607	(1.63)	0.398	(1.45)
id43	−2.164 *	(−2.45)	0.219	(0.39)	−0.0359	(-0.13)	0.0900	(0.35)
id44	−2.465 ***	(−−3.52)	−0.156	(−0.35)	0.336	(1.33)	0.468	(1.85)
id45	−2.110 **	(−3.26)	−1.315 **	(−3.20)	0.105	(0.64)	0.282	(1.18)
id46	0.132	(0.14)	−0.480	(−1.01)	−0.312 **	(−2.77)	−0.235	(−1.65)
id47	−2.957 ***	(−5.23)	−0.975	(−1.32)	−0.853 ***	(−4.77)	−0.656 ***	(−4.01)
id48	0.486	(0.78)	1.381 ***	(3.72)	0.829 ***	(5.36)	0.787 ***	(4.63)
id49	−25.28 ***	(−34.70)	−1.575 **	(−3.27)	−0.568 **	(−2.60)	−0.353	(−1.67)
id50	−2.556 ***	(−3.94)	−1.907 ***	(−3.87)	−0.413 *	(−2.43)	−0.331	(−1.80)
id51	−20.62 ***	(−20.07)	−20.01 ***	(−27.73)	1.123 ***	(3.56)	1.140 ***	(3.69)
id52	−21.73 ***	(−16.76)	−20.90 ***	(−19.35)	−0.354	(−1.10)	−0.952 ***	(−3.75)
id53	−20.65 ***	(−21.34)	−20.00 ***	(−31.74)	−0.133	(−0.72)	0.200	(1.01)
id54	−4.881 ***	(−5.39)	−1.082 **	(−2.66)	−0.686 ***	(−3.36)	−0.639 **	(−3.00)
id55	−4.290 ***	(−4.44)	−1.731 ***	(−3.90)	0.472	(1.80)	0.476	(1.37)
id56	−2.462 ***	(−3.72)	−0.0866	(−0.20)	0.119	(0.40)	0.377	(0.77)
id57	−21.96 ***	(−22.99)	−0.700	(−1.51)	0.110	(0.36)	0.719 *	(2.27)
id58	−1.877	(−1.66)	−0.692	(V0.83)	−0.344	(−0.77)	0.0660	(0.16)
id59	−38.77 ***	(−43.77)	−0.0709	(−0.10)	−0.726 *	(−2.11)	−0.587	(−1.68)
id60	−3.117 **	(−2.65)	−3.815 ***	(−4.00)	−0.711 *	(−2.07)	−0.565	(−1.69)
id61	0.821	(0.83)	1.078	(1.78)	−1.288 **	(−2.87)	−1.076 *	(−2.42)
id62	−0.465	(−0.61)	0.546	(1.46)	−0.670	(−1.58)	−0.473	(−1.15)
id63	−21.36 ***	(−28.39)	−20.68 ***	(−33.84)	1.393 ***	(10.47)	1.513 ***	(10.22)
id64	−2.529	(−1.39)	−1.707	(−1.18)	1.334 ***	(7.83)	1.339 ***	(6.13)
id65	−21.50 ***	(−34.71)	−20.76 ***	(−42.35)	−1.923 ***	(−5.19)	−1.288 ***	(−5.26)
id66	−1.389	(−1.49)	−1.510 ***	(−3.74)	0.504 **	(2.64)	0.971 ***	(5.69)
id67	−25.65 ***	(−34.83)	−3.400 ***	(−3.49)	−0.371 *	(−1.98)	−0.304	(−1.70)
id68	−19.09 ***	(−27.85)	−18.66 ***	(−35.37)	−1.286 **	(−2.97)	−1.660 ***	(−7.16)
id69	−24.40 ***	(−27.41)	−21.86 ***	(−33.37)	−0.589 **	(−2.94)	−0.625 *	(−2.48)
id70	−21.29 ***	(−22.38)	−3.693 ***	(−4.50)	−1.489 ***	(−3.60)	−1.501 ***	(−6.37)
id71	−31.22 ***	(−36.99)	−29.36 ***	(−53.20)	0.587 ***	(3.95)	1.212 ***	(7.82)
id72	−5.534 ***	(−5.68)	−1.058	(−1.96)	−0.516	(−1.41)	−0.643	(−1.84)
id73	−4.323 ***	(−4.06)	−2.863 ***	(−5.25)	−1.527 ***	(−6.22)	−1.523 ***	(−5.97)
id74	−30.92 ***	(−35.40)	−29.09 ***	(−51.97)	0.299	(1.03)	0.765 **	(3.02)
id75	−2.868 ***	(−3.45)	−1.677 ***	(−4.49)	−0.267	(−0.90)	−0.0911	(−0.31)
id76	−21.21 ***	(−22.83)	−21.18 ***	(−31.54)	−1.646 ***	(−3.80)	−1.903 ***	(−4.47)
id77	−19.83 ***	(−15.23)	−19.23 ***	(−21.15)	0.835 **	(2.84)	0.729 **	(2.71)
id78	−24.02 ***	(−33.62)	−3.260 ***	(−3.47)	−2.855 ***	(−10.67)	−2.759 ***	(−8.49)

Table A4. *Cont.*

Variable	Overspeed Coefficient	z	Highspeedbrake Coefficient	z	Harshacceleration Coefficient	z	Harshdeceleration Coefficient	z
id79	−3.449 **	(−2.81)	−0.618	(−1.43)	−0.232	(−1.23)	−0.110	(−0.57)
id80	−21.71 ***	(−15.28)	−20.69 ***	(−20.65)	−0.0149	(−0.05)	0.0509	(0.16)
id81	−33.98 ***	(−60.17)	−1.132 *	(−2.44)	−0.341 *	(−2.34)	−0.336 *	(−2.41)
id82	−1.391	(−1.64)	−0.541	(−0.76)	−0.312	(−1.38)	−0.326	(−1.55)
id83	−1.516 **	(−2.58)	0.157	(0.53)	−0.123	(−0.64)	−0.242	(−1.37)
id84	−23.95 ***	(−25.81)	−1.866 *	(−2.04)	−0.750 **	(−3.19)	−0.855 ***	(−4.50)
id85	−22.55 ***	(−23.68)	−3.844 ***	(−4.49)	−1.430 ***	(−5.83)	−1.318 ***	(−5.43)
id86	−29.06 ***	(−33.35)	−2.036 ***	(−3.87)	−1.272 ***	(−3.40)	−1.111 **	(−2.99)
id87	−1.851 *	(−2.47)	1.034 **	(2.59)	0.196	(1.03)	0.425 *	(2.03)
id88	−20.13 ***	(−31.73)	−19.83 ***	(−38.92)	−0.208	(−1.54)	−0.165	(−1.04)
id89	−25.42 ***	(−13.08)	−23.44 ***	(−19.03)	1.100 *	(2.18)	1.135 *	(2.49)
id90	−4.008 ***	(−4.24)	−0.841	(−1.18)	−0.972 ***	(−4.48)	−0.982 **	(−3.17)
id91	−19.51 ***	(−26.36)	−19.33 ***	(−31.18)	0.676 ***	(5.52)	0.818 ***	(4.54)
id92	−26.04 ***	(−14.88)	−24.17 ***	(−22.56)	0.848 *	(2.54)	0.663 *	(2.00)
id93	−23.79 ***	(−23.97)	−22.53 ***	(−27.69)	−0.290	(−1.41)	−0.300	(−1.26)
id94	−2.684 ***	(−3.53)	−1.034 **	(−3.05)	−0.157	(−0.99)	0.0139	(0.06)
id95	−24.75 ***	(−12.35)	−22.86 ***	(−19.16)	−0.503	(−0.78)	−0.670	(−1.26)
id96	−22.66 ***	(−28.80)	−21.77 ***	(−35.98)	1.374 ***	(7.81)	1.343 ***	(6.60)
id97	−20.83 ***	(−29.33)	−20.42 ***	(−32.43)	−0.464 *	(−2.30)	−0.282	(−1.24)
id98	−18.59 ***	(−24.60)	−18.56 ***	(−27.54)	−1.405 ***	(−4.98)	−0.887 *	(−2.38)
id99	−18.19 ***	(−27.16)	−18.21 ***	(−34.28)	−1.774 ***	(−5.28)	−1.369 ***	(−4.34)
id100	−4.226 **	(−3.13)	−21.52 ***	(−26.23)	0.802 *	(2.30)	0.824 *	(2.37)
id101	−22.60 ***	(−13.70)	−21.29 ***	(−22.32)	0.955 *	(2.21)	0.814	(1.96)
id102	−24.81 ***	(−34.96)	−23.71 ***	(−42.27)	0.0308	(0.14)	−0.0294	(−0.12)
id103	−17.82 ***	(−22.56)	−17.67 ***	(−27.25)	0.542 *	(2.27)	0.606 ***	(3.70)
id104	−20.05 ***	(−32.16)	−19.72 ***	(−37.06)	0.131	(0.73)	0.262 *	(1.98)
id105	−3.426 ***	(−4.29)	−0.430	(−0.94)	−0.464	(−0.90)	−0.925 *	(−2.01)
id106	−24.96 ***	(−26.36)	−23.73 ***	(−35.39)	0.317	(1.84)	0.252	(1.33)
id107	−21.01 ***	(−29.69)	−20.53 ***	(−36.62)	0.0144	(0.09)	0.147	(0.63)
id108	−23.28 ***	(−27.56)	−2.647 ***	(−5.35)	−0.532 *	(−2.40)	−0.635 ***	(−3.35)
id109	−20.89 ***	(−19.32)	−20.46 ***	(−21.38)	−0.347 **	(−2.97)	−0.782 ***	(−5.90)
id110	−20.70 ***	(−17.96)	−20.72 ***	(−20.15)	−1.801 ***	(−5.72)	−1.044 ***	(−5.99)
id111	−3.405 ***	(−4.19)	−1.278 ***	(−3.61)	0.173	(1.27)	0.198	(1.40)
id112	−19.65 ***	(−24.23)	−19.29 ***	(−25.89)	−1.453 ***	(−8.75)	−0.831 ***	(−7.19)
id113	−29.63 ***	(−42.75)	−2.998 ***	(−3.76)	−1.703 ***	(−6.85)	−1.296 ***	(−6.74)
id114	−23.43 ***	(−20.14)	−22.26 ***	(−23.54)	0.637 **	(2.80)	0.537 **	(3.22)
id115	−22.18 ***	(−29.89)	−21.49 ***	(−33.89)	−0.0179	(−0.14)	−0.109	(−0.69)
id116	−21.78 ***	(−24.48)	−3.753 ***	(−7.41)	−1.349 ***	(−4.54)	−1.135 **	(−2.79)
id117	−20.71 ***	(−26.29)	−20.08 ***	(−32.28)	−0.156	(−1.35)	−0.273 *	(−2.17)
id118	−18.97 ***	(−27.81)	−0.705	(−0.91)	0.116	(0.88)	0.00337	(0.02)
id119	−20.31 ***	(−30.19)	−19.89 ***	(−36.33)	−0.145	(−0.79)	−0.143	(−0.92)
id120	−27.27 ***	(−33.83)	−25.62 ***	(−41.44)	−0.0170	(−0.06)	0.133	(0.42)
id121	−28.62 ***	(−36.09)	−0.687	(−1.76)	0.239	(1.20)	0.387 *	(2.01)
id122	−21.69 ***	(−15.37)	−2.515 *	(−2.40)	0.623 *	(2.36)	0.653 *	(2.48)
id123	−23.18 ***	(−12.54)	−3.892 ***	(−4.57)	0.698	(1.51)	0.886 *	(2.13)
id124	−4.268 *	(−2.34)	−2.612 *	(−2.50)	0.698 **	(2.66)	0.361	(1.22)
id125	−3.828 *	(−2.35)	−22.52 ***	(−19.24)	0.296	(0.67)	0.619	(1.69)
id126	−2.023	(−1.24)	−2.183 *	(−2.54)	0.576	(1.84)	0.539	(1.73)
id127	−22.03 ***	(−12.96)	−20.58 ***	(−19.98)	1.158 ***	(4.23)	1.010 ***	(3.82)
id128	−20.90 ***	(−14.97)	−20.01 ***	(−22.10)	0.762 **	(2.84)	0.618 *	(2.18)
id129	−1.540 *	(−2.37)	0.776 *	(2.07)	0.0280	(0.17)	0.165	(0.76)
id130	−24.70 ***	(−30.90)	−23.31 ***	(−35.06)	−1.578 ***	(−5.17)	−1.635 ***	(−4.61)
id131	−1.659 *	(−2.10)	−0.403	(−1.01)	−0.980 ***	(−3.71)	−0.794 ***	(−4.13)
id132	−19.37 ***	(−26.07)	−18.83 ***	(−32.28)	−0.863 ***	(−3.40)	−0.435	(−1.85)
id133	−26.03 ***	(−21.40)	−2.904 ***	(−8.09)	−0.622 ***	(−4.18)	−0.691 ***	(−4.38)
id134	−31.36 ***	(−34.83)	−2.618 ***	(−5.23)	0.488 *	(2.40)	1.176 ***	(6.93)
id135	−23.37 ***	(−24.64)	−22.02 ***	(−35.69)	0.930 ***	(4.58)	1.350 ***	(7.18)
id136	3.358 ***	(3.79)	4.212 ***	(6.07)	2.661 ***	(8.38)	2.709 ***	(11.14)
id137	−23.49 ***	(−24.93)	−2.508 **	(−2.65)	0.0440	(0.26)	0.804 ***	(4.16)
id138	−19.02 ***	(−29.14)	−18.74 ***	(−37.30)	−0.827 ***	(−3.86)	−0.890 ***	(−3.54)
id139	−4.105 ***	(−4.05)	−1.187 *	(−2.54)	−0.922 **	(−3.06)	−0.677	(−1.61)
id140	−2.970 **	(−3.06)	−0.615	(−1.10)	−1.276 ***	(−3.79)	−1.035 ***	(−−3.61)
id141	−24.65 ***	(−33.01)	−23.40 ***	(−44.51)	−1.071 ***	(−4.15)	−1.100 ***	(−4.52)
id142	−37.04 ***	(−41.13)	−0.873	(−1.69)	0.0500	(0.22)	0.175	(0.76)
id143	−37.41 ***	(−40.91)	−0.397	(−0.82)	−0.368	(−1.15)	−0.157	(−0.52)
id144	−0.585	(−0.65)	0.551	(1.04)	0.0261	(0.09)	0.167	(0.58)
id145	−2.485 *	(−2.36)	−1.273 *	(−2.44)	−0.750	(−1.56)	−0.631	(−1.32)
id146	−22.93 ***	(−26.06)	−1.250 **	(−2.87)	−1.130 **	(−2.83)	−0.758 *	(−1.97)

Table A4. *Cont.*

Variable	Overspeed		Highspeedbrake		Harshacceleration		Harshdeceleration	
	Coefficient	z	Coefficient	z	Coefficient	z	Coefficient	z
id147	−2.851 ***	(−3.75)	−0.0796	(−0.17)	−1.021 **	(−2.90)	−0.896 **	(−2.59)
id148	−3.737 ***	(−4.05)	1.617 ***	(3.83)	−0.0483	(−0.14)	−0.0520	(−0.16)
id149	−3.202 ***	(−3.53)	−1.184 *	(−2.01)	−0.554 **	(−2.59)	−0.343	(−1.54)
id150	−3.616 **	(−3.24)	−0.905 *	(−2.24)	−1.167 ***	(−4.30)	−0.905 ***	(−3.46)
id151	−0.362	(−0.51)	−1.167 *	(−2.07)	−1.654 ***	(−5.15)	−1.677 ***	(−4.32)
id152	−32.89 ***	(−33.69)	−3.751 ***	(−6.31)	−1.421 ***	(−3.68)	−1.382 ***	(−4.40)
id153	−1.598 *	(−2.14)	−0.169	(−0.46)	−2.936 ***	(−5.34)	−3.067 ***	(−5.01)
id154	−21.84 ***	(−20.17)	−2.716 ***	(−3.85)	−1.703 ***	(−4.38)	−1.483 ***	(−4.26)
id155	−4.238 ***	(−3.67)	−2.441 *	(−2.46)	−0.759 **	(−3.12)	−0.814 **	(−2.63)
id156	−43.11 ***	(−52.10)	−1.456 **	(−3.27)	−0.590 **	(−2.64)	−0.429 *	(−2.03)
id157	−1.868 *	(−2.19)	0.337	(0.62)	−0.753 **	(−2.82)	−0.502 *	(−2.35)
id158	−19.28 ***	(−28.14)	−18.96 ***	(−35.46)	0.678 ***	(4.39)	0.744 ***	(3.87)
id159	−19.26 ***	(−28.40)	−19.02 ***	(−33.84)	0.827 ***	(4.09)	0.715 **	(3.07)
id160	−3.790 ***	(−4.60)	0.550	(1.25)	0.148	(0.72)	0.337	(1.90)
id161	−22.11 ***	(−30.73)	−21.31 ***	(−34.53)	0.608 ***	(4.58)	0.494 ***	(3.29)
id162	−20.15 ***	(−18.45)	−19.53 ***	(−23.73)	0.431	(1.86)	0.176	(0.52)
id163	−22.31 ***	(−22.00)	−21.43 ***	(−29.98)	1.844 ***	(9.58)	1.656 ***	(8.12)
id164	−2.923 **	(−2.69)	−2.557 ***	(−3.68)	−0.245	(−1.66)	−0.301	(−1.85)
id165	−20.82 ***	(−28.93)	−20.41 ***	(−32.12)	1.341 ***	(11.50)	1.447 ***	(10.42)
id166	−25.44 ***	(−20.51)	−2.439 ***	(−4.42)	−0.158	(−0.49)	−0.0534	(−0.15)
id167	−2.696 *	(−2.28)	−4.124 ***	(−4.21)	1.119 ***	(3.43)	1.089 ***	(3.53)
id168	−5.731 ***	(−6.49)	−1.940 ***	(−3.42)	0.0447	(0.23)	0.0115	(0.06)
id169	−26.04 ***	(−24.96)	−24.71 ***	(−35.44)	0.473 *	(2.06)	0.422	(1.82)
id170	−15.15 ***	(−15.82)	−15.11 ***	(−19.48)	−17.48 ***	(−24.79)	−0.684	(−0.99)
id171	−3.650 ***	(−3.49)	−1.497 **	(−2.86)	−0.344	(−1.62)	−0.313	(−1.47)
id172	−3.659 ***	(−3.93)	−1.951 ***	(−4.53)	−0.427	(−1.84)	−0.367	(−1.34)
id173	−3.036 **	(−2.98)	−3.500 ***	(−4.94)	−0.874 ***	(−3.64)	−0.888 ***	(−3.41)
id174	1.453	(1.39)	0.361	(0.38)	−1.484 ***	(−6.68)	−1.288 ***	(−4.64)
id175	−0.688	(−0.99)	1.615 ***	(3.66)	0.114	(0.57)	0.333	(1.64)
id176	−1.666	(−1.82)	−0.313	(−0.81)	0.530	(0.98)	−0.0614	(−0.13)
id177	−2.576 **	(−3.13)	−1.675 ***	(−3.66)	−0.245	(−1.10)	0.187	(0.75)
id178	−0.823	(−0.51)	0.510	(1.04)	0.213	(1.00)	0.0436	(0.18)
id179	−19.54 ***	(−27.56)	−1.071	(−1.13)	−1.386 ***	(−4.24)	−1.021	(−1.91)
id180	−4.457 ***	(−3.85)	−2.934 ***	(−3.90)	−0.402 *	(−2.17)	−0.277	(−1.44)
id181	−1.850 *	(−2.18)	−0.909 *	(−2.21)	−0.573	(−1.78)	−0.354	(−1.41)
id182	−4.754 **	(−3.03)	−2.082 **	(−2.75)	0.387	(1.35)	0.409	(1.60)
log-likelihood	−952.2391		−1519.954		−3479.969		−3488.38	
AIC	2292.478		3427.908		7347.937		7364.76	
BIC	3261.657		4397.086		8317.116		8333.939	
Observation	1092		1092		1092		1092	

*** $p < 0.01$, ** $p < 0.05$, * $p < 0.1$.

Table A5. Driving risk scores for four near-miss events after winsorizing and Min-Max scaling on regression coefficients.

Variable	Overspeed	Highspeedbrake	Harshacceleration	Harshdeceleration
id1	4.824741	4.344986	2.622834	2.52286
id2	1.242371	4.033808	3.753797	3.75
id3	2.476298	1.628204	4.413078	4.113936
id4	2.578824	1.749502	4.312131	4.315106
id5	1.133814	3.574272	1.557084	1.542612
id6	4.646467	4.258833	3.576023	3.377835
id7	4.434299	4.244682	3.208862	3.286394
id8	2.244711	3.976736	2.4531	2.636247
id9	4.685306	4.334427	2.398606	2.423189
id10	4.449618	4.107521	2.314633	2.288771
id11	4.580368	4.164127	2.478113	2.27414
id12	4.662871	4.224932	2.692603	2.612564
id13	4.542011	4.219333	2.499553	2.404901
id14	4.441416	4.413722	2.060925	2.1891
id15	4.673486	4.370957	2.542969	2.547549

Table A5. *Cont.*

Variable	Overspeed	Highspeedbrake	Harshacceleration	Harshdeceleration
id16	2.050515	1.186551	2.864928	2.924287
id17	2.12168	4.018102	2.444167	2.612747
id18	2.218175	4.2198	1.618724	1.364301
id19	4.704364	4.404081	3.147222	3.058705
id20	1.835814	3.761974	2.883688	2.607535
id21	2.124092	3.944234	2.910488	2.941661
id22	4.506067	4.309374	3.065928	3.014813
id23	4.729211	4.345159	2.199393	2.148866
id24	1.923866	1.063697	2.328926	2.428676
id25	2.207319	1.317181	1.834912	1.854426
id26	4.494367	4.189475	2.229766	1.912948
id27	1.957639	1.080804	2.581383	2.582846
id28	2.621041	1.695073	2.661426	2.805413
id29	4.687598	4.20938	3.150795	3.094367
id30	2.274866	1.419818	2.42362	2.42959
id31	4.77565	4.246703	1.597284	1.655084
id32	4.432128	3.890427	2.524567	2.653621
id33	2.476298	1.503794	2.464713	2.382955
id34	4.531518	4.10441	2.617715	2.482718
id35	4.362531	4.183099	2.080579	2.018105
id36	4.325984	3.970049	2.368233	2.209217
id37	0.021711	4.153396	2.194033	2.323519
id38	2.317082	4.132869	1.297123	1.497805
id39	0.024124	5	5	5
id40	4.664922	4.280294	2.374486	2.59519
id41	4.53007	4.186365	2.783634	2.664594
id42	4.768412	4.413722	3.165088	2.886796
id43	4.563723	4.379043	2.590763	2.605157
id44	4.527417	4.320727	2.922994	2.950805
id45	4.570236	4.140489	2.716634	2.780724
id46	4.840663	4.270341	2.344113	2.307974
id47	4.468072	4.193363	1.860818	1.923007
id48	4.883362	4.559747	3.363409	3.242502
id49	1.672979	4.100056	2.115419	2.200073
id50	4.51644	4.048426	2.253886	2.22019
id51	2.268835	1.384051	3.62605	3.565289
id52	2.192845	1.124347	2.306593	1.652341
id53	2.260391	1.348283	2.50402	2.705743
id54	4.236002	4.176723	2.010005	1.938552
id55	4.307288	4.075796	3.044488	2.95812
id56	4.527778	4.331519	2.729141	2.867593
id57	2.18802	4.236128	2.721101	3.180322
id58	4.59834	4.237372	2.315526	2.583211
id59	0	4.333961	1.974272	1.986101
id60	4.448773	3.752022	1.987672	2.006218
id61	4.923769	4.512628	1.472217	1.538954
id62	4.768654	4.429895	2.024299	2.090344
id63	2.165103	1.309405	3.86725	3.906364
id64	4.519697	4.079528	3.814544	3.747257
id65	2.171134	1.334287	0.904949	1.345099
id66	4.657202	4.110164	3.073075	3.410753
id67	1.641618	3.816248	2.291406	2.244879

Table A5. *Cont.*

Variable	Overspeed	Highspeedbrake	Harshacceleration	Harshdeceleration
id68	2.459412	1.607987	1.474004	1.004938
id69	1.78636	1.116571	2.096659	1.951353
id70	2.194051	3.770683	1.292657	1.150329
id71	0.937206	0	3.147222	3.631127
id72	4.157238	4.180455	2.161872	1.934894
id73	4.303307	3.899757	1.25871	1.130212
id74	0.979422	0	2.889941	3.222385
id75	4.478807	4.084194	2.384313	2.439557
id76	2.198876	0.898855	1.152403	0.782736
id77	2.366536	1.433814	3.368769	3.189466
id78	1.845464	3.838019	0.07236	0
id79	4.408728	4.24888	2.41558	2.422275
id80	2.130123	1.197437	2.609523	2.569404
id81	0.595856	4.168947	2.318206	2.215618
id82	4.656961	4.260855	2.344113	2.224762
id83	4.641884	4.369402	2.512953	2.301573
id84	1.823752	4.054802	1.952832	1.741039
id85	2.061371	3.747356	1.345364	1.317666
id86	1.21101	4.028365	1.486511	1.50695
id87	4.601476	4.505785	2.797927	2.911485
id88	2.341206	1.475802	2.43702	2.371982
id89	1.641618	0.869308	3.605503	3.560717
id90	4.341302	4.214201	1.754511	1.624909
id91	2.40634	1.387161	3.226729	3.270849
id92	1.571659	0.774446	3.380382	3.129115
id93	1.873206	1.03415	2.363766	2.248537
id94	4.501001	4.184188	2.48258	2.535571
id95	1.736907	0.925292	2.173486	1.910205
id96	1.999855	1.163225	3.850277	3.750914
id97	2.23868	1.320291	2.208326	2.264996
id98	2.526958	1.600211	1.367697	1.711778
id99	2.583649	1.687298	1.038056	1.271031
id100	4.315007	1.200547	3.339289	3.276335
id101	2.014329	1.197437	3.475969	3.267191
id102	1.72967	0.847537	2.650348	2.495977
id103	2.619835	1.71218	3.107022	3.076993
id104	2.34	1.460251	2.739861	2.762436
id105	4.411502	4.278116	2.208326	1.67703
id106	1.711577	0.852202	2.906021	2.753292
id107	2.215762	1.324956	2.635698	2.657279
id108	1.917835	3.933348	2.147579	1.942209
id109	2.236268	1.290744	2.312846	1.807791
id110	2.288134	1.175666	1.013936	1.568215
id111	4.414035	4.146398	2.777381	2.703914
id112	2.390659	1.522456	1.324817	1.762985
id113	1.147082	3.878919	1.101483	1.337783
id114	1.90336	1.068363	3.191889	3.013899
id115	2.062577	1.197437	2.606843	2.423189
id116	2.119268	3.761352	1.417724	1.485004
id117	2.253154	1.382496	2.483473	2.273226
id118	2.469061	4.235351	2.726461	2.525942

Table A5. *Cont.*

Variable	Overspeed	Highspeedbrake	Harshacceleration	Harshdeceleration
id119	2.30502	1.421373	2.4933	2.392099
id120	1.435361	0.55051	2.607647	2.644477
id121	1.255639	4.23815	2.836341	2.876737
id122	2.133742	3.953875	3.179382	3.119971
id123	1.987793	3.739736	3.246382	3.333029
id124	4.309941	3.938791	3.246382	2.852963
id125	4.363014	1.066808	2.887261	3.088881
id126	4.58073	4.005505	3.137395	3.015728
id127	2.083082	1.265862	3.657316	3.446416
id128	2.225412	1.262752	3.303555	3.087966
id129	4.638989	4.465819	2.647847	2.673738
id130	1.741732	0.892635	1.21315	1.027798
id131	4.624635	4.282315	1.747365	1.796818
id132	2.42202	1.567554	1.851885	2.125091
id133	1.568041	3.893381	2.067179	1.891002
id134	0.916701	3.937858	3.058781	3.598208
id135	1.906979	1.087024	3.453636	3.757315
id136	5	5	5	5
id137	1.893711	3.954964	2.66214	3.258047
id138	2.485948	1.635979	1.884045	1.709034
id139	4.329602	4.160394	1.799178	1.903804
id140	4.466504	4.249347	1.482937	1.576445
id141	1.759824	0.892635	1.666071	1.517008
id142	0.219526	4.209225	2.6675	2.682882
id143	0.171278	4.283248	2.294086	2.379298
id144	4.754179	4.430673	2.64615	2.675567
id145	4.525004	4.14702	1.952832	1.945867
id146	1.947989	4.150597	1.613364	1.829737
id147	4.480858	4.332608	1.710738	1.703548
id148	4.37399	4.596448	2.579686	2.475311
id149	4.438521	4.160861	2.127926	2.209217
id150	4.388585	4.204249	1.580311	1.695318
id151	4.781077	4.163505	1.145256	0.989393
id152	0.724917	3.761663	1.353404	1.259144
id153	4.631993	4.318705	0	0
id154	2.120474	3.922618	1.101483	1.166789
id155	4.31356	3.965383	1.944792	1.77853
id156	0	4.118562	2.095766	2.130578
id157	4.599426	4.397394	1.950152	2.063826
id158	2.434082	1.548893	3.228515	3.203182
id159	2.434082	1.480468	3.361622	3.176664
id160	4.367597	4.430518	2.755047	2.831017
id161	2.072226	1.250311	3.165982	2.974579
id162	2.325525	1.450921	3.007861	2.683797
id163	2.043278	1.16478	4.270145	4.037125
id164	4.472173	3.947344	2.403966	2.247623
id165	2.233855	1.31096	3.820797	3.846013
id166	1.624732	3.965694	2.481687	2.474031
id167	4.499554	3.703658	3.622476	3.518654
id168	4.133476	4.043294	2.662766	2.533376
id169	1.57769	0.707577	3.045381	2.908742
id170	2.962391	2.184934	0	1.897403
id171	4.384484	4.112186	2.315526	2.23665
id172	4.383398	4.041584	2.241379	2.187271

Table A5. *Cont.*

Variable	Overspeed	Highspeedbrake	Harshacceleration	Harshdeceleration
id173	4.458543	3.800697	1.842058	1.710863
id174	5	4.401126	1.297123	1.345099
id175	4.741756	4.596137	2.724674	2.827359
id176	4.623791	4.296311	3.096302	2.466715
id177	4.514028	4.084505	2.403966	2.693855
id178	4.725472	4.424297	2.813114	2.562729
id179	2.40634	4.178434	1.38467	1.589247
id180	4.287144	3.888716	2.263713	2.269568
id181	4.601597	4.203627	2.110952	2.199159
id182	4.2512	4.021212	2.968555	2.896854

References

1. Guillen, M.; Nielsen, J.P.; Pérez-Marín, A.M. Near-miss telematics in motor insurance. *J. Risk Insur.* **2021**, 1–21. [CrossRef]
2. Guillen, M.; Nielsen, J.P.; Pérez-Marín, A.M.; Elpidorou, V. Can automobile insurance telematics predict the risk of near-miss events? *N. Am. Actuar. J.* **2020**, *24*, 141–152. [CrossRef]
3. Litman, T. *Distance-Based Vehicle Insurance Feasibility, Costs and Benefits*; Comprehensive Technical Report; Victoria Transport Policy Institute: Victoria, BC, Canada, 2011.
4. Tselentis, D.I.; Yannis, G.; Vlahogianni, E.I. Innovative insurance schemes: Pay as/how you drive. *Transp. Res. Procedia* **2016**, *14*, 362–371. [CrossRef]
5. Paefgen, J.; Staake, T.; Thiesse, F. Evaluation and aggregation of pay-as-you-drive insurance rate factors: A classification analysis approach. *Decis. Support Syst.* **2013**, *56*, 192–201. [CrossRef]
6. Tselentis, D.I.; Yannis, G.; Vlahogianni, E.I. Innovative motor insurance schemes: A review of current practices and emerging challenges. *Accid. Anal. Prev.* **2017**, *98*, 139–148. [CrossRef]
7. Troncoso, C.; Danezis, G.; Kosta, E.; Balasch, J.; Preneel, B. Pripayd: Privacy-friendly pay-as-you-drive insurance. *IEEE Trans. Dependable Secur. Comput.* **2010**, *8*, 742–755. [CrossRef]
8. Pesantez-Narvaez, J.; Guillen, M.; Alcañiz, M. Predicting Motor Insurance Claims Using Telematics Data—XGBoost versus Logistic Regression. *Risks* **2019**, *7*, 70. [CrossRef]
9. Guillen, M.; Nielsen, J.P.; Ayuso, M.; Pérez-Marín, A.M. The use of telematics devices to improve automobile insurance rates. *Risk Anal.* **2019**, *39*, 662–672. [CrossRef]
10. Sun, S.; Bi, J.; Guillen, M.; Pérez-Marín, A.M. Assessing driving risk using internet of vehicles data: An analysis based on generalized linear models. *Sensors* **2020**, *20*, 2712. [CrossRef]
11. De Diego, I.M.; Siordia, O.S.; Crespo, R.; Conde, C.; Cabello, E. Analysis of hands activity for automatic driving risk detection. *Transp. Res. Part C Emerg. Technol.* **2013**, *26*, 380–395. [CrossRef]
12. Siordia, O.S.; de Diego, I.M.; Conde, C.; Cabello, E. Subjective traffic safety experts' knowledge for driving-risk definition. *IEEE Trans. Intell. Transp. Syst.* **2014**, *15*, 1823–1834. [CrossRef]
13. Charlton, S.G.; Starkey, N.J.; Perrone, J.A.; Isler, R.B. What's the risk? A comparison of actual and perceived driving risk. *Transp. Res. Part F Traffic Psychol. Behav.* **2014**, *25*, 50–64. [CrossRef]
14. Peng, J.; Shao, Y. Intelligent method for identifying driving risk based on V2V multisource big data. *Complexity* **2018**, *2018*. [CrossRef]
15. Wang, J.; Zheng, Y.; Li, X.; Yu, C.; Kodaka, K.; Li, K. Driving risk assessment using near-crash database through data mining of tree-based model. *Accid. Anal. Prev.* **2015**, *84*, 54–64. [CrossRef]
16. Yan, L.; Zhang, Y.; He, Y.; Gao, S.; Zhu, D.; Ran, B.; Wu, Q. Hazardous traffic event detection using Markov Blanket and sequential minimal optimization (MB-SMO). *Sensors* **2016**, *16*, 1084. [CrossRef]
17. Liao, Y.; Wang, M.; Duan, L.; Chen, F. Cross-regional driver–vehicle interaction design: An interview study on driving risk perceptions, decisions, and ADAS function preferences. *IET Intell. Transp. Syst.* **2018**, *12*, 801–808. [CrossRef]
18. Jiang, K.; Yang, D.; Xie, S.; Xiao, Z.; Victorino, A.C.; Charara, A. Real-time estimation and prediction of tire forces using digital map for driving risk assessment. *Transp. Res. Part C Emerg. Technol.* **2019**, *107*, 463–489. [CrossRef]
19. Yan, Y.; Dai, Y.; Li, X.; Tang, J.; Guo, Z. Driving risk assessment using driving behavior data under continuous tunnel environment. *Traffic Inj. Prev.* **2019**, *20*, 807–812. [CrossRef]
20. Lu, J.; Xie, X.; Zhang, R. Focusing on appraisals: How and why anger and fear influence driving risk perception. *J. Saf. Res.* **2013**, *45*, 65–73. [CrossRef]
21. Wang, J.; Huang, H.; Li, Y.; Zhou, H.; Liu, J.; Xu, Q. Driving risk assessment based on naturalistic driving study and driver attitude questionnaire analysis. *Accid. Anal. Prev.* **2020**, *145*, 105680. [CrossRef]
22. Handel, P.; Skog, I.; Wahlstrom, J.; Bonawiede, F.; Welch, R.; Ohlsson, J.; Ohlsson, M. Insurance telematics: Opportunities and challenges with the smartphone solution. *IEEE Intell. Transp. Syst. Mag.* **2014**, *6*, 57–70. [CrossRef]

23. Joubert, J.W.; De Beer, D.; De Koker, N. Combining accelerometer data and contextual variables to evaluate the risk of driver behaviour. *Transp. Res. Part F Traffic Psychol. Behav.* **2016**, *41*, 80–96. [CrossRef]

24. Verbelen, R.; Antonio, K.; Claeskens, G. Unravelling the predictive power of telematics data in car insurance pricing. *J. R. Stat. Soc. Ser. C Appl. Stat.* **2018**, *67*, 1275–1304. [CrossRef]

25. Ma, Y.L.; Zhu, X.; Hu, X.; Chiu, Y.C. The use of context-sensitive insurance telematics data in auto insurance rate making. *Transp. Res. Part A Policy Pract.* **2018**, *113*, 243–258. [CrossRef]

26. Jiang, Y.; Zhang, J.; Wang, Y.; Wang, W. Drivers' behavioral responses to driving risk diagnosis and real-time warning information provision on expressways: A smartphone app–based driving experiment. *J. Transp. Saf. Secur.* **2020**, *12*, 329–357. [CrossRef]

27. Jin, W.; Deng, Y.; Jiang, H.; Xie, Q.; Shen, W.; Han, W. Latent class analysis of accident risks in usage-based insurance: Evidence from Beijing. *Accid. Anal. Prev.* **2018**, *115*, 79–88. [CrossRef]

28. Carfora, M.F.; Martinelli, F.; Mercaldo, F.; Nardone, V.; Orlando, A.; Santone, A.; Vaglini, G. A "pay-how-you-drive" car insurance approach through cluster analysis. *Soft Comput.* **2019**, *23*, 2863–2875. [CrossRef]

29. Burton, A.; Parikh, T.; Mascarenhas, S.; Zhang, J.; Voris, J.; Artan, N.S.; Li, W. Driver identification and authentication with active behavior modeling. In Proceedings of the 2016 12th International Conference on Network and Service Management (CNSM), Montreal, QC, Canada, 31 October–4 November 2016; pp. 388–393.

30. Baecke, P.; Bocca, L. The value of vehicle telematics data in insurance risk selection processes. *Decis. Support Syst.* **2017**, *98*, 69–79. [CrossRef]

31. Guelman, L. Gradient boosting trees for auto insurance loss cost modeling and prediction. *Expert Syst. Appl.* **2012**, *39*, 3659–3667. [CrossRef]

32. Bian, Y.; Yang, C.; Zhao, J.L.; Liang, L. Good drivers pay less: A study of usage-based vehicle insurance models. *Transp. Res. Part A Policy Pract.* **2018**, *107*, 20–34. [CrossRef]

33. Jafarnejad, S.; Castignani, G.; Engel, T. Towards a real-time driver identification mechanism based on driving sensing data. In Proceedings of the 2017 IEEE 20th International Conference on Intelligent Transportation Systems (ITSC), Yokohama, Japan, 16–19 October 2017; pp. 1–7.

34. Paefgen, J.; Staake, T.; Fleisch, E. Multivariate exposure modeling of accident risk: Insights from Pay-as-you-drive insurance data. *Transp. Res. Part A Policy Pract.* **2014**, *61*, 27–40. [CrossRef]

35. Boucher, J.P.; Pérez-Marín, A.M.; Santolino, M. Pay-as-you-drive insurance: The effect of the kilometers on the risk of accident. In *Anales del Instituto de Actuarios Españoles*; Instituto de Actuarios Españoles: Madrid, Spain, 2013; Volume 19, pp. 135–154.

36. Sun, S.; Bi, J.; Ding, C. Cleaning and Processing on the Electric Vehicle Telematics Data. In Proceedings of the INFORMS International Conference on Service Science, Nanjing, China, 27–29 June 2019; Springer: Berlin/Heidelberg, Germany, 2019; pp. 1–6.

37. Gao, G.; Wüthrich, M.V.; Yang, H. Evaluation of driving risk at different speeds. *Insur. Math. Econ.* **2019**, *88*, 108–119. [CrossRef]

38. Gao, G.; Wang, H.; Wüthrich, M.V. Boosting Poisson regression models with telematics car driving data. *Mach. Learn.* **2021**, 1–30. [CrossRef]

39. So, B.; Boucher, J.P.; Valdez, E.A. Synthetic Dataset Generation of Driver Telematics. *Risks* **2021**, *9*, 58. [CrossRef]

Article

A Proposal of a Motion Measurement System to Support Visually Impaired People in Rehabilitation Using Low-Cost Inertial Sensors

Karla Miriam Reyes Leiva [1,2,*]**, Milagros Jaén-Vargas** [1]**, Miguel Ángel Cuba** [1]**, Sergio Sánchez Lara** [1] **and José Javier Serrano Olmedo** [1,3]

1 Escuela Superior Técnica de Ingenieros de Telecomunicación, Universidad Politécnica de Madrid, 28013 Madrid, Spain; milagros.jaen@ctb.upm.es (M.J.-V.); ma.cuba@alumnos.upm.es (M.Á.C.); sergio.sanchez.lara@alumnos.upm.es (S.S.L.); josejavier@ctb.upm.es (J.J.S.O.)
2 Engineering Faculty, Universidad Tecnológica Centroamericana UNITEC, San Pedro Sula 211001, Honduras
3 Networking Center of Biomedical Research for Bioengineering Biomaterials and Nanomedicine, Instituto de Salud Carlos III, 28029 Madrid, Spain
* Correspondence: karla.reyes@ctb.upm.es

Abstract: The rehabilitation of a visually impaired person (VIP) is a systematic process where the person is provided with tools that allow them to deal with the impairment to achieve personal autonomy and independence, such as training for the use of the long cane as a tool for orientation and mobility (O&M). This process must be trained personally by specialists, leading to a limitation of human, technological and structural resources in some regions, especially those with economical narrow circumstances. A system to obtain information about the motion of the long cane and the leg using low-cost inertial sensors was developed to provide an overview of quantitative parameters such as sweeping coverage and gait analysis, that are currently visually analyzed during rehabilitation. The system was tested with 10 blindfolded volunteers in laboratory conditions following constant contact, two points touch, and three points touch travel techniques. The results indicate that the quantification system is reliable for measuring grip rotation, safety zone, sweeping amplitude and hand position using orientation angles with an accuracy of around 97.62%. However, a new method or an improvement of hardware must be developed to improve gait parameters' measurements, since the step length measurement presented a mean accuracy of 94.62%. The system requires further development to be used as an aid in the rehabilitation process of the VIP. Now, it is a simple and low-cost technological aid that has the potential to improve the current practice of O&M.

Keywords: absolute orientation; inertial sensors; orientation and mobility; visually impaired rehabilitation

Citation: Reyes Leiva, K.M.; Jaén-Vargas, M.; Cuba, M.Á.; Lara, S.S.; Olmedo, J.J.S. A Proposal of a Motion Measurement System to Support Visually Impaired People in Rehabilitation Using Low-Cost Inertial Sensors. *Entropy* **2021**, *23*, 848. https://doi.org/10.3390/e23070848

Academic Editors: Felipe Ortega and Emilio López Cano

Received: 28 April 2021
Accepted: 29 June 2021
Published: 1 July 2021

Publisher's Note: MDPI stays neutral with regard to jurisdictional claims in published maps and institutional affiliations.

1. Introduction

People with visual impairments face many daily challenges that limit their quality of life. These challenges include basic life activities such as finding and keeping a job, mobility, and displacement, using public transport, among others. When a person is born with a visual disability or suffers from a traumatism or disease that leads to a visual impairment, they must be assisted trough a rehabilitation process. During this rehabilitation process, the person is provided with tools to help them deal with their visual impairments with greater independence and self-confidence. Tools as learning braille, learning how to use a long cane, sightless feeding, also to optimize the use of residual vision and teaching skills in order to improve visual functioning in daily life as well as other daily activities as O&M trained by specialists [1–6]. This process of rehabilitation is specialized according to the cognitive capacities of each user, the regular rehabilitation programs worldwide, as reported by the World Blind Union, which includes several stages, such as activities of daily living services, career exploration services, travel-training services/O&M, and others [7]. In several references [6,8–12] the emphasis and importance of the O&M service and training in order to improve the quality of life, is widely emphasized [5,13]. Therefore, there is a specific health discipline in charge of the study, development, and improvement of the O&M training in VIP [14–16]. The latest report of the international approaches to rehabilitation programs from the World Blind Union [7] presents two important challenges on which this project was motivated: (1) the limitation of resources to provide basic rehabilitation services and (2) transportation and geographic limitations, where many VIP must displace themselves to other cities in order to access the rehabilitation services which, in some cases, is impossible for some VIP.

A fundamental part of the mobility training is the use of the long cane, the VIP should learn how to hold it correctly, how to grip it, how to walk with it and sweep it in order to detect obstacles, different techniques of exploration, and other parameters according to the complexity of the environment in which the VIP will navigate [17,18]. This training is usually done in person with an O&M specialist, which, as mentioned before, leads to an accessibility problem in rural communities, also it compromises the rehabilitation duration, as well as the number of VIP that can be rehabilitated at the same time. In this training, depending on the scenario there is a recommended technique and according to the complexity and advances of the training, the scenarios will change [19]. However, the parameters for evaluation of the correct use, regardless of the change of scenario, will remain the same; this allows the possibility to register parameters and quantitative values of the motion of the person and the long cane [20], in order to support the O&M training in the rehabilitation processes.

According to the literature, a diversity of technological proposals have been designed for orientation and mobility, such as ETAS (Electronic Travel Aid Systems) [2], focused on obtaining information from the environment and providing it to the visually impaired in order to assist them in autonomous navigation. There have been many attempts to enhance the long cane with technology [21–25]. These systems are developed from technologies such as Global Positioning System, BLE beacons, RFID or radio frequency identification, to obtain information on position and displacement and optical sensors (RGB-D cameras, laser), inertial sensors, speed sensors among others for obtaining information regarding object detection [26–32]. However, the use of any of these ETAS requires previous O&M training [33,34], leading to an existing gap, which is the development of assistive technologies specifically focused on evaluation and assistance of the training process, so it can be more accessible for users.

Three articles of assistive rehabilitation tools for O&M were found in the literature; Schloerb et al. [35] developed a virtual environment system named BlindAid, created in order to enhance the O&M training. This is a software with haptic and auditory feedback in which the user can virtually visit different unknown places in order to create cognitive mental maps of the representation of these places. Oliveira et al. [36] created a programming language named GoDonnie, to be used as a tool to aid in the resolution of spatial problems involving O&M. This programming language was developed considering the criteria of accessibility and usability for VIP, with the assumption that by using GoDonnie, the user could improve programming and O&M skills, since the users are able to create mental maps of the environments and related objects. On the other hand, Gong et al. [37] developed HeliCoach an O&M training system created to help VIP to train the ability of audio orientation. This training environment is composed of a drone, which moves through 3D space and is used as a sound source. It is composed of a belt with a set of vibration motors for haptic feedback, the belt also contains an BNO055 IMU and six vibration motors controlled by an Arduino DFRobot Leonardo + Xbee. In this system high accuracy indoor localization system is needed for the perspective-driven interaction. For this goal, Ultra-Wide Bandwidth Microwave is used: the system uses four base stations and two tracking tags which are embedded into the drone and the cap of the user, respectively.

In comparison to the mentioned developed technologies, the aim of this research was to develop a simple-architecture hardware system using low-cost inertial sensors for data acquisition and test its reliability in the quantitative analysis of the parameters evaluated in the rehabilitation process of VIP by obtaining metrics that the O&M specialists personally examined to aid the rehabilitators during current practice of O&M while training travel techniques.

The system can provide information about the hand grip rotation, the safety zone, the hand height during the travel techniques, amplitude and patterns of the sweeping, and gait parameters with a high accuracy using only two inertial sensors.

Technologies based on inertial measurement unit sensors (IMU) are used in a large and ever-growing number of applications, such as intelligence guidance, self-driving robots [38,39], full body motion tracking [40–43] and navigation [26,44–46]. An accelerometer measures the external specific force acting on the sensor, which consists of both the sensor's acceleration and the acceleration due to the earth's gravity. A gyroscope measures angular velocity: the rate of change of the sensor's orientation. Thus, the integration of gyroscope measurements provides information about the orientation on the sensor. Magnetometers complement accelerometers by providing sensor heading (orientation around the gravity vector), which is information that accelerometers or gyroscopes cannot provide. With the fusion of accelerometer, gyroscope and magnetometers, the orientation is estimated based on the direction of the magnetic field [39,44]. In the system presented in this paper, the parameters of O&M are calculated using absolute orientation values of the sensor fusion provided by the BNO055 IMU module. Note that the present article is an extended version of [47], where the algorithms to measure amplitude of the sweeping techniques and the orientation of the long cane were tested with 97% and 98% accuracy, respectively.

2. Materials and Methods

A tool was developed to evaluate the rehabilitation parameters during the experimental procedure. For the data acquisition an Arduino MKR1010 microprocessor was used with two 9DOF BNO055 IMU Bosch sensors. One sensor placed on the outer side of the leg of each participant and the other on the higher part of a 117 cm long cane. Serial communication was done via I^2C protocol at a sample rate of 0.01 s. In order to remove noise components from the signal, a low pass filtering was performed, with a cutoff frequency of 20 Hz. The microprocessor was wired to a SD card module via SPI protocol and to two push buttons settled as input parameters to control the acquisitions manually. With the use of the Euler roll angle θ_{leg} and the interpretation of step detection according to the values of the filtered absolute orientation, an algorithm was developed to calculate step length using the local coordinates of the sensor placed in the leg. Additionally, to obtain the sweeping metrics with the local coordinates of sensor placed in the cane, the Euler roll φ_{cane}, pitch θ_{cane} and yaw γ_{cane} angles were used to provide the grip rotation, the safety zone metrics and sweeping characteristics consecutively.

For the experimental procedure, the acquisitions were performed with 10 blindfolded volunteers. First, the volunteers were instructed and trained for each travel technique while sighted. A floor carrel was marked for the sweep training with an amplitude of around 1 m, they were asked to train each technique walking 20 steps three times. After that, they were blindfolded and asked to perform the travel techniques when displacing around 20 steps in the indicated direction, as described in Table 1. Each acquisition was repeated blindfolded three times, obtaining nine comparative metrics for each participant. The total time and displacement were measured using a 50 m measuring tape and a chronometer. This value served as references values to evaluate the accuracy of the measured gait parameters.

Table 1. Description and representation of the top view of the travel techniques for the experimental evaluation of the developed system.

Constant Contact Technique (CCT)	Two Points Touch Technique (2PT)	Three Points Touch Technique (3PT)
	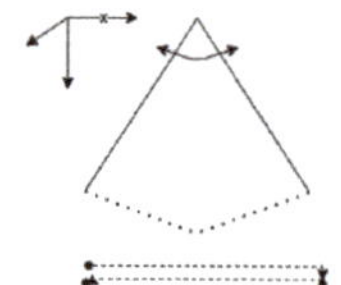	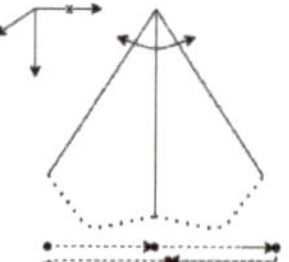
The CCT travel technique consisted of sweeping the long cane on the floor between two points with constant contact with an approximate amplitude of 1 m in order to provide coverage of the walking path.	The 2PT travel technique consisted of sweeping the long cane on the floor between two points taking the cane off the ground and creating an arc of around 5 cm, with an approximate amplitude of 1 m.	The 3PT travel technique consisted of sweeping the long cane on the floor between three points. One point on the left, one on the center and one on the right. Taking the cane off the ground in each point and creating an arc of around 5 cm.

3. Results

3.1. Measurement of the Hand Height and the Safety Zone

The Hand Height (HH) and Safety Zone (SZ) are reference parameters to evaluate the reaction distance in O&M, which refers to the warning distance provided by the cane of an object in one's path, the time that is provided by the cane to be warmed about an object or danger [48]. By implementing trigonometrical ratios and using the local coordinates of the sensor, the pitch angle θ_{cane} (which is the transversal axis, equivalent to the angle produced between the floor plane and the long cane) was continuously measured to obtain the height of the hand during and the distance between the tip of the cane and the leg, in the repetitions of the three different travel techniques. Being the HH, the opposite leg of the θ_{cane}, the SZ then is the adjacent leg from the θ_{cane}, as shown in Figure 1.

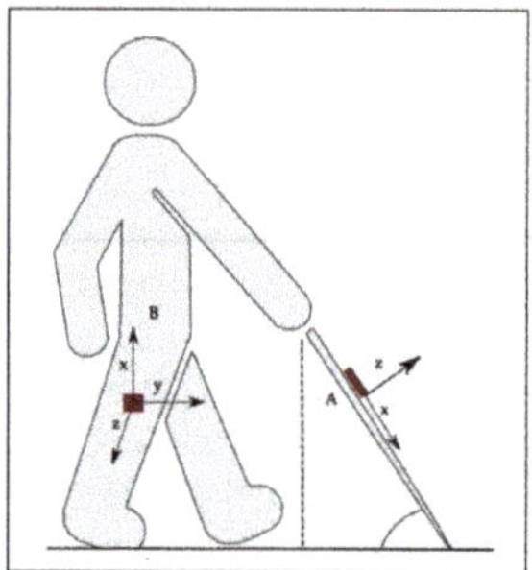

Figure 1. Local coordinate system of the sensor placed on the cane (A) and local coordinate system of the sensor placed in the leg (B).

An extract of the measured values for HH and SZ for each subject is presented in Table 2. This value is compared with the real value (RV), which is the self-reported HH and the calculated SZ according to Pythagoras theorem. The mean value is the calculated media of the HH and SZ measurements within the nine travel technique acquisitions. The values of standard deviation (SD) and %Error vary for each subject. The major precision and accuracy obtained was with S01, being the standard deviation of only 1.39 cm, which represents 1.46% of the mean HH and 1.95 cm which represents 2.83% of the mean SZ and the %Error of 0.63% and 1.37%, respectively. Additionally, S09 presented a very low %Error, however a high SD (6.15) which together with S05 presented the less precision on repeatability, the SD being 5.03% of the mean HH and 8.60% of the SZ. On the other hand, the lower accuracy was shown by S06, followed by S07 and S08 with a %Error of 4.62% in the HH and 4.86% in the SZ measurement. Finally, a media accuracy of around 97.62% was obtained by joining all the subjects in both measurements proving that the algorithm applied is reliable to measure these O&M parameters using absolute orientation angles.

Table 2. Extract of the measured Hand Height and Safety Zone and statistic characteristics.

	RV cm	Mean cm	SD cm	%Error	RV cm	Mean cm	SD cm	%Error
	Hand Height (HH)				Safety Zone (SZ)			
S01	94.00	94.59	1.39	0.63	69.66	68.71	1.95	1.37
S02	89.00	90.17	3.17	1.31	76.00	74.34	3.77	2.18
S03	95.00	94.28	3.37	0.76	68.29	69.35	5.03	1.55
S04	86.00	82.64	2.60	3.90	79.32	82.68	2.59	4.24
S05	94.00	92.56	4.66	1.53	68.66	71.04	6.11	3.47
S06	86.00	82.03	3.75	4.62	79.32	83.17	3.54	4.86
S07	87.00	90.85	2.38	4.43	77.10	73.50	2.86	4.68
S08	88.00	84.13	2.51	4.40	78.23	80.21	4.35	2.54
S09	82.00	81.86	6.15	0.17	83.46	83.08	6.32	0.46
S10	83.00	81.15	2.50	2.23	82.46	84.19	2.39	2.09

3.2. Measurement of the Grip Rotation

A proper grip was one of the first parameters to be observed by the rehabilitators during the very first stage of the O&M training. With the inertial sensors, is not possible to analyze all the characteristics of the grip, but it is possible to determine the variation of the rotation of the cane which is the consequence of the grip rotation by analyzing the absolute orientation angles, as shown in Table 3. In this table, the SD in degrees for each travel technique by subject was calculated and presented. For this, it was taken into account the total raw data of the roll angle φ_{cane}, which according to the local coordinates of the placed sensor represents the rotation of the grip of the user during the development of the travelling techniques. As shown in the Table 3, this value can be representative for technical analysis of the performance of the traveling techniques independently of the stage of and scene in which the user is being rehabilitated. It can also provide a numerical representation to establish what is considered as adequate and acceptable grip rotation according to each travel technique.

Note that the variation of the values represents the percentage of rotation of the grip during each experiment which means that each column represents how much variation in the rotation of the hand occurred during the experimental acquisition. In the results, it can be observed that S04 and S10 present less grip rotation in the 2P and 3P techniques, which is an indication of a better execution than for instance for S03 and S09. This is direct indication for the specialist to determine which is the acceptable percentage of rotation for each travelling technique and which technique is more appropriate for the visually impaired; it can also allow to have a tracking of the performance during the rehabilitation stages.

Table 3. Standard deviation of the measured grip rotation for each subject in the acquisitions of the different travelling techniques.

SD in Degrees														
S01			S02			S03			S04			S05		
CCT	2PT	3PT	CCT	2PT	3PT	CCT	2PT	3PT	CCT	2PT	3PT	CCT	2PT	3PT
4.55	7.21	5.34	3.44	3.05	2.82	4.34	4.43	2.88	3.72	3.12	2.13	4.45	4.28	3.01
5.68	6.47	4.97	2.76	4.12	3.84	1.88	4.45	3.06	1.93	2.11	2.03	3.67	3.49	3.01
5.31	6.29	4.25	6.14	4.93	3.09	4.84	5.18	2.9	2.66	2.12	2.22	3.43	3.42	3.28

S06			S07			S08			S09			S10		
CCT	2PT	3PT	CCT	2PT	3PT	CCT	2PT	3PT	CCT	2PT	3PT	CCT	2PT	3PT
3.1	3.92	3.92	8.37	5.61	3.92	5.68	3.63	3.5	7.43	7.43	6.19	2.17	2.88	2.47
3.47	3.06	3.06	7.4	6.55	4.72	6.8	4.16	2.91	8.84	7.12	4.4	6.15	2.98	2.41
2.97	2.84	2.84	7.97	6.24	4.5	6.18	5.1	3.11	6.13	7.15	5.75	4.35	3.1	2.21

3.3. Representation of the Sweeping

In [47], it was clearly demonstrated that using absolute orientation angles was reliable to measure the amplitude of the sweepings with the long cane. As described by Blasch and LaGrow [48], the performance of the O&M rehabilitation can be evaluated in terms of "coverage" provided by the long cane, where a full coverage includes, for instance, object preview: the capacity to identify objects in the path of travel with a correct sweeping of the long cane. As the carried out traveling techniques consists of sweeping oscillatory movements, by extracting the motion of the yaw angle γ_{cane}, it is possible to graphically represent the movement of the long cane beside the value of the sweeping amplitude, as shown in Figure 2.

This graphical representation is indispensable in order to have an estimation of the performance of the travelling techniques while the user is in training, since it is a detailed characterization of the movement of the cane in each millisecond for the dynamic conditions. Additionally, it can help the rehabilitators to evaluate the coverage that is being provided in that moment of the execution of the travelling techniques. As well, for the user to self-correct any lack of coverage with immediate feedback to prevent an accident while correcting the amplitude and execution of the sweeping during the training. It can also help the user and the rehabilitator to quantitatively determinate which is the most appropriate travelling technique for the user. As shown in Figure 2, many differences are observed in the development of the traveling techniques for two subjects (A and B) with the same characteristics. This brings us to one last advantage of this tool, which is the possibility to register the performance of each user during the entire rehabilitation process for future data analysis.

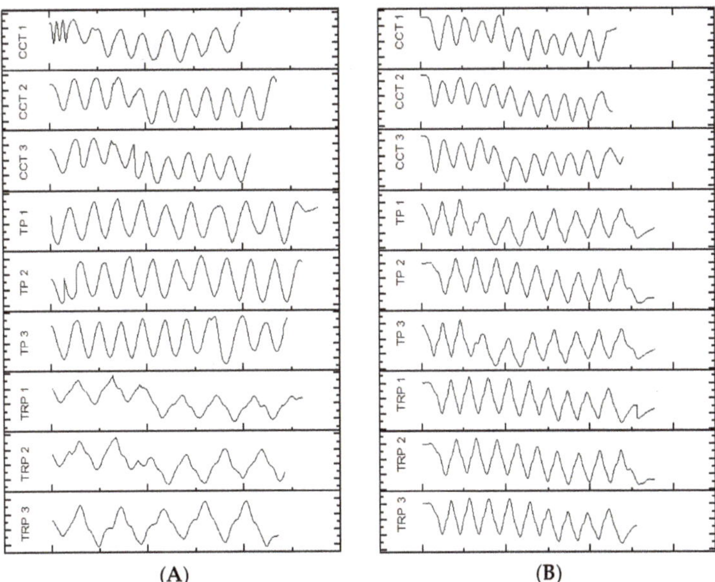

(A)   (B)

Figure 2. Sweeping preview (γ_{cane}) in the 20-step displacement for each travelling technique, S05 (**A**) and S06 (**B**).

3.4. Measurement of Gait Parameters

In terms of coverage, an appropriate gait is crucial for the development of O&M abilities [49], therefore during the O&M training, the gait velocity and the stride length is being constantly visually evaluated by the rehabilitator. With this tool, the method to evaluate the step length for calculating the gait parameters (Stride Length and Gait Velocity) was developed using also absolute orientation angles. With the inertial sensor placed on the outer side of the leg, with the same local coordinates as the sensor placed in the long cane, the pitch angle was used to calculate the step length in a walking cycle and two of the travelling techniques (see Figure 1). The step length was calculated in an algorithm averaging the estimation of the displacement of the leg during the gait cycle following the difference of each peak-to-peak representation of the oscillatory movements of the pitch angle, where each peak represents the higher value of each phase in the gait cycle. Therefore, by knowledge of the leg length of each user, and constantly laying up the values of θ_{legmax} and θ_{legmin}, the step length could be calculated using the following equation:

$$SL = 2 \times \sin\left(\frac{\theta_{legmax} - \theta_{legmin}}{2}\right) \times LL \tag{1}$$

where SL is the length of the step and LL is the length of the leg of the user. The algorithm is capable of detecting if a step is being executed with the θ_{legmax} and θ_{legmin} thresholds. Table 4 summarizes the measurements obtained in each experiment. Note that the value of the measurement of the SL is an average of the three measurements obtained for each repetition and the mean difference (MD) in centimeters is measured with the resulting three values of the average.

Table 4. Step length measurement analysis.

Activity	SL m	RSL m	MD cm	SL m	RSL m	MD cm
		S01			S06	
W	0.553	0.500		0.560	0.546	
CCT	0.517	0.405	7.046	0.490	0.443	2.769
3PT	0.577	0.539		0.500	0.478	
		S02			S07	
W	0.553	0.551		0.678	0.625	
CCT	0.590	0.601	2.704	0.731	0.716	4.224
3PT	0.577	0.583		0.664	0.607	
		S03			S08	
W	0.447	0.432		0.567	0.502	
CCT	0.530	0.534	4.937	0.581	0.536	12.370
3PT	0.483	0.542		0.572	0.540	
		S04			S09	
W	0.603	0.592		0.520	0.502	
CCT	0.563	0.603	3.333	0.500	0.536	3.047
3P	0.580	0.550		0.536	0.540	
		S05			S10	
W	0.637	0.498		0.538	0.505	
CCT	0.603	0.567	9.800	0.534	0.566	5.152
3P	0.673	0.554		0.515	0.425	

W = walking.

The difference in centimeters between the actual value and the measured value is very low in most of the cases (2.704 cm–12.370 cm), which indicates that the system is also reliable to estimate the step length, however, in order to calculate traveled distances using this value, it is necessary to set the measurement error and thus dismiss the accumulated errors. This was not possible because there is an extended variation of the mean %Error of the measurement from one subject to other, from 1.07% to 15.06%. The reason for this variation is unknown, perhaps so the proposed method does not estimate hip displacement in the gait cycle. Another reason could be the reliance on the sensor decalibration, however, the accuracy of the absolute angles sensed varies very little with calibration but, as the step length values lie in the order of centimeters, this can be a factor affecting the variation of the %Error, which has a mean of 94.62%.

4. Discussion

Kim et al. [20], presented a quantification of the characteristics of long cane usage. In this work, similar parameters are evaluated in terms of the coverage of the travelling techniques in relation to the rotation angles of the movement of the long cane. However, to develop this study, optical tracking cameras were needed in addition to an inertial sensor placed in the long cane. The presented tool allows the dynamic quantification of the characteristics of the movement of the long cane with a lower cost dispositive and complexity and with high precision. With the inertial sensors and the presented metrics, it will be possible to obtain outcome measures as stride rate, gate velocity (meters per minute), and grip characteristics. Additionally, the provided coverage and long mechanics will allow interpretations of the sweeping characteristics as amplitude, frequency and the ability to detect obstacles in the path, as it has been done previously either with more complex acquisition systems [15,50,51], simulated [52] or in some cases manually [53].

The presented measure of the SL can be considered for the estimation of the gait parameters in O&M. Considering the limitations of the method, the most remarkable element of this tool is the fact that the system brings a measurement with the simplicity of one inertial sensor placed in the leg, using only one absolute orientation angle. Most of the algorithms found in the literature considered, beside orientation angles, acceleration values for step detection and calculation of displacement [54], as addition of at least another sensing method, which brings many other limitations and complexity in the development of the algorithm [34]. This article presents a simple method for computing clinically relevant gait parameters, with acceptable precision and accuracy, as in [55]. However, it is a fact that more precision can be obtained implementing a new method considering the details of the swing of the gait cycle or implementing artificial intelligence for instance [55,56].

Currently, a motion analysis device able to evaluate the percentage of coverage provided by a travelling technique according to specific parameters of a user cannot be found in the literature. The RoboCane software [48] was not successfully adopted by the O&M research community in the last decade. This software was designed to calculate the coverage according to direct measurement (manual) of the specific variables of the user. On the other hand, the proposed tool will allow the O&M specialists to have a real estimation of the coverage that the users are providing to themselves in dynamic conditions, which will also help them to be more objective in the evaluation of the O&M training. Among O&M specialists and researchers, it is known that there is no standardization in training methods, and these methods may vary according to the experience of each specialist. That is why research in O&M can also benefit from this tool. The development of the presented tool permits evaluating these mobility parameters independently of the environment complexity in which the training is gradually subjected [57], as it is a low cost portable device. Moreover, further development must be done to obtain more quantification characteristics of the O&M performance of the VIP. By adding one more sensor to the body, for instance, parameters of postural stability and balance analysis can be obtained.

5. Conclusions

This article proposed a system able to overview the quantitative parameters of O&M for VIP, which are currently visually analyzed by O&M specialist during rehabilitation, such as sweeping coverage and gait analysis. The proposed tool provides motion analysis of the long cane and the leg by using placed low-cost inertial measurement unit sensors (IMU). The system was tested in laboratory conditions by six blindfolded volunteers following three travel techniques trained by VIP during rehabilitation. The experimental results indicate that this system is reliable for measuring grip rotation, safety zone, sweeping amplitude and hand position using orientation angles with an accuracy of 97%. In terms of future work, a further development is required for the system to be implemented as a rehabilitation aid. Thereby, a more precise method for step length must be obtained, since the mean %Error varies between 1.07% and 15.06% among experiments. Also, more parameters of O&M can be analyzed using IMU's absolute angles, including postural stability and balance analysis. Finally, as the main purpose, the proposed system is a new, simple and low-cost technological aid that that has the potential to improve the current practice of O&M.

Author Contributions: Conceptualization, K.M.R.L. and J.J.S.O.; methodology, K.M.R.L.; software, K.M.R.L., M.Á.C., S.S.L.; validation, K.M.R.L., M.J.-V. and J.J.S.O.; formal analysis, K.M.R.L. and J.J.S.O.; investigation, K.M.R.L.; resources, J.J.S.O.; data curation, K.M.R.L., M.Á.C., S.S.L.; writing—original draft preparation, K.M.R.L.; writing—review and editing, M.J.-V., M.Á.C., S.S.L. and J.J.S.O.; visualization, K.M.R.L., M.J.-V.; supervision, J.J.S.O. All authors have read and agreed to the published version of the manuscript.

Funding: This work was partially financed by the Ministerio de Ciencia, Innovación y Universidades, Ref.: PGC2018-097531-B-I00.

Institutional Review Board Statement: The study was conducted according to the guidelines of the Declaration of Helsinki, and approved by the Institutional Ethics Committee of Universidad Politécnica de Madrid (Ref. ID 2020000224, 30 October 2020).

Informed Consent Statement: Informed consent was obtained from all subjects involved in the study.

Data Availability Statement: The data supporting reported results can be found in the following link: https://docs.google.com/spreadsheets/d/1-4NMnHSUrvO-NimwVJImX7G-dZibNwdqc_E2CO-AR00/edit?usp=sharing, accessed on 1 July 2021.

Acknowledgments: The author Karla Miriam Reyes acknowledges scholarship support from the Fundación Carolina FC and the Universidad Tecnológica Centroamericana UNITEC. The author Milagros Jaén-Vargas would like to thank the Secretaria Nacional de Ciencia y Tecnología SENACYT for her scholarship in the IFARHU-SENACYT program.

Conflicts of Interest: The authors declare no conflict of interest.

References

1. Brady, E.; Morris, M.R.; Zhong, Y.; White, S.; Bigham, J.P. Visual challenges in the everyday lives of blind people. *Conf. Hum. Factors Comput. Syst. Proc.* **2013**, 2117–2126. [CrossRef]
2. Real, S.; Araujo, A. Navigation systems for the blind and visually impaired: Past work, challenges, and open problems. *Sensors* **2019**, *19*, 3404. [CrossRef]
3. Aciem, T.M.; Mazzotta, M.J.d. Personal and social autonomy of visually impaired people who were assisted by rehabilitation services. *Rev. Bras. Oftalmol.* **2013**, *72*, 261–267. [CrossRef]
4. Kacorri, H.; Kitani, K.M.; Bigham, J.P.; Asakawa, C. People with visual impairment training personal object recognizers: Feasibility and challenges. *Conf. Hum. Factors Comput. Syst. Proc.* **2017**, 5839–5849. [CrossRef]
5. Stelmack, J. Quality of life of low-vision patients and outcomes of low-vision rehabilitation. *Optom. Vis. Sci.* **2001**, *78*, 335–342. [CrossRef]
6. Lopera, G.; Aguirre, Á.; Parada, P.; Baquet, J. Manual Tecnico De Servicios De Rehabilitacion Integral Para Personas Ciegas O Con Baja Vision En America Latina Unión Latinoamericana De Ciegos -Ulac. 2010. Available online: http://www.ulacdigital.org/downloads/manual_de_rehabilitacion.pdf (accessed on 20 April 2021).
7. American Foundation for the Blind. International Approaches to Rehabilitation Programs for Adults who are Blind or Visually Impaired: Delivery Models. In *Services, Challenges, and Trends*; American Foundation for the Blind: Arlington, VA, USA, 2016; Available online: https://www.foal.es/es/content/international-approaches-rehabilitation-programs-adults-who-are-blind-or-visually-impaired (accessed on 13 February 2021).
8. National Rehabilitation Center for the disabled Japan. Rehabilitation Manual, tactile ground surface indicators for blind persons. 2003. Available online: http://www.rehab.go.jp/english/whoclbc/pdf/E13.pdf (accessed on 1 February 2021).
9. Welsh, R.L.; Blasch, B.B. Manpower needs in orientation and mobility. *New Outlook Blind* **1974**, *68*, 433–443. [CrossRef]
10. Blasch, B.; Gallimore, D. Back to the Future: Expanding the Profession—O&M for People with Disabilities. *Int. J. Orientat. Mobil.* **2013**, *6*, 21–33. [CrossRef]
11. Zijlstra, G.A.R.; Ballemans, J.; Kempen, G.I.J.M. Orientation and mobility training for adults with low vision: A new standardized approach. *Clin. Rehabil.* **2013**, *27*, 3–18. [CrossRef]
12. Szabo, J.; Panikkar, R.K. Bridging the gap between physical therapy and orientation and mobility in schools: Using a collaborative team approach for students with visual impairments. *J. Vis. Impair. Blind.* **2017**, *111*, 495–510. [CrossRef]
13. Cuturi, L.F.; Aggius-Vella, E.; Campus, C.; Parmiggiani, A.; Gori, M. From science to technology: Orientation and mobility in blind children and adults. *Neurosci. Biobehav. Rev.* **2016**, *71*, 240–251. [CrossRef]
14. Teskeredžić, A. The significance of orientantion of blind pupuls to ther body in regard to mobility and space orientation. *Human* **2018**, *8*, 10–16. [CrossRef]
15. Ramsey, V.K.; Blasch, B.B.; Kita, A. Effects of Mobility Training on Gait and Balance. *J. Vis. Impair. Blind.* **2003**, *97*, 720–726. [CrossRef]
16. Scott, B.S. Opening Up the World: Early Childhood Orientation and Mobility Intervention as Perceived by Young Children Who are Blind, Their Parents, and Specialist Teachers. 2015. Available online: https://search.proquest.com/docview/1925329675?accountid=14548%0Ahttps://julac.hosted.exlibrisgroup.com/openurl/HKU_ALMA/SERVICES_PAGE??url_ver=Z39.88-2004&rft_val_fmt=info:ofi/fmt:kev:mtx:dissertation&genre=dissertations+%26+theses&sid=ProQ:Australian+Ed (accessed on 13 February 2021).
17. Blasch, B.B.; la Grow, S.; Penrod, W. Environmental Rating Scale for Orientation and Mobility. *Int. J. Orientat. Mobil.* **2008**, *1*, 9–16. [CrossRef]
18. Pissaloux, E.; Velázquez, R. Mobility of visually impaired people: Fundamentals and ICT assistive technologies. *Mobil. Vis. Impair. People Fundam. ICT Assist. Technol.* **2017**, 1–652. [CrossRef]
19. Organización Nacional de Ciegos Españoles. Discapacidad Visual y Autonomía Personal. Enfoque Práctico de la Rehabilitación. 2011. Available online: https://sid.usal.es/idocs/F8/FDO26230/discap_visual.pdf (accessed on 1 February 2021).
20. Kim, Y.; Moncada-Torres, A.; Furrer, J.; Riesch, M.; Gassert, R. Quantification of long cane usage characteristics with the constant contact technique. *Appl. Ergon.* **2016**, *55*, 216–225. [CrossRef]
21. Fan, K.; Lyu, C.; Liu, Y.; Zhou, W.; Jiang, X.; Li, P.; Chen, H. Hardware implementation of a virtual blind cane on FPGA. In Proceedings of the 2017 IEEE International Conference on Real-time Computing and Robotics (RCAR), Okinawa, Japan, 14–18 July 2017; pp. 344–348. [CrossRef]

22. Dastider, A.; Basak, B.; Safayatullah, M.; Shahnaz, C.; Fattah, S.A. Cost efficient autonomous navigation system (e-cane) for visually impaired human beings. In Proceedings of the 2017 IEEE region 10 humanitarian technology conference (R10-HTC), Dhaka, Bangladesh, 21–23 December 2017; pp. 650–653. [CrossRef]

23. Meshram, V.V.; Patil, K.; Meshram, V.A.; Shu, F.C. An Astute Assistive Device for Mobility and Object Recognition for Visually Impaired People. *IEEE Trans. Human Mach. Syst.* **2019**, *49*, 449–460. [CrossRef]

24. Zhang, H.; Ye, C. A Visual Positioning System for Indoor Blind Navigation. In Proceedings of the 2020 IEEE International Conference on Robotics and Automation (ICRA), Paris, France, 31 May–31 August 2020; pp. 9079–9085. [CrossRef]

25. Bernieri, G.; Faramondi, L.; Pascucci, F. Augmenting white cane reliability using smart glove for visually impaired people. In Proceedings of the 37th Annual International Conference of the IEEE Engineering in Medicine and Biology Society (EMBC), Milan, Italy, 25–29 August 2015; pp. 8046–8049. [CrossRef]

26. Islam, M.M.; Sadi, M.S.; Zamli, K.Z.; Ahmed, M.M. Developing Walking Assistants for Visually Impaired People: A Review. *IEEE Sens. J.* **2019**, *19*, 2814–2828. [CrossRef]

27. Biswas, M.; Dhoom, T.; Pathan, R.K.; Chaiti, M.S. Shortest Path Based Trained Indoor Smart Jacket Navigation System for Visually Impaired Person. In Proceedings of the 2020 IEEE International Conference on Smart Internet of Things (SmartIoT), Beijing, China, 14–16 August 2020; pp. 228–235. [CrossRef]

28. Ferrand, S.; Alouges, F.; Aussal, M. An Augmented Reality Audio Device Helping Blind People Navigation. In *International Conference on Computers Helping People with Special Needs*; Springer International Publishing: Cham, Switzerland, 2018; Volume 10897, Lecture Notes in Computer Science.

29. Ferrand, S.; Alouges, F.; Aussal, M. An electronic travel aid device to help blind people playing sport. *IEEE Instrum. Meas. Mag.* **2020**, *23*, 14–21. [CrossRef]

30. Jabbar, M.S.; Hussain, G.; Cho, J. Indoor Positioning System: Improved deep learning approach based on LSTM and multi-stage activity classification. In Proceedings of the 2020 IEEE International Conference on Consumer Electronics-Asia (ICCE-Asia), Seoul, Korea, 1–3 November 2020; pp. 15–18. [CrossRef]

31. Guerreiro, J.; Sato, D.; Asakawa, S.; Dong, H.; Kitani, K.M.; Asakawa, C. Cabot: Designing and evaluating an autonomous navigation robot for blind people. In Proceedings of the 21st International ACM SIGACCESS Conference on Computers and Accessibility, Pittsburgh, PA, USA, 28–30 October 2019; pp. 68–82. [CrossRef]

32. Paredes, N.E.G.; Cobo, A.; Martín, C.; Serrano, J.J. Methodology for building virtual reality mobile applications for blind people on advanced visits to unknown interior spaces. In Proceedings of the 14th International Conference on Mobile Learning, Lisbon, Portugal, 14–16 April 2018; pp. 3–14.

33. Davies, T.C.; Burns, C.M.; Pinder, S.D. Mobility interfaces for the visually impaired: What's missing? *ACM Int. Conf. Proc. Ser.* **2007**, *254*, 41–47. [CrossRef]

34. Kandalan, R.N.; Namuduri, K. Techniques for Constructing Indoor Navigation Systems for the Visually Impaired: A Review. *IEEE Trans. Hum. Mach. Syst.* **2020**, *50*, 492–506. [CrossRef]

35. Schloerb, D.W.; Lahav, O.; Desloge, J.G.; Srinivasan, M.A. BlindAid: Virtual environment system for self-reliant trip planning and orientation and mobility training. In Proceedings of the 2010 IEEE Haptics Symposium, Waltham, MA, USA, 25–26 March 2010; pp. 363–370. [CrossRef]

36. Oliveira, J.D.; Campos, M.D.; Bordini, R.H.; Amory, A. Godonnie: A robot programming language to improve orientation and mobility skills in people who are visually impaired. In Proceedings of the 21st International ACM SIGACCESS Conference on Computers and Accessibility, Pittsburgh, PA, USA, 28–30 October 2019; pp. 679–681. [CrossRef]

37. Gong, J.; Ding, Q.; Xu, P.; Zhang, Y.; Zhang, L.; Wang, Q. HeliCoach: An Adaptive Multimodal Orientation and Mobility Training System in a Drone-Based Simulated 3D Audio Space. *Jisuanji Fuzhu Sheji Yu Tuxingxue Xuebao/J. Comput. Des. Comput. Graph.* **2020**, *32*, 1129–1136. [CrossRef]

38. Zheng, Y. Miniature Inertial Measurement Unit. *Space Microsyst. Micro/Nano Satell.* **2018**, 233–293. [CrossRef]

39. Kok, M.; Hol, J.D.; Schön, T.B. Using Inertial Sensors for Position and Orientation Estimation. *Found. Trends Signal Process.* **2017**, *11*, 1–153. [CrossRef]

40. Filippeschi, A.; Schmitz, N.; Miezal, M.; Bleser, G.; Ruffaldi, E.; Stricker, D. Survey of motion tracking methods based on inertial sensors: A focus on upper limb human motion. *Sensors* **2017**, *17*, 1257. [CrossRef]

41. Ligorio, G.; Zanotto, D.; Sabatini, A.M.; Agrawal, S.K. A novel functional calibration method for real-time elbow joint angles estimation with magnetic-inertial sensors. *J. Biomech.* **2017**, *54*, 106–110. [CrossRef]

42. Roetenberg, D.; Luinge, H.; Slycke, P. Xsens MVN: Full 6DOF human motion tracking using miniature inertial sensors. *Xsens Motion Technol. BV Tech. Rep.* **2009**. Available online: http://human.kyst.com.tw/upload/pdfs120702543998066.pdf (accessed on 2 February 2021).

43. Zhu, R.; Zhou, Z. A real-time articulated human motion tracking using tri-axis inertial/magnetic sensors package. *IEEE Trans. Neural Syst. Rehabil. Eng.* **2004**, *12*, 295–302. [CrossRef]

44. Shaeffer, D.K. MEMS inertial sensors: A tutorial overview. *IEEE Commun. Mag.* **2013**, *51*, 100–109. [CrossRef]

45. Simdiankin, A.; Byshov, N.; Uspensky, I. A method of vehicle positioning using a non-satellite navigation system. *Transp. Res. Procedia* **2018**, *36*, 732–740. [CrossRef]

46. Mahida, P.; Shahrestani, S.; Cheung, H. Deep learning-based positioning of visually impaired people in indoor environments. *Sensors* **2020**, *20*, 6238. [CrossRef]

47. Leiva, K.M.R.; Lara, S.S.; Olmedo, J.J.S. Development of a motion measurement system of a white cane for Visually Impaired People rehabilitation. In Proceedings of the XXXVIII Congreso Anual de la Sociedad Española de Ingeniería Biomédica (CASEIB 2020), Virtual Congress, 25–27 November 2020.
48. Blasch, B.B.; LaGrow, S.J.; de L'Aune, W.R. Three aspects of coverage provided by the long cane: Object, surface, and foot-placement preview. *J. Vis. Impair. Blind.* **1996**, *90*, 295–301. [CrossRef]
49. Sankako, A.N.; Marília, P.; Lucareli, P.R.G.; de Carvalho, S.M.R.; Braccialli, L.M.P. Temporal spatial parameters analysis of the gait in children with vision impairment. *Int. J. Orientat. Mobil.* **2016**, *8*, 90–100. [CrossRef]
50. Ramsey, V.K.; Blasch, B.B.; Kita, A.; Johnson, B.F. A biomechanical evaluation of visually impaired persons' gait and lone-cane mechanics. *J. Rehabil. Res. Dev.* **1999**, *36*, 323–332. [PubMed]
51. Emerson, R.W.; Kim, D.S.; Naghshineh, K.; Myers, K.R. Biomechanics of Long Cane Use. *J. Vis. Impair. Blind.* **2019**, *113*, 235–247. [CrossRef]
52. Blasch, B.B.; de L'aune, W.R.; Coombs, F.K. Computer Simulation of Cane Techniques Used by People with Visual Impairments for Accessibility Analysis. In *Enabling Environments. Plenum Series in Rehabilitation and Health*; Springer: Boston, MA, USA, 1999.
53. LaGrow, S.J.; Blasch, B.B.; de L'Aune, W. Efficacy of the touch technique for surface and foot-placement preview. *J. Vis. Impair. Blind.* **1997**, *91*, 47–52. [CrossRef]
54. Rampp, A.; Barth, J.; Schülein, S.; Gaßmann, K.G.; Klucken, J.; Eskofier, B.M. Inertial Sensor-Based Stride Parameter Calculation From Gait Sequences in Geriatric Patients. *IEEE Trans. Biomed. Eng.* **2015**, *62*, 1089–1097. [CrossRef] [PubMed]
55. Flores, G.H.; Manduchi, R. WeAllWalk. *ACM Trans. Access. Comput.* **2018**, *11*, 1–28. [CrossRef]
56. Xing, H.; Li, J.; Hou, B.; Zhang, Y.; Guo, M. Pedestrian Stride Length Estimation from IMU Measurements and ANN Based Algorithm. *J. Sens.* **2017**, *2017*, 6091261. [CrossRef]
57. Finger, R.P.; Ayton, L.N.; Deverell, L.; O'Hare, F.; McSweeney, S.C.; Luu, C.D.; Fenwick, E.K.; Keeffe, J.E.; Guymer, R.H.; Bentley, S.A. Developing a very low vision orientation and mobility test battery (O&M-VLV). *Optom. Vis. Sci.* **2016**, *93*, 1127–1136. [PubMed]

Article

Optimal 3D Angle of Arrival Sensor Placement with Gaussian Priors

Rongyan Zhou [1,2], **Jianfeng Chen** [1,*], **Weijie Tan** [3], **Qingli Yan** [4] **and Chang Cai** [1]

[1] School of Marine Science and Technology, Northwestern Polytechnical University, Xi'an 710072, China; zhoury@mail.nwpu.edu.cn or zhoury619@163.com (R.Z.); caichang@mail.nwpu.edu.cn (C.C.)

[2] School of Information Engineering, Nanyang Institute of Technology, Nanyang 473004, China

[3] State Key Laboratory of Public Big Data, Guizhou University, Guiyang 550025, China; wjtan@gzu.edu.cn

[4] School of Computer Science & Technology, Xi'an University of Posts & Telecommunications, Xi'an 710121, China; yql@xupt.edu.cn

* Correspondence: chenjf@nwpu.edu.cn

Abstract: Sensor placement is an important factor that may significantly affect the localization performance of a sensor network. This paper investigates the sensor placement optimization problem in three-dimensional (3D) space for angle of arrival (AOA) target localization with Gaussian priors. We first show that under the A-optimality criterion, the optimization problem can be transferred to be a diagonalizing process on the AOA-based Fisher information matrix (FIM). Secondly, we prove that the FIM follows the invariance property of the 3D rotation, and the Gaussian covariance matrix of the FIM can be diagonalized via 3D rotation. Based on this finding, an optimal sensor placement method using 3D rotation was created for when prior information exists as to the target location. Finally, several simulations were carried out to demonstrate the effectiveness of the proposed method. Compared with the existing methods, the mean squared error (MSE) of the maximum a posteriori (MAP) estimation using the proposed method is lower by at least 25% when the number of sensors is between 3 and 6, while the estimation bias remains very close to zero (smaller than 0.15 m).

Keywords: 3D angle of arrival (AOA) localization; Cramér–Rao lower bound (CRLB); optimal sensor placement; covariance matrix; fisher information matrix (FIM)

Citation: Zhou, R.; Chen, J.; Tan, W.; Yan, Q.; Cai, C. Optimal 3D Angle of Arrival Sensor Placement with Gaussian Priors. *Entropy* **2021**, *23*, 1379. https://doi.org/10.3390/e23111379

Academic Editors: Felipe Ortega and Emilio López Cano

Received: 14 September 2021
Accepted: 18 October 2021
Published: 21 October 2021

Publisher's Note: MDPI stays neutral with regard to jurisdictional claims in published maps and institutional affiliations.

1. Introduction

Tracking and localization using sensor networks have a wide range of applications in radar, sonar, and wireless sensor networks [1,2]. There are several types of localization techniques that have been developed in recent years: time difference of arrival (TDOA) or time of arrival (TOA) [3,4], angle of arrival (AOA) [5–7], and received signal strength (RSS) [8,9].

AOA target localization has been an active research area during the past two decades. It does not require synchronization with the signal target or among the different distributed sensors, unlike TOA and TDOA localization. Many estimators have been developed for AOA-based localization. A 3D one-step pseudolinear estimator (PLE) with a bias compensation strategy was proposed in [10]. An asymptotically unbiased weight instrumental variable (WIV) technique was presented in [11] to solve the bias problem, and then a 3D, improved WIV estimator was derived to break down the correlation between the instrumental variable (IV) matrix and the error vector in [12]. Furthermore, a closed-form solution for 3D AOA localization, which can handle the presence of sensor location errors, was presented in [13]. Recently, an approximately unbiased estimator was proposed by approximating the bias and subtracting it from the weighted least squares (WLS) solution obtained using semidefinite relaxation (SDR) in [14].

Apart from the above localization methods, generating the target–sensor geometry for localization is also a non-trivial task and attracts great interest in the localization area. The optimization problem for sensor placement was usually formulated to minimize the

Cramer–Rao lower bound (CRLB) or maximize the Fisher information matrix (FIM) [15–18], and the differences between the above two methods were reported in [19]. In [20], the trace of CRLB was adopted to find the optimal geometric configuration, which yielded the minimum possible covariance of any unbiased target estimator in a constrained 3D space. The optimal placement analysis for 3D AOA target localization using the A-optimality criterion (minimize the trace of CRLB) appeared in [21]. In addition, a frame theory was also presented that can handle the optimal sensor placement with three types of sensor placement strategy in [22] as an identical parameter optimization problem in two-dimensional (2D) and 3D space. In [23], the frame theory was used to derive an evaluation function for optimal placement with random numbers of newly added sensors in AOA target localization.

The majority of previous work on optimal sensor placement assumed that the target location was known perfectly, which is impossible in actual scenarios. Therefore, it is beneficial to solve the optimal sensor placement problem when the target location is uncertain. The optimal sensor placement algorithm for TDOA localization with an unknown target location was proposed in [24]. An equivalence between minimizing the estimation mean squared error and minimizing the area of the estimation uncertainty ellipse was established for the geometry optimization problem of target localization with Bayesian priors in [25], which makes the optimal geometry conditions algebraically simple and easy to be computed. However, the above proposed algorithms can only be used in 2D space. In addition, an analysis of the performance measures of covariance and information matrices in resource management for target state estimation was provided in [26]. Then the analysis results were extended in [27] to find the optimal placement of heterogeneous sensors for the target with Gaussian priors. Furthermore, the updated FIM was used to derive optimal placement conditions for heterogeneous sensors tracking the unknown number of targets in [28]. Nevertheless, the solutions in [27,28] were complicated, particularly in the case of more than two sensors.

Several valuable conclusions have been obtained about the coordinate system rotation, which provides a new path to solving target localization and optimal sensor placement. As pointed out in [29], local coordinate translations and rotations do not influence the PLE and maximum likelihood estimator (MLE) performance of the bearings-only target localization algorithm. Furthermore, it was demonstrated that the trace of CRLB was invariant in XY-coordinates and the AOA-based FIM was invariant to flipping a sensor about the target in [21]. Lately, a TOA-based FIM invariant to sensor rotation about the target in 3D space was shown in [30].

In this paper, we address the optimal 3D AOA sensor placement problem with Gaussian priors. The key contributions of this paper are summarized as follows:

- A detailed 3D AOA optimal sensor placement problem with Gaussian priors is analyzed using the A-optimality criterion (minimizing the trace of the inverse FIM). We show analytically that the problem can be transformed to diagonalize the AOA-based FIM under the A-optimality criterion.
- The invariance property of the 3D rotation for the AOA-based FIM with Gaussian priors is deduced. Thus, the Gaussian covariance matrix of the FIM can be diagonalized via 3D rotation.
- An optimal sensor placement method using 3D rotation is proposed for when prior information exists as to the target location using the invariance property of the AOA-based FIM and the A-optimality criterion.
- Simulation studies are presented to demonstrate the analytical findings. The comparison results show that the proposed method significantly improves the localization performance.

The rest of the paper is organized as follows: The 3D AOA sensor placement with Gaussian priors optimization problem is formulated in Section 2. Section 3 derives the FIM with Gaussian priors after the 3D rotation and then exploits the invariance property for the 3D AOA-based FIM. Section 4 presents the optimal sensor-target geometric solutions with the help of a resistor network analogy. The main results are presented with simulation examples in Section 5, and the conclusion and discussion of future work are in Section 6.

2. Problem Formulation

We consider a 3D AOA configuration with N sensors localizing a stationary target, as depicted in Figure 1, and each sensor is assumed to be omnidirectional. $\mathbf{s} = (x, y, z)^T$ is the unknown location of the target with ^{T} denoting matrix transpose, $\mathbf{p}_k = \left(p_{xk}, p_{yk}, p_{zk} \right)^T$, $k = 1, 2, \cdots N$ is the location of the sensors. It is assumed that $\mathbf{s}$ is a Gaussian random variable with a distribution as $\mathbf{s} \sim \mathcal{N}(\mathbf{s}_0, \mathbf{P}_0)$, where $\mathbf{s}_0$ and $\mathbf{P}_0$ represent the mean and the covariance matrix of $\mathbf{s}$. Note that the gray ellipse in Figure 1 illustrates the confidence region corresponding to the Gaussian priors, and $\{\theta_k, \phi_k\}$ denotes the bearing measurement with the azimuth and elevation angle in spherical coordinates. Using $\mathbf{s}_0 = (x_0, y_0, z_0)^T$ as a reference, the AOA measurement of the kth sensor can be expressed as

$$
\begin{aligned}
\theta_k &= \tan^{-1} \frac{y_0 - p_{yk}}{x_0 - p_{xk}}, \quad -\pi < \theta \le \pi, \\
\phi_k &= \sin^{-1} \frac{z_0 - p_{zk}}{r_k}, \quad -\frac{\pi}{2} < \phi \le \frac{\pi}{2},
\end{aligned}
\tag{1}
$$

where $r_k = \|\mathbf{s}_0 - \mathbf{p}_k\|$, $\tan^{-1}$ is the fourth quadrant arctangent, and $\|\cdot\|$ denotes the Euclidean norm. In terms of azimuth and elevation angles, the unit bearing vector $\mathbf{g}_k^0$ can be given by

$$
\mathbf{g}_k^0 = \begin{bmatrix} \cos \phi_k \cos \theta_k \\ \cos \phi_k \sin \theta_k \\ \sin \phi_k \end{bmatrix},
\tag{2}
$$

In the 3D localization system, the AOA measurements are always affected by multi-path effects, the propagation environment, the transmitted power, and other unfavorable factors. In order to focus our study on the sensor placement optimization problem itself, in our paper, although we do not consider these inference factors explicitly, we take them into account, as a whole, by modeling them as the additive Gaussian white noise on the true angle measurements $\{\tilde{\theta}_k, \tilde{\phi}_k\}$ as

$$
\begin{aligned}
\tilde{\theta}_k &= \theta_k + n_{\theta_k}, \, n_{\theta_k} \sim \mathcal{N}\left(0, \sigma_{\theta_k}^2\right), \\
\tilde{\phi}_k &= \phi_k + n_{\phi_k}, \, n_{\phi_k} \sim \mathcal{N}\left(0, \sigma_{\phi_k}^2\right).
\end{aligned}
\tag{3}
$$

where $\sigma_{\theta_k}^2$ and $\sigma_{\phi_k}^2$ are sensor-dependent noise variances [31].

The sensor measurement covariance matrix can be expressed as

$$
\Sigma = \begin{bmatrix} \mathbf{P}_0 & \mathbf{0}_{2N \times 3} \\ \mathbf{0}_{3 \times 2N} & \Sigma_0 \end{bmatrix},
\tag{4}
$$

with

$$
\Sigma_0 = \mathrm{diag}\left\{ \sigma_{\theta_1}^2, \sigma_{\phi_1}^2, \ldots, \sigma_{\theta_N}^2, \sigma_{\phi_N}^2 \right\},
\tag{5}
$$

Here we define $\mathbf{e}(\mathbf{s})$ and $\mathbf{r}(\mathbf{s})$

$$
\begin{aligned}
\mathbf{r}(\mathbf{s}) &= \mathbf{s} - \mathbf{s}_0, \\
\mathbf{e}(\mathbf{s}) &= [\tilde{\theta}_1 - \theta_1(\mathbf{s}), \tilde{\phi}_1 - \phi_1(\mathbf{s}), \ldots, \tilde{\theta}_N - \theta_N(\mathbf{s}), \tilde{\phi}_N - \phi_N(\mathbf{s})]^T.
\end{aligned}
\tag{6}
$$

The Jacobian matrix of measurement errors evaluated at the mean location $\mathbf{s}_0$ can be written as

$$
\mathbf{J} = \begin{bmatrix} \mathbf{J}_1 & \mathbf{J}_2 \end{bmatrix}^T,
\tag{7}
$$

where $\mathbf{J}_1$ is the 3×3 Jacobian of $\mathbf{r}(\mathbf{s})$, given by

$$\mathbf{J}_1 = \mathbf{I}_{3\times3} \, , \tag{8}$$

The Jacobian vector of the kth sensor measurement error evaluated at the true target location $\mathbf{s} = (x, y, z)^T$ as

$$
\mathbf{J}'_k = \left[\frac{\partial\theta_k}{\partial\mathbf{s}^T}, \frac{\partial\phi_k}{\partial\mathbf{s}^T} \right]^T \Bigg|_{\mathbf{s}} =
\begin{bmatrix}
\dfrac{\partial\theta_k}{\partial x} & \dfrac{\partial\theta_k}{\partial y} & \dfrac{\partial\theta_k}{\partial z} \\[2mm]
\dfrac{\partial\phi_k}{\partial x} & \dfrac{\partial\phi_k}{\partial y} & \dfrac{\partial\phi_k}{\partial z}
\end{bmatrix}\Bigg|_{\mathbf{s}}
$$
$$
= \begin{bmatrix}
-\dfrac{\sin\theta_k}{r_k \cos\phi_k} & \dfrac{\cos\theta_k}{r_k \cos\phi_k} & 0 \\[3mm]
-\dfrac{\sin\phi_k \cos\theta_k}{r_k} & -\dfrac{\sin\phi_k \sin\theta_k}{r_k} & \dfrac{\cos\phi_k}{r_k}
\end{bmatrix}, \tag{9}
$$

Therefore, we can obtain the Jacobian matrix of the $2N$ measurements as

$$
\mathbf{J}_2 =
\begin{bmatrix}
-\dfrac{\sin\theta_1}{r_1 \cos\phi_1} & \dfrac{\cos\theta_1}{r_1 \cos\phi_1} & 0 \\[3mm]
-\dfrac{\sin\phi_1 \cos\theta_1}{r_1} & -\dfrac{\sin\phi_1 \sin\theta_1}{r_1} & \dfrac{\cos\phi_1}{r_1} \\[2mm]
\vdots & \vdots & \vdots \\[2mm]
-\dfrac{\sin\theta_N}{r_N \cos\phi_N} & \dfrac{\cos\theta_N}{r_N \cos\phi_N} & 0 \\[3mm]
-\dfrac{\sin\phi_N \cos\theta_N}{r_N} & -\dfrac{\sin\phi_N \sin\theta_N}{r_N} & \dfrac{\cos\phi_N}{r_N}
\end{bmatrix}, \tag{10}
$$

The FIM for 3D AOA localization with Gaussian problem yields

$$\mathbf{\Phi} = \mathbf{J}^T \mathbf{\Sigma}^{-1} \mathbf{J}. \tag{11}$$

For simplification, $\mathbf{J}$ is expressed as the following three vectors:

$$
\mathbf{a} = \left[-\dfrac{\sin\theta_1}{r_1 \cos\phi_1}, -\dfrac{\sin\phi_1 \cos\theta_1}{r_1}, \cdots, -\dfrac{\sin\theta_N}{r_N \cos\phi_N}, -\dfrac{\sin\phi_N \cos\theta_N}{r_N} \right]^T,
$$
$$
\mathbf{b} = \left[\dfrac{\cos\theta_1}{r_1 \cos\phi_1}, -\dfrac{\sin\phi_1 \sin\theta_1}{r_1}, \cdots, \dfrac{\cos\theta_N}{r_N \cos\phi_N}, -\dfrac{\sin\phi_N \sin\theta_N}{r_N} \right]^T, \tag{12}
$$
$$
\mathbf{c} = \left[0, \dfrac{\cos\phi_1}{r_1}, \cdots, 0, \dfrac{\cos\phi_N}{r_N} \right]^T,
$$

Thus,

$$\mathbf{J} = \begin{bmatrix} \mathbf{a} & \mathbf{b} & \mathbf{c} \end{bmatrix}_{(2N+3)\times3}, \tag{13}$$

Hence, the FIM is

$$
\mathbf{\Phi} = \begin{bmatrix} \mathbf{a}^T \\ \mathbf{b}^T \\ \mathbf{c}^T \end{bmatrix} \mathbf{\Sigma}^{-1} \begin{bmatrix} \mathbf{a} & \mathbf{b} & \mathbf{c} \end{bmatrix} =
\begin{bmatrix}
\widehat{\mathbf{a}}^T\widehat{\mathbf{a}} & \widehat{\mathbf{a}}^T\widehat{\mathbf{b}} & \widehat{\mathbf{a}}^T\widehat{\mathbf{c}} \\
\widehat{\mathbf{b}}^T\widehat{\mathbf{a}} & \widehat{\mathbf{b}}^T\widehat{\mathbf{b}} & \widehat{\mathbf{b}}^T\widehat{\mathbf{c}} \\
\widehat{\mathbf{c}}^T\widehat{\mathbf{a}} & \widehat{\mathbf{c}}^T\widehat{\mathbf{b}} & \widehat{\mathbf{c}}^T\widehat{\mathbf{c}}
\end{bmatrix}, \tag{14}
$$

where $\widehat{\mathbf{a}} = \mathbf{\Sigma}^{-1/2}\mathbf{a}$, $\widehat{\mathbf{b}} = \mathbf{\Sigma}^{-1/2}\mathbf{b}$, $\widehat{\mathbf{c}} = \mathbf{\Sigma}^{-1/2}\mathbf{c}$, and $\mathbf{\Sigma}^{-1/2}\mathbf{\Sigma}^{-1/2} = \mathbf{\Sigma}^{-1}$. Given $\widehat{\mathbf{a}}$, $\widehat{\mathbf{b}}$, and $\widehat{\mathbf{c}}$ in $\Re^{2n}$, then $|\widehat{\mathbf{a}}|^2 = \langle \widehat{\mathbf{a}}, \widehat{\mathbf{a}} \rangle$ and $\langle \widehat{\mathbf{a}}, \widehat{\mathbf{b}} \rangle = |\widehat{\mathbf{a}}||\widehat{\mathbf{b}}|\cos(\theta_{\widehat{\mathbf{a}\mathbf{b}}})$, from which it follows that the angle

$\theta_{\widehat{\mathbf{a}}\widehat{\mathbf{b}}}$ between vector $\widehat{\mathbf{a}}$ and $\widehat{\mathbf{b}}$ is given by $\theta_{\widehat{\mathbf{a}}\widehat{\mathbf{b}}} = \cos^{-1}\left(\langle\widehat{\mathbf{a}}, \widehat{\mathbf{b}}\rangle / \left(|\widehat{\mathbf{a}}||\widehat{\mathbf{b}}|\right)\right)$, $\theta_{\widehat{\mathbf{a}}\widehat{\mathbf{c}}}$, $\theta_{\widehat{\mathbf{b}}\widehat{\mathbf{c}}}$ are the angle defined by vectors $\widehat{\mathbf{a}}$; and $\widehat{\mathbf{c}}$, $\widehat{\mathbf{b}}$, and $\widehat{\mathbf{c}}$ [32]. With this notion, the FIM becomes

$$\boldsymbol{\Phi} = \begin{bmatrix} |\widehat{\mathbf{a}}|^2 & |\widehat{\mathbf{a}}||\widehat{\mathbf{b}}|\cos(\theta_{\widehat{\mathbf{a}}\widehat{\mathbf{b}}}) & |\widehat{\mathbf{a}}||\widehat{\mathbf{c}}|\cos(\theta_{\widehat{\mathbf{a}}\widehat{\mathbf{c}}}) \\ |\widehat{\mathbf{a}}||\widehat{\mathbf{b}}|\cos(\theta_{\widehat{\mathbf{a}}\widehat{\mathbf{b}}}) & |\widehat{\mathbf{b}}|^2 & |\widehat{\mathbf{b}}||\widehat{\mathbf{c}}|\cos(\theta_{\widehat{\mathbf{b}}\widehat{\mathbf{c}}}) \\ |\widehat{\mathbf{a}}||\widehat{\mathbf{c}}|\cos(\theta_{\widehat{\mathbf{a}}\widehat{\mathbf{c}}}) & |\widehat{\mathbf{b}}||\widehat{\mathbf{c}}|\cos(\theta_{\widehat{\mathbf{b}}\widehat{\mathbf{c}}}) & |\widehat{\mathbf{c}}|^2 \end{bmatrix}, \tag{15}$$

The determinant of $\boldsymbol{\Phi}$ is

$$|\boldsymbol{\Phi}| = |\widehat{\mathbf{a}}|^2 |\widehat{\mathbf{b}}|^2 |\widehat{\mathbf{c}}|^2 \lambda, \tag{16}$$

where

$$\lambda = 1 - \cos^2(\theta_{\widehat{\mathbf{a}}\widehat{\mathbf{b}}}) - \cos^2(\theta_{\widehat{\mathbf{a}}\widehat{\mathbf{c}}}) - \cos^2(\theta_{\widehat{\mathbf{b}}\widehat{\mathbf{c}}}) + 2\cos(\theta_{\widehat{\mathbf{a}}\widehat{\mathbf{b}}})\cos(\theta_{\widehat{\mathbf{a}}\widehat{\mathbf{c}}})\cos(\theta_{\widehat{\mathbf{b}}\widehat{\mathbf{c}}}). \tag{17}$$

Thus, the trace of CRLB is

$$\begin{aligned} \text{tr}(\text{CRLB}) &= \text{tr}\left(\boldsymbol{\Phi}^{-1}\right) \\ &= \frac{|\widehat{\mathbf{b}}|^2|\widehat{\mathbf{c}}|^2(1 - \cos^2(\theta_{\widehat{\mathbf{b}}\widehat{\mathbf{c}}}))}{|\boldsymbol{\Phi}|} + \frac{|\widehat{\mathbf{a}}|^2|\widehat{\mathbf{c}}|^2(1 - \cos^2(\theta_{\widehat{\mathbf{a}}\widehat{\mathbf{c}}}))}{|\boldsymbol{\Phi}|} + \frac{|\widehat{\mathbf{a}}|^2|\widehat{\mathbf{b}}|^2(1 - \cos^2(\theta_{\widehat{\mathbf{a}}\widehat{\mathbf{b}}}))}{|\boldsymbol{\Phi}|} \\ &= \frac{(1 - \cos^2(\theta_{\widehat{\mathbf{b}}\widehat{\mathbf{c}}}))}{|\widehat{\mathbf{a}}|^2\lambda} + \frac{(1 - \cos^2(\theta_{\widehat{\mathbf{a}}\widehat{\mathbf{c}}}))}{|\widehat{\mathbf{b}}|^2\lambda} + \frac{(1 - \cos^2(\theta_{\widehat{\mathbf{a}}\widehat{\mathbf{b}}}))}{|\widehat{\mathbf{c}}|^2\lambda}, \end{aligned} \tag{18}$$

Thus, we can get

$$\text{tr}(\text{CRLB}) \geq \frac{1}{|\widehat{\mathbf{a}}|^2} + \frac{1}{|\widehat{\mathbf{b}}|^2} + \frac{1}{|\widehat{\mathbf{c}}|^2}. \tag{19}$$

The $\text{tr}(\text{CRLB})$ is minimum when $\cos(\theta_{\widehat{\mathbf{a}}\widehat{\mathbf{b}}}) = \cos(\theta_{\widehat{\mathbf{a}}\widehat{\mathbf{c}}}) = \cos(\theta_{\widehat{\mathbf{b}}\widehat{\mathbf{c}}}) = 0$. Note that when $\text{tr}(\text{CRLB})$ becomes minimum, the FIM becomes diagonal, so the optimal sensor placement is obtained by diagonalizing the FIM [33].

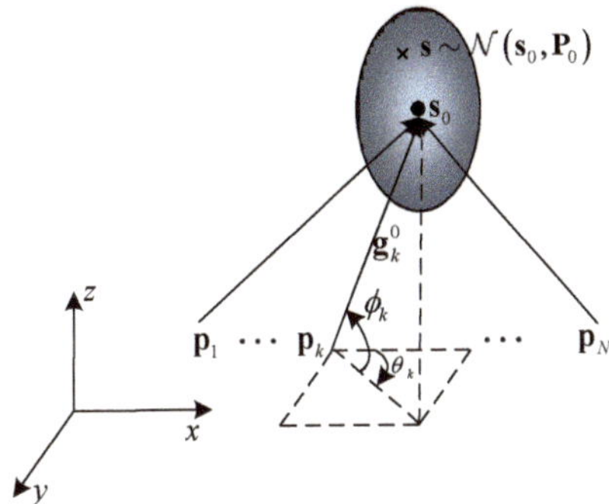

Figure 1. 3D AOA localization sensor placement with Gaussian priors.

3. The Proposed Method

Under the Gaussian assumption, the prior covariance matrix $\mathbf{P}_0$ may be a diagonal or non-diagonal matrix, which physically represents an ellipsoid bounding the uncertain target measurement estimators. Since the rotation does not affect the size of the ellipsoid, the covariance $\mathbf{P}_0$ should be invariant to any similarity transform $\mathbf{U}\mathbf{P}_0\mathbf{U}^T$, where $\mathbf{U}$ is a unitary matrix. Therefore, a proper 3D rotation provides a solution for diagonalizing the non-diagonal matrix $\mathbf{P}_0$. Additionally, in this section, we derive the FIM for 3D AOA localization with Gaussian priors after the 3D rotation, and then the invariance property of the 3D AOA-based FIM is exploited.

3.1. 3D Rotation Matrix

First, we define rotation matrices of the AOA measurement as follows:

$$\mathbf{R}_x = \begin{pmatrix} 1 & 0 & 0 \\ 0 & \cos\alpha & -\sin\alpha \\ 0 & \sin\alpha & \cos\alpha \end{pmatrix}, \mathbf{R}_y = \begin{pmatrix} \cos\beta & 0 & \sin\beta \\ 0 & 1 & 0 \\ -\sin\beta & 0 & \cos\beta \end{pmatrix}, \mathbf{R}_z = \begin{pmatrix} \cos\gamma & -\sin\gamma & 0 \\ \sin\gamma & \cos\gamma & 0 \\ 0 & 0 & 1 \end{pmatrix}. \tag{20}$$

Here α, β, and γ are counterclockwise rotation angles around the x, y, and z axes, respectively, which is depicted in Figure 2. The rotation matrix is

$$\mathbf{R} = \mathbf{R}_x \mathbf{R}_y \mathbf{R}_z. \tag{21}$$

and satisfies $\mathbf{R}\mathbf{R}^T = \mathbf{R}\mathbf{R}^{-1} = \mathbf{I}$.

Next, when the rotation happens in the 3D space, we can get

$$\mathbf{s}^r = \mathbf{R}\mathbf{s}, \quad \mathbf{s}_0^r = \mathbf{R}\mathbf{s}_0, \quad \mathbf{p}^r = \mathbf{R}\mathbf{p}, \quad \mathbf{P}_0^r = \mathbf{R}\mathbf{P}_0\mathbf{R}^T. \tag{22}$$

where $\mathbf{s}^r$, $\mathbf{s}_0^r$, $\mathbf{p}^r$, and $\mathbf{P}_0^r$ are the new measurements compared with $\mathbf{s}$, $\mathbf{s}_0$, $\mathbf{p}$, and $\mathbf{P}_0$ after rotation.

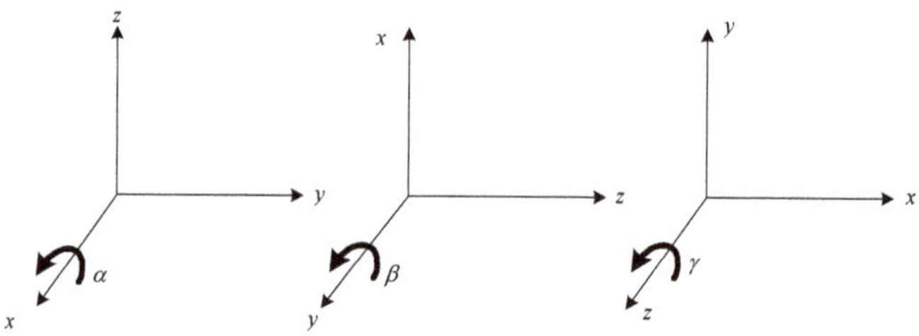

Figure 2. The rotation angles α, β, γ around the x, y, and z axes.

3.2. Invariance to 3D Rotation for AOA-Based FIM

When the 3D AOA measurements are assumed to be corrupted by additive white Gaussian noise with zero mean, the k-th sensor bearing unit vector in (2) is modified as

$$\mathbf{g}_k = \begin{bmatrix} \cos\tilde{\phi}_k \cos\tilde{\theta}_k \\ \cos\tilde{\phi}_k \sin\tilde{\theta}_k \\ \sin\tilde{\phi}_k \end{bmatrix}, \tag{23}$$

From (20) and (22), the bearing unit vector after rotation is

$$\mathbf{g}_k^r = \mathbf{R}\mathbf{g}_k = \begin{bmatrix} \cos\tilde{\phi}_k^r \cos\tilde{\theta}_k^r \\ \cos\tilde{\phi}_k^r \sin\tilde{\theta}_k^r \\ \sin\tilde{\phi}_k^r \end{bmatrix}, \tag{24}$$

Therefore, the azimuth and elevation angles are given by

$$\tilde{\theta}_k^r = \tan^{-1}\left(\frac{\mathbf{g}_k^r(2)}{\mathbf{g}_k^r(1)}\right), \quad \tilde{\phi}_k^r = \sin^{-1}(\mathbf{g}_k^r(3)). \tag{25}$$

Here we define

$$\tilde{\theta}_k^r = g(\tilde{\theta}_k, \tilde{\phi}_k), \quad \tilde{\phi}_k^r = h(\tilde{\theta}_k, \tilde{\phi}_k), \tag{26}$$

To compute the covariance matrix after rotation, we can adopt the First-order Taylor series approximation for the rotated noisy angles using (θ_k, ϕ_k) in $(\tilde{\theta}_k, \tilde{\phi}_k)$ with respect to the noise variables $n_{\theta k}$ and $n_{\phi k}$. Therefore, (26) can be rewritten as

$$
\begin{aligned}
\tilde{\theta}_k^r &= g(\tilde{\theta}_k, \tilde{\phi}_k) = g(\theta_k + n_{\theta_k}, \phi_k + n_{\phi_k}) = g(\theta_k, \phi_k) + \left[\frac{\partial g(\theta_k, \phi_k)}{\partial \theta_k} \quad \frac{\partial g(\theta_k, \phi_k)}{\partial \phi_k} \right] \begin{bmatrix} n_{\theta_k} \\ n_{\phi_k} \end{bmatrix}, \\
\tilde{\phi}_k^r &= h(\tilde{\theta}_k, \tilde{\phi}_k) = h(\theta_k + n_{\theta_k}, \phi_k + n_{\phi_k}) = h(\theta_k, \phi_k) + \left[\frac{\partial h(\theta_k, \phi_k)}{\partial \theta_k} \quad \frac{\partial h(\theta_k, \phi_k)}{\partial \phi_k} \right] \begin{bmatrix} n_{\theta_k} \\ n_{\phi_k} \end{bmatrix}.
\end{aligned}
\tag{27}
$$

According to the error propagation law [34], the noise covariance matrix for the k-th sensor after 3D rotation can be written as

$$
\mathbf{K}_k^r = \begin{bmatrix} \dfrac{\partial g(\theta_k, \phi_k)}{\partial \theta_k} & \dfrac{\partial g(\theta_k, \phi_k)}{\partial \phi_k} \\ \dfrac{\partial h(\theta_k, \phi_k)}{\partial \theta_k} & \dfrac{\partial h(\theta_k, \phi_k)}{\partial \phi_k} \end{bmatrix} \times \begin{bmatrix} \sigma_\theta^2 & 0 \\ 0 & \sigma_\phi^2 \end{bmatrix} \times \begin{bmatrix} \dfrac{\partial g(\theta_k, \phi_k)}{\partial \theta_k} & \dfrac{\partial g(\theta_k, \phi_k)}{\partial \phi_k} \\ \dfrac{\partial h(\theta_k, \phi_k)}{\partial \theta_k} & \dfrac{\partial h(\theta_k, \phi_k)}{\partial \phi_k} \end{bmatrix}^T,
\tag{28}
$$

By substituting (26) into (27), the maximum likelihood (ML) covariance matrix of the bearing measurement noise can be expressed as

$$
\Sigma_0^r = \begin{bmatrix} \mathbf{K}_1^r & 0 & 0 \\ 0 & \ddots & 0 \\ 0 & 0 & \mathbf{K}_N^r \end{bmatrix},
\tag{29}
$$

Using the prior covariance matrix after rotation $\mathbf{P}_0^r$ given in (22) and the above equation, the covariance matrix after rotation is given by

$$
\Sigma^r = \begin{bmatrix} \mathbf{P}_0^r & \mathbf{0}_{2N \times 3} \\ \mathbf{0}_{3 \times 2N} & \Sigma_0^r \end{bmatrix}.
\tag{30}
$$

By substituting (22) into (8) and (10), $\mathbf{J}_1^r$ and $\mathbf{J}_2^r$ after rotation are computed. We thus obtain

$$
\mathbf{J}^r = \begin{bmatrix} \mathbf{J}_1^r & \mathbf{J}_2^r \end{bmatrix}^T,
\tag{31}
$$

Hence, the FIM after three rotations becomes

$$
\hat{\boldsymbol{\Phi}} = \mathbf{J}^{r^T} (\Sigma^r)^{-1} \mathbf{J}^r.
\tag{32}
$$

After the 3D rotations, the FIM becomes

$$
\hat{\boldsymbol{\Phi}} = \mathbf{R} \boldsymbol{\Phi} \mathbf{R}^{-1},
\tag{33}
$$

Substituting (21) into the above equation yields

$$
\hat{\boldsymbol{\Phi}} = \mathbf{R}_x \mathbf{R}_y \mathbf{R}_z \boldsymbol{\Phi} \mathbf{R}_z^{-1} \mathbf{R}_y^{-1} \mathbf{R}_x^{-1},
\tag{34}
$$

By using $(\mathbf{AB})^{-1} = \mathbf{B}^{-1} \mathbf{A}^{-1}$, $\mathbf{A}$ and $\mathbf{B}$ are full rank square matrices. The inverse of the new FIM $\hat{\boldsymbol{\Phi}}$ is

$$
\hat{\boldsymbol{\Phi}}^{-1} = \mathbf{R}_x \mathbf{R}_y \mathbf{R}_z \boldsymbol{\Phi}^{-1} \mathbf{R}_z^{-1} \mathbf{R}_y^{-1} \mathbf{R}_x^{-1},
\tag{35}
$$

Based on the properties of the rotation matrix and the above expression, it can be seen that $\hat{\boldsymbol{\Phi}}^{-1}$ and $\boldsymbol{\Phi}^{-1}$ are similarity matrices. Thus,

$$
\mathrm{tr}\left(\hat{\boldsymbol{\Phi}}^{-1} \right) = \mathrm{tr}\left(\boldsymbol{\Phi}^{-1} \right).
\tag{36}
$$

Thus, we can conclude that 3D rotations do not affect the $\mathrm{tr}\left(\mathbf{\Phi}^{-1}\right)$ calculated from the AOA-based FIM. In the next section, we will derive the optimal sensor placement with Gaussian priors using the invariance of the trace of FIM to 3D rotations.

4. Optimal Sensor Placement with Gaussian Priors

In this section, we investigate the optimal sensor placement with Gaussian priors. First, the FIM for 3D AOA localization with Gaussian priors is derived, and the solution of minimizing the trace of CRLB is developed. Moreover, Section 3 provided a solution for diagonalizing $\mathbf{P}_0$ with proper 3D rotation. The invariance property for 3D rotation of the AOA-based $\mathrm{tr}\left(\mathbf{\Phi}^{-1}\right)$ is used to diagonalize the non-diagonal covariance. Therefore, we suppose that the coordinate system is rotated such that the covariance matrix is diagonal $\mathbf{P}_0 = \mathrm{diag}([a, b, c])$.

Based on (11), the FIM for the 3D AOA target localization problem is

$$\mathbf{\Phi} = \mathbf{P}_0^{-1} + \mathbf{J}_2^T \mathbf{\Sigma}^{-1} \mathbf{J}_2 = \mathbf{P}_0^{-1} + \sum_{k=1}^{N} \frac{1}{r_k^2 \sigma_{\theta_k}^2 \cos^2\phi_k} \mathbf{u}_k \mathbf{u}_k^T + \sum_{k=1}^{N} \frac{1}{r_k^2 \sigma_{\phi_k}^2} \mathbf{v}_k \mathbf{v}_k^T, \tag{37}$$

where $\mathbf{u}_k$ and $\mathbf{v}_k$ are unit vectors orthogonal to the 2D azimuth vector and 3D range vector, respectively,

$$\mathbf{u}_k = \begin{bmatrix} -\sin\theta_k \\ \cos\theta_k \\ 0 \end{bmatrix}, \quad \mathbf{v}_k = \begin{bmatrix} -\sin\phi_k \cos\theta_k \\ -\sin\phi_k \sin\theta_k \\ \cos\phi_k \end{bmatrix}. \tag{38}$$

Following (19), we aim to determine optimal sensor locations, and the optimality criterion is to minimize the trace of CRLB, which is also known as the optimality criterion [35]. This section first investigates the optimal palcement of one sensor and then expands to multiple sensors.

4.1. Optimal Sensor Placement for One Sensor

Let us discuss the optimal placement for one sensor with Gaussian priors. Substitute (38) into (37) and then use (19). Then we can see that the trace of CRLB satisfies

$$\mathrm{tr}(\mathrm{CRLB}) = \mathrm{tr}(\mathbf{\Phi}^{-1}) \geq \left(a^{-1} + \frac{1}{r^2} \left(\frac{\sin^2\theta}{\sigma_\theta^2 \cos^2\phi} + \frac{1}{\sigma_\phi^2} \sin^2\phi \cos^2\theta \right) \right)^{-1}$$
$$+ \left(b^{-1} + \frac{1}{r^2} \left(\frac{\cos^2\theta}{\sigma_\theta^2 \cos^2\phi} + \frac{1}{\sigma_\phi^2} \sin^2\phi \sin^2\theta \right) \right)^{-1} + \left(c^{-1} + \frac{\cos^2\phi}{\sigma_\phi^2 r^2} \right)^{-1}, \tag{39}$$

with equality if

$$-\frac{\sin 2\theta}{\sigma_\theta^2 \cos^2\phi} + \frac{1}{\sigma_\phi^2} \sin^2\phi \sin 2\theta = 0,$$
$$\frac{1}{\sigma_\phi^2} \sin 2\phi \cos\theta = 0, \quad \frac{1}{\sigma_\phi^2} \sin 2\phi \sin\theta = 0. \tag{40}$$

To satisfy the above expression, we compute the azimuth and elevation angle as follows:

$$\{\theta, \phi\} \in \{\{\pm\pi/2, 0\}, \{\pm\pi/2, \pm\pi/2\}, \{0, 0\}, \{0, \pm\pi/2\}\}. \tag{41}$$

Substituting the optimal angle $\{\theta, \phi\}$ into (39), we can obtain different configurations, as listed in Table 1. We set $R1 = a$, $R2 = b$, $R3 = c$, $R4 = r^2\sigma_\theta^2$, and $R5 = r^2\sigma_\phi^2$, then adopt the resistor network model to find the minimum $\mathrm{tr}(\mathrm{CRLB})$, which depends on the prior covariance matrices, the angle noise variances σ_θ and σ_ϕ, and the sensor-target ranges r. The resistor network model for optimal sensor placement with different configurations is shown in Figure 3.

Table 1. Trace of CRLB with different optimal angles and configurations.

Configuration	θ	ϕ	tr(CRLB)
1	$\pm\pi/2$	0	$\left(a^{-1}+\dfrac{1}{r^2\sigma_\theta^2}\right)^{-1}+b+\left(c^{-1}+\dfrac{1}{r^2\sigma_\phi^2}\right)^{-1}$
2	$\pm\pi/2$	$\pm\pi/2$	$\left(b^{-1}+\dfrac{1}{r^2\sigma_\theta^2}+\dfrac{1}{r^2\sigma_\phi^2}\right)^{-1}+c$
3	0	0	$a+\left(b^{-1}+\dfrac{1}{r^2\sigma_\theta^2}\right)^{-1}+\left(c^{-1}+\dfrac{1}{r^2\sigma_\phi^2}\right)^{-1}$
4	0	$\pm\pi/2$	$\left(a^{-1}+\dfrac{1}{r^2\sigma_\theta^2}+\dfrac{1}{r^2\sigma_\phi^2}\right)^{-1}+c$

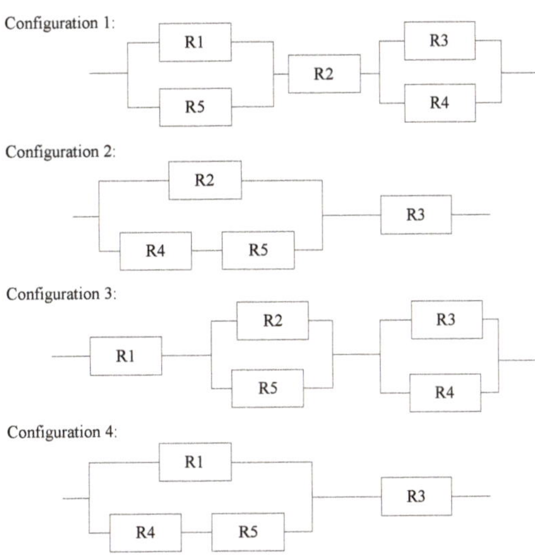

Figure 3. Resistor network model for optimal sensor placement for one sensor.

Furthermore, the resistor networks can help determine the optimal geometry rapidly using the analysis of different configurations, and the value of a, b, c with the prior covariance matrix $\mathbf{P}_0$ mainly decides the optimal placement when $r^2\sigma_\phi^2$ and $r^2\sigma_\theta^2$ are fixed by using the parallel resistor equation. The explanation of configurations in Table 1:

- Configuration 1: The values of resistors $R1$ and $R2$ can be reduced owing to the parallel resistors $R4$ and $R5$. Thus, the angle is suited for $a > c > b$ and $c > a > b$.
- Configuration 2: The value of resistor $R1$ is eliminated, so the angle is suited for $a > b > c$.
- Configuration 3: The value of resistor $R2$, $R3$ can be reduced owing to the parallel resistors $R4$ and $R5$. Thus, the angle is suited for $b > c > a$, $c > b > a$.
- Configuration 4: The value of resistor $R2$ is eliminated, so the angle is suited for $b > a > c$.

In conclusion, when the maximum value is a, the optimal angle of $\{\theta, \phi\}$ is $\{\pm\pi/2, 0\}$, $\{\pm\pi/2, \pm\pi/2\}$, and the line of sight (LOS) $\left\{[0, 1, 0]^T, [0, 0, 1]^T\right\}$ is orthogonal to the largest eigenvector of $\mathbf{P}_0$. A similar conclusion can be derived when the maximum value is b or c, which has the same results as [26]. Moreover, the non-diagonal covariance placement can easily be attained using the above analytical finding. This method is much simpler than the sensor update method in [26].

4.2. Optimal Sensor Placement for $N = 2$

In this subsection, we consider the case of two sensors and use the resistor network model to determine the optimal sensor placement. Substituting $N = 2$ into (37), the trace of inverse of FIM is written as

$$
\begin{aligned}
\text{tr(CRLB)} = \text{tr}(\mathbf{\Phi}^{-1}) \geq & \left(a^{-1} + \sum_{k=1}^{2}\frac{1}{r_k^2}\left(\frac{\sin^2\theta_k}{\sigma_{\theta_k}^2\cos^2\phi_k} + \frac{1}{\sigma_{\phi_k}^2}\sin^2\phi_k\cos^2\theta_k\right)\right)^{-1} \\
& + \left(b^{-1} + \sum_{k=1}^{2}\frac{1}{r_k^2}\left(\frac{\cos^2\theta_k}{\sigma_{\theta_k}^2\cos^2\phi_k} + \frac{1}{\sigma_{\phi_k}^2}\sin^2\phi_k\sin^2\theta_k\right)\right)^{-1} + \left(c^{-1} + \sum_{k=1}^{2}\left(\frac{\cos^2\phi_k}{\sigma_{\phi_k}^2 r_k^2}\right)\right)^{-1},
\end{aligned}
\tag{42}
$$

with equality if

$$
\begin{aligned}
& \sum_{k=1}^{2}\frac{1}{r_k^2}\left(\frac{1}{\sigma_{\phi_k}^2}\sin^2\phi_k\sin 2\theta_k - \frac{\sin 2\theta_k}{\sigma_{\theta_k}^2\cos^2\phi_k}\right) = 0, \\
& \sum_{k=1}^{2}\frac{1}{r_k^2\sigma_{\phi_k}^2}\sin 2\phi_k\cos\theta_k = 0, \quad \sum_{k=1}^{2}\frac{1}{r_k^2\sigma_{\phi_k}^2}\sin 2\phi_k\sin\theta_k = 0.
\end{aligned}
\tag{43}
$$

For azimuth angles, the two-sensor optimal placement in the 2D plane that minimizes the tr(CRLB) is given by $|\theta_1 - \theta_2| = \pi/2$, regardless of noise variance and sensor ranges [23]. Since we set $\{\theta_1, \theta_2\} = \{0, \pm\pi/2\}$, and the above equations can be satisfied when

$$
\{\phi_1, \phi_2\} \in \{\{0,0\}, \{0,\pm\pi/2\}, \{\pm\pi/2,0\}, \{\pm\pi/2,\pm\pi/2\}\}.
\tag{44}
$$

By substituting (44) into (42), we can obtain the tr(CRLB) for $\{\theta_1, \theta_2\} = \{0, \pm\pi/2\}$ with different elevation angles that listed in Table 2. Besides, we set $R1 = a$, $R2 = b$, $R3 = c$, $R4 = r_1^2\sigma_{\theta_1}^2$, $R5 = r_2^2\sigma_{\theta_2}^2$, $R6 = r_1^2\sigma_{\phi_1}^2$, and $R7 = r_2^2\sigma_{\phi_2}^2$. The minimum trace of CRLB depends on the prior covariance matrix, the angle noise variances, and the sensor-target ranges. The resistor network model for optimal sensor placement with the different configurations is shown in Figure 4.

Table 2. Trace of CRLB for $\{\theta_1, \theta_2\} = \{0, \pm\pi/2\}$ and different elevation-angles.

Configuration	ϕ_1	ϕ_2	tr(CRLB)
1	0	0	$\left(a^{-1} + \dfrac{1}{r_2^2\sigma_{\theta_2}^2}\right)^{-1} + \left(b^{-1} + \dfrac{1}{r_1^2\sigma_{\theta_1}^2}\right)^{-1} + \left(c^{-1} + \dfrac{1}{r_1^2\sigma_{\phi_1}^2} + \dfrac{1}{r_2^2\sigma_{\phi_2}^2}\right)^{-1}$
2	0	$\pm\pi/2$	$\left(b^{-1} + \dfrac{1}{r_1^2\sigma_{\theta_1}^2} + \dfrac{1}{r_2^2\sigma_{\phi_2}^2}\right)^{-1} + \left(c^{-1} + \dfrac{1}{r_1^2\sigma_{\phi_1}^2}\right)^{-1}$
3	$\pm\pi/2$	0	$\left(a^{-1} + \dfrac{1}{r_1^2\sigma_{\phi_1}^2} + \dfrac{1}{r_2^2\sigma_{\theta_2}^2}\right)^{-1} + \left(c^{-1} + \dfrac{1}{r_2^2\sigma_{\phi_2}^2}\right)^{-1}$
4	$\pm\pi/2$	$\pm\pi/2$	c

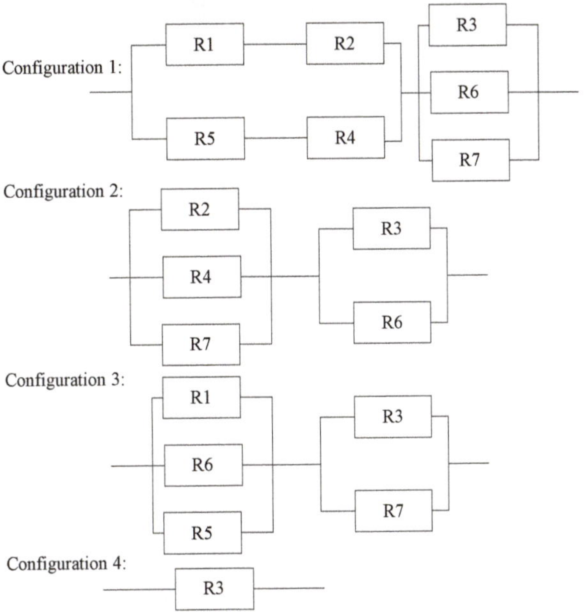

Figure 4. Resistor network model for optimal sensor placement for $N = 2$.

4.3. Optimal Sensor Placement for $N \geq 3$

In this section, we consider the optimal placement of N sensors in 3D space with different angle noises and distances. The trace of inverse of FIM is written as

$$
\mathrm{tr}(\mathrm{CRLB}) = \mathrm{tr}(\boldsymbol{\Phi}^{-1}) \geq \left(a^{-1} + \sum_{k=1}^{N} \frac{1}{r_k^2} \left(\frac{\sin^2\theta_k}{\sigma_{\theta_k}^2 \cos^2\phi_k} + \frac{1}{\sigma_{\phi_k}^2} \sin^2\phi_k \cos^2\theta_k \right) \right)^{-1}
$$

$$
+ \left(b^{-1} + \sum_{k=1}^{N} \frac{1}{r_k^2} \left(\frac{\cos^2\theta_k}{\sigma_{\theta_k}^2 \cos^2\phi_k} + \frac{1}{\sigma_{\phi_k}^2} \sin^2\phi_k \sin^2\theta_k \right) \right)^{-1} + \left(c^{-1} + \sum_{k=1}^{N} \left(\frac{\cos^2\phi_k}{\sigma_{\phi_k}^2 r_k^2} \right) \right)^{-1},
\tag{45}
$$

subject to

$$
\sum_{k=1}^{N} \frac{1}{r_k^2} \left(\frac{1}{\sigma_{\phi_k}^2} \sin^2\phi_k \sin 2\theta_k - \frac{\sin 2\theta_k}{\sigma_{\theta_k}^2 \cos^2\phi_k} \right) = 0,
$$

$$
\sum_{k=1}^{N} \frac{1}{r_k^2 \sigma_{\phi_k}^2} \sin 2\phi_k \cos\theta_k = 0, \quad \sum_{k=1}^{N} \frac{1}{r_k^2 \sigma_{\phi_k}^2} \sin 2\phi_k \sin\theta_k = 0.
\tag{46}
$$

To diagonalize FIM, the azimuth and elevation angle can be shown to obey the following equality [21]:

$$
\sin 2\theta_k = 0, k = 1, \ldots, N, \quad \sin 2\phi_k = 0, k = 1, \ldots, N.
\tag{47}
$$

Define the subset of $\mathbb{C}$ as the optimal azimuth angles, which is given by

$$
\mathbb{C} = \{ \{\theta_1, \theta_2, \ldots, \theta_N\} | \theta_k \in \{0, \pm\pi/2\}, k = 1, \ldots, N \},
\tag{48}
$$

The elevation angles satisfy (45) form a set defined as

$$
\mathbb{Z} = \{ \{\phi_1, \phi_2, \ldots \phi_N\} | \phi_k \in \{0, \pm\pi/2\}, k = 1, \ldots, N \}.
\tag{49}
$$

Thus, we can get the minimum trace of CRLB with the angle combination of $\mathbb{C}$ and $\mathbb{Z}$.

$$
\text{tr}\left(\boldsymbol{\Phi}_{\text{opt}}^{-1}(\theta_1,\ldots,\theta_N,\phi_1,\ldots,\phi_N)\right) = \left(a^{-1} + \sum_{k=1}^{N}\frac{1}{r_k^2}\left(\frac{\sin^2\theta_k}{\sigma_{\theta_k}^2\cos^2\phi_k} + \frac{1}{\sigma_{\phi_k}^2}\sin^2\phi_k\cos^2\theta_k\right)\right)^{-1}
$$

$$
+ \left(b^{-1} + \sum_{k=1}^{N}\frac{1}{r_k^2}\left(\frac{\cos^2\theta_k}{\sigma_{\theta_k}^2\cos^2\phi_k} + \frac{1}{\sigma_{\phi_k}^2}\sin^2\phi_k\sin^2\theta_k\right)\right)^{-1} + \left(c^{-1} + \sum_{k=1}^{N}\left(\frac{\cos^2\phi_k}{\sigma_{\phi_k}^2 r_k^2}\right)\right)^{-1}. \tag{50}
$$

Therefore, (48) and (49) can be used to determine the optimal sensor placement $N \geq 3$.

Based on the analysis above, we can get the optimal azimuth and elevation angles subset. This conclusion is consistent with the literature [21]. In addition, it can be seen that the parameters of $\mathbf{P}_0$ also affect the sensor placement with the analysis of the resistor network models. Therefore, the minimum trace of CRLB depends on the angle noise variances, the sensor-target distance, and the value of $\mathbf{P}_0$.

5. Simulation Studies

5.1. Gradient Descent Alogorithm Simulations

In this subsection, we adopt a gradient descent algorithm to verify the optimal sensor placement conditions derived in the above section. Assume that the distribution of target is given, and $\mathbf{s}_0 = (0,0,0)^T$. The minimum distances between the target and sensors are represented by d_k. A group of mobile sensors is moving to minimize the trace of CRLB in 3D space [21]. This exact gradient descent simulation was run 10,000 steps.

- Example 1: For optimal sensor placement with one sensor

Case A: We used these simulation parameters: $\mathbf{P}_0 = \begin{bmatrix} 500 & 0 & 0 \\ 0 & 200 & 0 \\ 0 & 0 & 100 \end{bmatrix}, d = 150\text{ m}$, $\sigma_\theta^2 = \sigma_\phi^2 = 1°$, and the initial sensor location was $\begin{bmatrix} 200 & -100 & -100\sqrt{2} \end{bmatrix}^T$. The sensor trajectory is shown in Figure 5a, and the final angles were $\theta = -91.34°$ and $\phi = -89.53°$, which matches Configuration 2 ($a > b > c$) in Table 1, and the LOS was orthogonal to the largest eigenvector of $\mathbf{P}_0$.

Case B: The simulation parameters were as follows: $\mathbf{P}_0 = \begin{bmatrix} 100 & 0 & 0 \\ 0 & 200 & 0 \\ 0 & 0 & 500 \end{bmatrix}, d = 200\text{ m}$, $\sigma_\theta^2 = 1°$, $\sigma_\phi^2 = 2°$, and the initial sensor location was $\begin{bmatrix} 100 & 200 & 100\sqrt{2} \end{bmatrix}^T$. The sensor trajectory is shown in Figure 5b and the final angles were $\theta = -0.03°$ and $\phi = -0.02°$, which matches Configuration 3 ($c > b > a$) in Table 1, and the LOS was orthogonal to the largest eigenvector of $\mathbf{P}_0$. Moreover, although the initial sensor location and d were different in Cases A and B, it is shown that the final optimal sensor placement also matches the analysis results in Figure 5a,b. The simulation results also can prove the proposed method without any restriction on the sensor-target range and initial sensor locations.

Case C: We used the parameters of Case B except $\mathbf{P}_0 = \begin{bmatrix} 100 & 50 & 20 \\ 50 & 200 & 30 \\ 20 & 30 & 500 \end{bmatrix}$. The rotation angles were computed using (20) and (21), i.e., $\alpha = 7.10°$, $\beta = 358.88°$, $\gamma = 22.26$, and $\mathbf{P}_0$ can be rewritten as $\mathbf{P}_0^r = \begin{bmatrix} 79.17 & 0 & 0 \\ 0 & 216.31 & 0 \\ 0 & 0 & 504.52 \end{bmatrix}$. The rest of simulation parameters can be obtained from (22), and the tr(CRLB) was computed using (33). The sensor trajectory is shown in Figure 5c, and the final angles were $\theta = -0.03°$ and $\phi = -0.01°$, which matches Configuration 3 ($c > b > a$) in Table 1. The LOS was orthogonal to the largest eigenvector of $\mathbf{P}_0$.

Case D: We used the parameters of Case B except $\mathbf{P}_0 = \begin{bmatrix} 300 & 10 & 20 \\ 10 & 500 & 15 \\ 20 & 15 & 100 \end{bmatrix}$, and

$$\mathbf{P}_0^r = \begin{bmatrix} 301.46 & 0 & 0 \\ 0 & 501.13 & 0 \\ 0 & 0 & 97.41 \end{bmatrix}$$ after the 3D rotation. As in Case C, the final angles were $\theta = -0.16°$ and $\phi = -89.67°$, which matches Configuration 4 ($b > a > c$) in Table 1; and the LOS was orthogonal to the largest eigenvector of $\mathbf{P}_0$, and the sensor trajectory is shown in Figure 5d.

More specifically, the prior covariance matrices $\mathbf{P}_0$ in Cases A and B were diagonal covariance matrices $\mathbf{P}_0$. We could quickly obtain the optimal placement through the gradient simulation, and the results of Figure 5a,b match the findings in Section 4.1. Besides, the prior covariance matrices $\mathbf{P}_0$ in Cases C and D were non-diagonal covariance matrices, and the invariance property for 3D rotation of the AOA-based trace of CRLB was used to diagonalize the non-diagonal covariance. Then, we obtained the optimal placement using the gradient simulation, and the results of Figure 5c,d also match the findings in Section 4.1.

- Example 2: Optimal sensor placement for two and three sensors:

Case A: The simulation parameters were as follows: $\mathbf{P}_0 = \begin{bmatrix} 200 & 0 & 0 \\ 0 & 600 & 0 \\ 0 & 0 & 900 \end{bmatrix}$, $d_1 = d_2 = 200$ m, $\sigma_{\theta_1}^2 = \sigma_{\theta_2}^2 = 0.5°, \sigma_{\phi_1}^2 = \sigma_{\phi_2}^2 = 1°$, and the initial sensor locations were $\begin{bmatrix} 200 & -100 & -100\sqrt{2} \end{bmatrix}^T, \begin{bmatrix} 100 & -100 & 200 \end{bmatrix}^T$. The sensors' trajectories are shown in Figure 6a, and the final angles were $\theta_1 = -37.24°$, $\theta_2 = -130.29°$, $\phi_1 = -0.03°$, and $\phi_2 = 88.91°$, which matches Configuration 2 in Table 2.

Case B: We used the parameters of Case A except $\mathbf{P}_0 = \begin{bmatrix} 200 & 20 & 15 \\ 20 & 600 & 50 \\ 15 & 50 & 900 \end{bmatrix}$, and

$$\mathbf{P}_0^r = \begin{bmatrix} 198.52 & 0 & 0 \\ 0 & 596.23 & 0 \\ 0 & 0 & 905.25 \end{bmatrix}$$ after rotation. The sensors' trajectories are shown in Figure 6b, and the final angles were $\theta_1 = -27.95°$, $\theta_2 = -116.51°$, $\phi_1 = -0.06°$, and $\phi_2 = 88.65°$, which also matches Configuration 2 in Table 2.

In Cases A and B, we adopted the same parameters except for the covariance matrix $\mathbf{P}_0$. Similarly, the non-diagonal covariance matrix in Case B was diagonalized by the 3D rotation. It is shown that the sensor trajectories and the final optimal sensor-target geometries were almost identical in Figure 6 a,b, which satisfies the results of Section 4.2.

Case C: For three sensors, we used the simulation parameters as follows: $\mathbf{P}_0 = \begin{bmatrix} 300 & 0 & 0 \\ 0 & 800 & 0 \\ 0 & 0 & 900 \end{bmatrix}$, $d_1 = d_2 = d_3 = 200$ m, $\sigma_{\theta_1}^2 = \sigma_{\theta_2}^2 = \sigma_{\theta_3}^2 = 0.5°, \sigma_{\phi_1}^2 = \sigma_{\phi_2}^2 = \sigma_{\phi_3}^2 = 0.5°$, and the initial sensor locations were $\begin{bmatrix} -100\sqrt{2} & 100 & -200 \end{bmatrix}^T, \begin{bmatrix} 100 & -100\sqrt{2} & 0 \end{bmatrix}^T, \begin{bmatrix} -100\sqrt{2} & 100 & 200 \end{bmatrix}^T$. The sensors' trajectories are shown in Figure 6c and the final angles were $\theta_1 = 118.51°$, $\theta_2 = -61.49°$, $\theta_3 = -155.64°$, $\phi_1 = -0.01°$, $\phi_2 = 0.01°$, and $\phi_3 = 88.91°$.

Case D: We used the parameters of Case C except $\mathbf{P}_0 = \begin{bmatrix} 300 & 15 & 20 \\ 15 & 800 & 30 \\ 20 & 30 & 900 \end{bmatrix}$, and

$$\mathbf{P}_0^r = \begin{bmatrix} 299.18 & 0 & 0 \\ 0 & 795.77 & 0 \\ 0 & 0 & 905.05 \end{bmatrix}$$ after rotation. The sensors' trajectories are shown in Figure 6d, and the final angles were $\theta_1 = 120.51°$, $\theta_2 = -59.49°$, $\theta_3 = -146.41°$, $\phi_1 = 0.05°$, $\phi_2 = 0.01°$, and $\phi_3 = 88.33°$.

Similarly, we used the same parameters except for the covariance matrix $\mathbf{P}_0$ in Cases C and D. The non-diagonal covariance matrix in Case D was diagonalized by the 3D rotation. It is shown that the sensor trajectories and the final optimal sensor-target geometries were almost identical in Figure 6c,d, which also satisfies the results of Section 4.3.

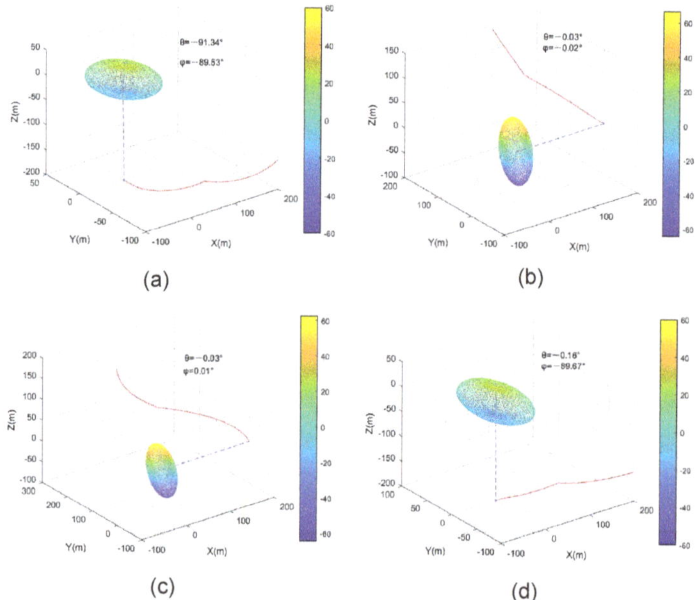

Figure 5. Optimal sensor placement for one sensor. (**a**) $\mathbf{P}_0$ is a diagonal matrix with $a > b > c$, (**b**) $\mathbf{P}_0$ is a diagonal matrix with $c > b > a$, (**c**) $\mathbf{P}_0$ is a non-diagonal matrix with $c > b > a$, (**d**) $\mathbf{P}_0$ is a non-diagonal matrix with $b > a > c$.

For Cases A and B in Example 2, the $\mathrm{tr}\left(\boldsymbol{\Phi}_{\mathrm{opt}}^{-1}\right)$ computed by the gradient descent algorithm were approximately the same; besides, we could obtain the theoretical minimum trace of CRLB using (42) with the optimal sensor placement. The $\mathrm{tr}(\boldsymbol{\Phi}^{-1})$ from Case A and $\mathrm{tr}(\hat{\boldsymbol{\Phi}}^{-1})$ from Case B were equal, which is in agreement with the analytical result of (36). Table 3 lists the $\mathrm{tr}\left(\boldsymbol{\Phi}_{\mathrm{opt}}^{-1}\right)$, $\mathrm{tr}(\boldsymbol{\Phi}^{-1})$ and $\mathrm{tr}(\hat{\boldsymbol{\Phi}}^{-1})$ for different cases of Example 2. It is clear that the same conclusion was obtained for $N = 3$ in Example 2 for Cases C and D. Furthermore, the $\mathrm{tr}\left(\boldsymbol{\Phi}_{\mathrm{opt}}^{-1}\right)$ is close to the theoretical minimum trace; i.e., $\mathrm{tr}(\boldsymbol{\Phi}^{-1})$ and $\mathrm{tr}(\hat{\boldsymbol{\Phi}}^{-1})$.

Table 3. Trace of CRLB for Example 2.

Example 2	$\mathrm{tr}(\boldsymbol{\Phi}_{\mathrm{opt}}^{-1})$ (m^2)	$\mathrm{tr}(\boldsymbol{\Phi}^{-1})$ (m^2)	$\mathrm{tr}(\hat{\boldsymbol{\Phi}}^{-1})$ (m^2)
Case A	5.4678	5.4620	/
Case B	5.5156	/	5.4620
Case C	2.5389	2.5310	/
Case D	2.5680	/	2.5310

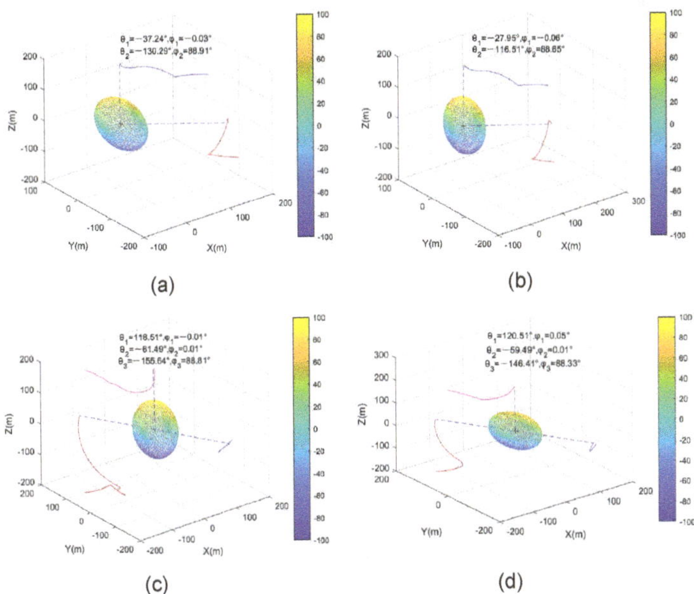

Figure 6. Optimal sensor placement with two and three sensors. (**a**) $\mathbf{P}_0$ is a diagonal matrix with $N = 2$, (**b**) $\mathbf{P}_0$ is a non-diagonal matrix with $N = 2$, (**c**) $\mathbf{P}_0$ is a diagonal matrix with $N = 3$, (**d**) $\mathbf{P}_0$ is a non-diagonal matrix with $N = 3$.

5.2. The Comparison Results

This subsection demonstrates the optimal sensor placement with the maximum a posteriori (MAP) estimation simulations, and the MAP is deduced in Appendix A. In the example, the method in [21] and the method in [22] using the D-optimality criterion are compared with the proposed method. In this paper, we use "the method in [21]" and "the method in [22]" to denote the optimal placement methods in [21,22], respectively. The parameters were as follows: $\mathbf{s}_0 = (0,0,0)^T$, $\mathbf{P}_0 = \begin{bmatrix} 100 & 0 & 0 \\ 0 & 200 & 0 \\ 0 & 0 & 800 \end{bmatrix}$, and the initial sensor locations were $\begin{bmatrix} -200 & -100 & -100\sqrt{2} \end{bmatrix}^T$, $\begin{bmatrix} 100 & -100\sqrt{2} & -200 \end{bmatrix}^T$, $\begin{bmatrix} -100\sqrt{2} & 100 & 200 \end{bmatrix}^T$, $\begin{bmatrix} 100\sqrt{2} & 200 & -100\sqrt{2} \end{bmatrix}^T$. We added different noise levels and show the theoretical minimum trace of CRLBs and MSEs; i.e., $\sigma_{\theta_1}^2 = \sigma_{\theta_2}^2 = \sigma_{\theta_3}^2 = \sigma_{\theta_4}^2 = 0.5°$, and $\sigma_{\phi_1}^2 = \sigma_{\phi_2}^2 = \sigma_{\phi_3}^2 = \sigma_{\phi_4}^2$, the value of σ_{ϕ}^2 from $0.2°$ to $1.8°$.

The theoretical trace of CRLBs and MSEs of different sensor placements are shown in Figure 7. The MSEs of MAP were estimated using 10,000 Monte Carlo simulations. The MAP estimator was implemented using the Gauss–Newton method and initialized to the prior mean target location $\mathbf{s}_0$. The results showed that the optimal sensor placement can always provide better MSEs than the other existing methods.

Next, we compare the localization accuracies of different methods. We fixed $N = 3$, $\sigma_{\theta}^2 = 0.5°$ and increased the value of σ_{ϕ}^2 from $0.1°$ to $1°$. The settings of others parameters were the same as in Case C of Example 2. The optimal angles in [21] are $\theta_1 = 0°$, $\theta_2 = 90°$, $\theta_3 = -90°$, $\phi_1 = 0°$, $\phi_2 = 0°$, and $\phi_3 = 0°$. The correspondingly optimal angles were adopted in Case C of Example 2 as $\theta_1 = 118°$, $\theta_2 = -62°$, $\theta_3 = -152°$, $\phi_1 = 0°$, $\phi_2 = 0°$, and $\phi_3 = 90°$. Figure 8 shows the comparison of tr(CRLB)s computed by the method in [21], the method in [22], and the final sensor locations in Case C of Example 2.

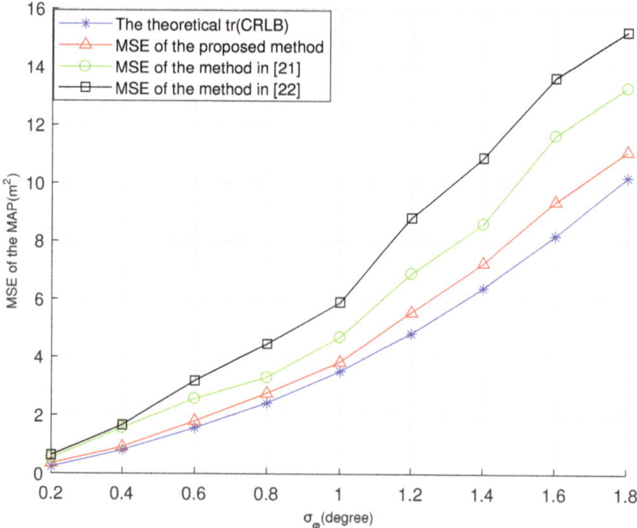

Figure 7. Estimation comparison with $\sigma_\theta^2 = 0.5°$ and $\sigma_\phi^2 = 0.2°$ to $1.8°$.

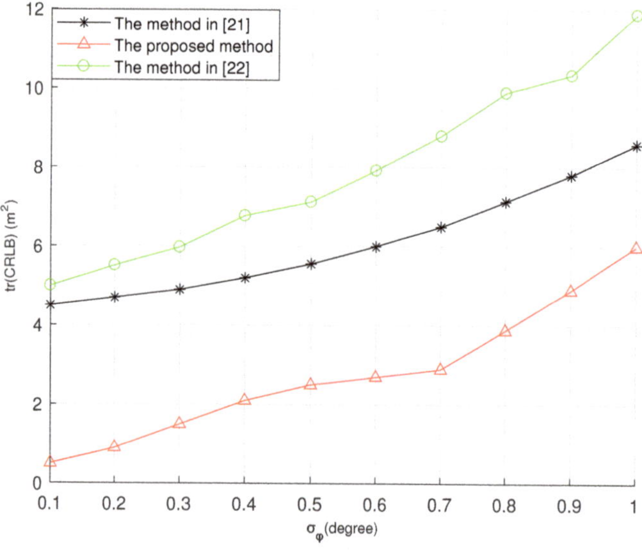

Figure 8. The comparison results with $\sigma_\theta^2 = 1°$ and $\sigma_\phi^2 = 0.1°$ to $1°$.

From Figure 8, it can be seen that the proposed method in this paper had better estimation performance than the existing methods, even if both the proposed method and the method in [21] contained optimal azimuth and elevation angles subsets. This result also can confirm the analytical optimal sensor placement in Section 4.

Finally, we compare the method in [21,22] in terms of estimation performance for different sensor numbers. The sensors started from different original locations, and we set $\mathbf{P}_0 = \begin{bmatrix} 200 & 0 & 0 \\ 0 & 500 & 0 \\ 0 & 0 & 700 \end{bmatrix}$, $d = 200$ m, $\sigma_\theta^2 = \sigma_\phi^2 = 1°$. Table 4 lists the MSEs and bias norms when the number of sensors is $N = 3, 4, 5, 6$. Due to the effect of the prior covariance matrix, the performance of the existing methods was worse than that of the proposed method. The MSEs of our proposed method were much smaller than those of the existing methods with the different sensor numbers. From Figure 8 and Table 4, we conclude that the proposed method can achieve the optimal estimation performance.

Table 4. MAP estimation performances of three different methods with $N = 3, 4, 5, 6$.

Number	Method	MSE (m^2)	Bias Norm (m)
$N = 3$	The proposed method	6.12	0.1472
	The method in [21]	12.35	0.8225
	The method in [22]	14.67	1.3557
$N = 4$	The proposed method	4.32	0.0925
	The method in [21]	9.97	0.4634
	The method in [22]	11.43	0.8143
$N = 5$	The proposed method	1.54	0.055
	The method in [21]	4.81	0.2415
	The method in [22]	5.94	0.5468
$N = 6$	The proposed method	0.48	0.0123
	The method in [21]	1.61	0.1022
	The method in [22]	2.58	0.3967

6. Conclusions

In this paper, an optimal sensor placement method for an uncertain target with Gaussian priors was presented. Our analysis was conducted based on minimizing the trace of the inverse FIM. The invariance property for the 3D rotation of the AOA-based FIM was provided, which can be used to diagonalize the non-diagonal covariance matrix. An optimal sensor placement analysis for the 3D space with the diagonal covariance matrix of the target was presented, and a resistor network was used to represent the optimal sensor placement strategy. It was demonstrated that the optimal localization placements have a similar geometric configuration, regardless of the diagonality of the covariance matrix. Finally, the analytical results were verified via a series of numerical simulations. The analytical and numerical findings coincide with the simulation results.

For future work, we will consider a case with multiple uncertain targets with different Gaussian priors, which changes the optimization problem to a convex combination of FIMs. In addition, the optimal trajectories also can be developed for the uncertain moving target with Gaussian priors.

Author Contributions: Conceptualization, R.Z. and J.C.; methodology, R.Z.; validation, R.Z. and W.T.; writing—original draft preparation, R.Z.; writing—review and editing, R.Z., J.C. and C.C.; supervision, J.C.; funding acquisition, W.T. and Q.Y. All authors have read and agreed to the published version of the manuscript.

Funding: This research was funded by the National Natural Science Foundation of China under grant 62071383 and grant 6210012056.

Data Availability Statement: Not applicable.

Conflicts of Interest: The authors declare no conflict of interest.

Appendix A. The Deduction of MAP

The MAP estimation of the target was obtained from maximizing $\hat{\mathbf{s}}_{\mathrm{MAP}} = (\hat{x}, \hat{y}, \hat{z})^{T}$ to maximize the posterior probability density function (PDF) and can be written as

$$\hat{\mathbf{s}}_{\mathrm{MAP}} = \arg \max_{\mathbf{s}} p(\mathbf{s}|\tilde{\mathbf{q}}), \tag{A1}$$

where $\tilde{\mathbf{q}} = [\tilde{\theta}_1, \tilde{\phi}_1, \tilde{\theta}_2, \tilde{\phi}_2, \cdots, \tilde{\theta}_N, \tilde{\phi}_N]^T$ is the $2N \times 1$ vector of noisy angle measurements. In maximizing $p(\mathbf{s}|\tilde{\mathbf{q}})$, we observe that

$$p(\mathbf{s}|\tilde{\mathbf{q}}) = \frac{p(\tilde{\mathbf{q}}|\mathbf{s})p(\mathbf{s})}{p(\tilde{\mathbf{q}})}, \tag{A2}$$

Note that (A1) is equivalent to the maximization of $p(\tilde{\mathbf{q}}|\mathbf{s})p(\mathbf{s})$. This is reminiscent of the maximum likelihood estimation (MLE) except for the presence of the prior PDF [36]. Hence, the MAP estimation can be rewritten as

$$\hat{\mathbf{s}}_{\text{MAP}} = \arg\max_{\mathbf{s}} p(\tilde{\mathbf{q}}|\mathbf{s})p(\mathbf{s}), \tag{A3}$$

Assuming that the target location has a prior distribution as $\mathbf{s} \sim \mathcal{N}(\mathbf{s}_0, \mathbf{P}_0)$, the prior PDF for the target is given by

$$\begin{aligned} p(\mathbf{s}) =& \frac{1}{(2\pi)^{N/2} \det{(\mathbf{P}_0)}^{1/2}} \\ &\times \exp\left[-\frac{1}{2}(\mathbf{s} - \mathbf{s}_0)^T \mathbf{P}_0^{-1}(\mathbf{s} - \mathbf{s}_0)\right], \end{aligned} \tag{A4}$$

The maximum likelihood function of $\mathbf{s}$ is given by

$$\begin{aligned} p(\tilde{\mathbf{q}}|\mathbf{s}) =& \frac{1}{(2\pi)^{N/2} \det{(\mathbf{K})}^{1/2}} \\ &\times \exp\left[-\frac{1}{2}(\tilde{\mathbf{q}} - \mathbf{q}(\mathbf{s}))^T \mathbf{K}^{-1}(\tilde{\mathbf{q}} - \mathbf{q}(\mathbf{s}))\right], \end{aligned} \tag{A5}$$

where $\mathbf{K} = \text{diag}(\sigma_{\theta_1}^2, \sigma_{\phi_1}^2, \sigma_{\theta_2}^2, \sigma_{\phi_2}^2, \ldots, \sigma_{\theta_N}^2, \sigma_{\phi_N}^2)$ is the $2N \times 2N$ diagonal covariance matrix of the angle noise.

By substituting (A4) and (A5) into (A3), $\hat{\mathbf{s}}_{\text{MAP}}$ is obtained by the log-likelihood function $\ln p(\tilde{\mathbf{q}}|\mathbf{s})p(\mathbf{s})$ over $\mathbf{s}$, which is equivalent to

$$\hat{\mathbf{s}}_{\text{MAP}} = \arg\min_{\mathbf{s}} J_{\text{MAP}}(\mathbf{s}), \tag{A6}$$

with

$$J_{\text{MAP}}(\mathbf{s}) = \mathbf{e}(\mathbf{s})^T \mathbf{K}^{-1} \mathbf{e}(\mathbf{s}) + \mathbf{r}(\mathbf{s})^\mathbf{T} \mathbf{P}_0^{-1} \mathbf{r}(\mathbf{s}), \tag{A7}$$

and the $J_{\text{MAP}}(\mathbf{s})$ is the maximum A posterior cost function.

Here $\mathbf{e}(\mathbf{s})$ and $\mathbf{r}(\mathbf{s})$ are defined by

$$\begin{aligned} \mathbf{e}(\mathbf{s}) &= \tilde{\mathbf{q}} - \mathbf{q}(\mathbf{s}) \\ &= [\tilde{\theta}_1 - \theta_1(\mathbf{s}), \tilde{\phi}_1 - \phi_1(\mathbf{s}), \ldots, \tilde{\theta}_N - \theta_N(\mathbf{s}), \tilde{\phi}_N - \phi_N(\mathbf{s})]^T, \\ \mathbf{r}(\mathbf{s}) &= \mathbf{s} - \mathbf{s}_0, \end{aligned} \tag{A8}$$

and the residual can be written as

$$\boldsymbol{\Gamma}(\mathbf{s}) = [\mathbf{e}(\mathbf{s}); \mathbf{r}(\mathbf{s})], \tag{A9}$$

Note that the error covariance matrix of $\hat{\mathbf{s}}_{\text{MAP}}$ is given by (A4) and (A5)

$$\mathbf{Q} = \begin{bmatrix} \mathbf{K} & \mathbf{0}_{2N \times 3} \\ \mathbf{0}_{3 \times 2N} & \mathbf{P}_0 \end{bmatrix}. \tag{A10}$$

$\mathbf{J}_{1i}$ is the $2N \times 3$ Jacobian of $\mathbf{e}(\mathbf{s})$ with respect to $\mathbf{s}$ evaluated at $\mathbf{s} = \hat{\mathbf{s}}_i$, which can be expressed as

$$\mathbf{J}_{1i} = \begin{bmatrix} -\dfrac{\sin\theta_1^{(\hat{\mathbf{s}}_i)}}{\hat{d}_{i1}} & \dfrac{\cos\theta_1^{(\hat{\mathbf{s}}_i)}}{\hat{d}_{i1}} & 0 \\ -\dfrac{\sin\phi_1^{(\hat{\mathbf{s}}_i)}\cos\theta_1^{(\hat{\mathbf{s}}_i)}}{\hat{r}_{i1}} & -\dfrac{\sin\phi_1^{(\hat{\mathbf{s}}_i)}\sin\theta_1^{(\hat{\mathbf{s}}_i)}}{\hat{r}_{i1}} & \dfrac{\cos\phi_1^{(\hat{\mathbf{s}}_i)}}{\hat{r}_{i1}} \\ \vdots & \vdots & \vdots \\ -\dfrac{\sin\theta_N^{(\hat{\mathbf{s}}_i)}}{\hat{d}_{iN}} & \dfrac{\cos\theta_N^{(\hat{\mathbf{s}}_i)}}{\hat{d}_{iN}} & 0 \\ -\dfrac{\sin\phi_N^{(\hat{\mathbf{s}}_i)}\cos\theta_N^{(\hat{\mathbf{s}}_i)}}{\hat{r}_{iN}} & -\dfrac{\sin\phi_N^{(\hat{\mathbf{s}}_i)}\sin\theta_N^{(\hat{\mathbf{s}}_i)}}{\hat{r}_{iN}} & \dfrac{\cos\phi_N^{(\hat{\mathbf{s}}_i)}}{\hat{r}_{iN}} \end{bmatrix}, \tag{A11}$$

In the above expression

$$\begin{aligned} \hat{r}_{ik} &= \|\hat{\mathbf{s}}_i - \mathbf{p}_k\|, \\ \hat{d}_{ik} &= \hat{r}_{ik}\cos\phi_k(\hat{\mathbf{s}}_i), \end{aligned} \tag{A12}$$

$\mathbf{J}_{2i}$ is the 3×3 Jacobian of $\mathbf{r}(\mathbf{s})$ is given by

$$\mathbf{J}_{2i} = \mathbf{I}_{3\times3}, \tag{A13}$$

Combining (A11) and (A13), $\mathbf{J}_i$ is the Jacobian of (A9) defined by

$$\mathbf{J}_i = -[\mathbf{J}_{i1}; \mathbf{J}_{i2}], \tag{A14}$$

The MAP is calculated by the Gauss–Newton (GN) algorithm, as stated in [36], which is defined as

$$\hat{\mathbf{t}}_{i+1} = \hat{\mathbf{t}}_i - \left(\mathbf{J}_i^T \mathbf{Q}^{-1} \mathbf{J}_i\right)^{-1}\mathbf{J}_i^T \mathbf{Q}^{-1}\mathbf{\Gamma}(\hat{\mathbf{t}}_i). \tag{A15}$$

References

1. Sayed, A.H.; Tarighat, A.; KandKhajehnouri, N. Network-based wireless location: Challenges faced in developing techniques for accurate wireless location information. *IEEE Signal Process.* **2005**, *22*, 24–40. [CrossRef]
2. Akyildiz, F.; Su, W.; Sankarasubramaniam, Y.; Cayirci, E. A survey on sensor networks. *IEEE Commun. Mag.* **2002**, *40*, 102–114. [CrossRef]
3. Shen, J.; Molisch, A.F.; Salmi, J. Accurate passive location estimation using TOA measurements. *IEEE Trans. Wirel. Commun.* **2012**, *11*, 2182–2192. [CrossRef]
4. Chan, Y.T.; Ho, K.C. A simple and efficient estimator for hyperbolic location. *IEEE Trans. Signal Process.* **1994**, *42*, 1905–1915. [CrossRef]
5. Kułakowski, P.; Vales-Alonso, J.; Egea-Lopez, E.; Ludwin, W.; García-Harob, J. Angle-of-arrival localization based on antenna arrays for wless sensor. *Comput. Elect. Eng.* **2010**, *36*, 1181–1186. [CrossRef]
6. Peng, R.; Sichitiu, M.L. Angle of arrival localization for wireless sensor networks. In Proceedings of the 2006 3rd Annual IEEE Communications Society on Sensor and Ad Hoc Communications and Networks, Reston, VA, USA, 28 September 2006; pp. 374–382.
7. Wang, C.; Qi, F.; Shi, G.; Wang, X. Convex combination based target localization with noisy angle of arrival measurements. *IEEE Commun. Lett.* **2014**, *3*, 14–17. [CrossRef]
8. Li, X. Performance study of RSS-based location estimation techniques for wireless sensor networks. *Proc. IEEE Mil. Commun. Conf.* **2005**, *2*, 1064–1068.
9. Bishop, A.N.; Jensfelt, P. An optimality analysis of sensor-target geometries for signal strength based localization. In Proceedings of the 2009 International Conference on Intelligent Sensors, Sensor Networks and Information Processing (ISSNIP), Melbourne, Australia, 7–10 December 2009; pp. 127–132.
10. Doğançay, K. Bias compensation for the bearings-only pseu-dolinear target track estimator. *IEEE Trans. Signal Process.* **2006**, *54*, 59–68. [CrossRef]
11. Shao, H.J.; Zhang, X.P.; Wang, Z. Efficient closed-form algorithms for AOA based self-localization of sensor nodes using auxiliary variables. *IEEE Trans. Signal Process.* **2014**, *62*, 2580–2594. [CrossRef]

12. Doğançay, K. 3D pseudolinear target motion analysis from angle measurements. *IEEE Trans. Signal Process.* **2015**, *63*, 1570–1580. [CrossRef]
13. Wang, Y.; Ho, K.C. An asymptotically efficient estimator in closed-form for 3-D AOA localization using a sensor network. *IEEE Trans. Wirel. Commun.* **2015**, *14*, 6524–6535. [CrossRef]
14. Chen, X.; Gang, W.; Ho, K.C. Semidefinite relaxation method for unified near-Field and far-Field localization by AOA—ScienceDirect. *Signal Process.* **2021**, *181*, 107916. [CrossRef]
15. Doğançay, K.; Hmam, H. Optimal angular sensor separation for AOA localization. *Signal Process.* **2008**, *88*, 1248–1260. [CrossRef]
16. Bishop, A.N.; Fidan, B.; Anderson, B.; Doğançay, K.; Pathirana, P.N. Optimality analysis of sensor-target localization geometries. *Automatica* **2010**, *46*, 479–492. [CrossRef]
17. Xu, S.; Doğançay, K. Optimal sensor deployment for 3D AOA target localization. In Proceedings of the 2015 IEEE International Conference on Acoustics, Speech and Signal Processing (ICASSP), South Brisbane, Australia, 19–24 April 2015; pp. 2544–2548.
18. Nguyen, N.H.; Doğançay, K. Optimal Geometry Analysis for Multistatic TOA Localization. *IEEE Trans. Signal Process.* **2016**, *64*, 4180–4193. [CrossRef]
19. Ucinski, D. *Optimal Measurement Methods for Distributed Parameter System Identification*; CRC Press: Boca Raton, FL, USA, 2004.
20. Moreno-Salinas, D.; Pascoal, A.; Aranda, J. Sensor networks for optimal target localization with bearings-only measurements in constrained three-dimensional scenarios. *Sensors* **2013**, *13*, 10386–10417. [CrossRef]
21. Xu, S.; Doğançay, K. Optimal sensor placement for 3-D angle-of-arrival target localization. *IEEE Trans. Aerosp. Electron. Syst.* **2017**, *53*, 1196–1211. [CrossRef]
22. Zhao, S.; Chen, B.M.; Lee, T.H. Optimal sensor placement for target localisation and tracking in 2D and 3D. *Int. J. Control* **2013**, *86*, 1687–1704. [CrossRef]
23. Fang, X.; Li, J. Frame Theory for Optimal Sensor Augmentation Problem of AOA Localization. *IEEE Signal Process. Lett.* **2018**, *25*, 1310–1314. [CrossRef]
24. Isaacs, J.T.; Klein, D.J.; Hespanha, J.P. Optimal sensor placement for time difference of arrival localization. In Proceedings of the Proceedings of the 48h IEEE Conference on Decision and Control (CDC) Held Jointly with 2009 28th Chinese Control Conference, Shanghai, China, 15–18 December 2009; pp. 7878–7884.
25. Nguyen, N.H. Optimal geometry analysis for target localization with bayesian priors. *IEEE Access* **2021**, *9*, 33419–33437. [CrossRef]
26. Yang, C.; Kaplan, L.; Blasch, E. Performance measures of covariance and information matrices for resource management for target state estimation. *IEEE Trans. Aero. Electron. Syst.* **2012**, *48*, 2594–2613. [CrossRef]
27. Yang, C.; Kaplan, L.; Blasch, E.; Bakich, M. Optimal placement of heterogeneous sensors for targets with Gaussian priors. *IEEE Trans. Aerosp. Electron. Syst.* **2013**, *49*, 1637–1653. [CrossRef]
28. Yang, C.; Kaplan, L.; Blasch, E.; Bakich, M. Optimal placement of heterogeneous sensors in target tracking. In Proceedings of the 14th International Conference on Information Fusion, Chicago, IL, USA, 5–8 July 2011; pp. 1–8.
29. Doğançay, K. Relationship between geometric translations and TLS estimation bias in bearings-only target localization. *IEEE Trans. Signal Process.* **2008**, *56*, 1005–1017. [CrossRef]
30. Xu, S.; Ou, Y.; Wu, X. Optimal sensor placement for 3-D time-of-arrival target localization. *IEEE Trans. Signal Process.* **2019**, *67*, 5018–5031. [CrossRef]
31. Luo, J.; Zhang, X.; Wang, Z.; Lai, X. On the accuracy of passive source localization using acoustic sensor array networks. *IEEE Sens. J.* **2017**, *17*, 1795–1809. [CrossRef]
32. Xu, S. Optimal sensor placement for target localization using hybrid RSS, AOA and TOA measurements. *IEEE Commun. Lett.* **2020**, *24*, 1966–1970.
33. Luo, J.A.; Shao, X.H.; Peng, D.L.; Zhang, X.P. A novel subspace approach for bearing-only target localization. *IEEE Sens. J.* **2019**, *19*, 8174–8182. [CrossRef]
34. Kay, S.M. *Fundamentals of Statistical Signal Processing: Estimation Theory*; Prentice-Hal. Press: Englewood Cliffs, NJ, USA, 1993.
35. Zhang, F.; Sun, Y.; Zou, J.; Zhang, D.; Wan, Q. Closed-form localization method for moving target in passive multistatic radar network. *IEEE Sens. J.* **2020**, *20*, 980–990. [CrossRef]
36. Nguyen, N.H.; Doğançay, K. Closed-form algebraic solutions for Angle-of-Arrival source localization with Bayesian priors. *IEEE Trans. Wirel. Commun.* **2019**, *18*, 3827–3842. [CrossRef]

Article

Toward Accelerated Training of Parallel Support Vector Machines Based on Voronoi Diagrams

Cesar Alfaro, Javier Gomez *, Javier M. Moguerza, Javier Castillo and Jose I. Martinez

Department of Computer Science, University Rey Juan Carlos, 28933 Móstoles, Spain; cesar.alfaro@urjc.es (C.A.); javier.moguerza@urjc.es (J.M.M.); javier.castillo@urjc.es (J.C.); joseignacio.martinez@urjc.es (J.I.M.)
* Correspondence: javier.gomez@urjc.es

Abstract: Typical applications of wireless sensor networks (WSN), such as in Industry 4.0 and smart cities, involves acquiring and processing large amounts of data in federated systems. Important challenges arise for machine learning algorithms in this scenario, such as reducing energy consumption and minimizing data exchange between devices in different zones. This paper introduces a novel method for accelerated training of parallel Support Vector Machines (pSVMs), based on ensembles, tailored to these kinds of problems. To achieve this, the training set is split into several Voronoi regions. These regions are small enough to permit faster parallel training of SVMs, reducing computational payload. Results from experiments comparing the proposed method with a single SVM and a standard ensemble of SVMs demonstrate that this approach can provide comparable performance while limiting the number of regions required to solve classification tasks. These advantages facilitate the development of energy-efficient policies in WSN.

Keywords: classification; machine learning; Support Vector Machines; sensor networks; distributed algorithms

Citation: Alfaro, C.; Gomez, J.; Moguerza, J.M.; Castillo, J.; Martinez, J.I. Toward Accelerated Training of Parallel Support Vector Machines Based on Voronoi Diagrams. *Entropy* **2021**, *23*, 1605. https://doi.org/10.3390/e23121605

Academic Editor: Sotiris Kotsiantis

Received: 23 October 2021
Accepted: 25 November 2021
Published: 29 November 2021

Publisher's Note: MDPI stays neutral with regard to jurisdictional claims in published maps and institutional affiliations.

1. Introduction

Machine learning applications are radically changing our world as a key asset of an Information Society. New algorithms and methods for data processing and analysis, along with the capacity to deal with large and complex datasets, has led to the rise of a new industry. Over the next decades, data science and machine learning are expected to transform the way in which we interact with our surrounding environment.

One of the main challenges is to effectively prepare and analyze vast and distributed datasets. Classical algorithms for classification, such as convolutional neural networks (CNN) [1,2] or SVMs [3], are being pushed to their limits. Therefore, it is essential to develop efficient parallel architectures and techniques that can cope with massive data in distributed systems. As a result, algorithm parallelization is taking a key role, as it enables the exploitation of computing power available in large data centers, especially in cloud computing environments, to train and deploy these algorithms.

More classical machine learning algorithms, such as SVMs, can also be used as a viable alternative for classification of large datasets. Nonetheless, one of their main disadvantages is that, unlike CNN and other intrinsically parallel algorithms, SVMs lack from such property. For this reason, several proposals have been presented for their parallelization [4–6]. In general, parallel Support Vector Machines (pSVMs) are based on algorithm modifications to execute some code sections simultaneously. As well, alternative approaches consider incremental executions deployed on distributed architectures, such as MapReduce [7].

In this article, we present an alternative method for machine learning classification via SVMs, specially designed for structures similar to a federated network of sensors, such as wireless sensor networks [8]. These networks are characterized by the fact that it is necessary to discern between two classes in each region. These tasks arise in many contexts, such as decentralized intrusion detection systems [9], controlling environmental conditions in smart buildings [10], or emergency alert networks [11], among others. Processing large datasets acquired by WSN devices can be challenging. Specific goals in this context are to minimize communication between nodes or groups of nodes, to optimize energy consumption, as well as to attain conservative management of limited storage capacity [12].

The main contribution of our algorithm, in comparison to other similar approaches, is that it takes advantage of this kind of spatial distribution. Roughly speaking, our method works as a guided ensemble-type approach. In practice, this spatial distribution can be emulated by dividing the dataset into Voronoi regions [13]. In the case of sensor networks, data subregions contain almost complete Voronoi regions, with a rather empty intersection to other regions. At this point, it is important to remark that, in cases in which the spatial distribution of the data is known in advance and made up of small groups, this process could be avoided by using the known groups as approximated Voronoi-type regions. However, under the presence of large groups of data, the use of Voronoi regions will still be of help from a computational point of view. For the sake of completeness, in this paper, we describe the full process of building the Voronoi regions although, as commented above, it could be skipped in some situations.

In the same way, each task can be independently solved, using a standard SVM implementation such as libSVM [14], already available in popular programming languages such as Python, R, or C. As a result, any system already based on SVM can take advantage of this method, not only to reduce execution time but to also increase its processing capacity.

To this aim, we create a set of small SVMs that work as an ensemble of classifiers [15]. The key point is that members of the ensemble can be trained following a parallelization scheme. The success of these kinds of ensembles based on SVMs have already been proved in [16]. In related work, however, the SVM used for the selection of the ensemble does not admit parallelization.

The rest of the paper is organized as follows. We review previous related work in Section 2. In Section 3, the proposed algorithm is presented. Then, Section 4 describes the experimental setup to validate the proposed algorithm and presents the discussion of the results. Finally, the main conclusions and future lines of work are presented in Section 5.

2. Related Works

Nowadays, machine learning algorithms play a central role in wireless sensor networks [17]. In particular, SVMs are involved in diverse applications in this context such as localization techniques, anomaly and fault detection, or congestion control, among others. The new method introduced in this paper is based on the parallel implementation of SVM algorithms in Voronoi regions, efficiently combining selected results from some of these regions, following ensemble learning principles. In this section, we review the main background machine learning concepts and tools related to this work.

2.1. Support Vector Machines

SVMs are one of the most popular supervised learning methods that is used for both classification and regression tasks. They appeared by the end of the last century as optimal margin classifiers in the context of Vapnik's statistical learning theory [18]. The goal of the SVM algorithm is to find a hyperplane that optimally separates a higher dimensional space into different categories. SVM training consists of solving an optimization problem whose objective function gives a tradeoff between margin and misclassification error over the training dataset [19]. An advantage of the support vector method is that only a few training samples are involved in the determination of the prediction functions, facilitating the application of SVM to data mining problems with a huge amount of data. The whole formulation and some discussion can be found at [20].

SVM has been widely used in real application due to its efficient performance in machine learning problems. In the last years, different SVM methods have been successfully applied to solve the practical problems. In [21], a hybrid of k-means and SVM methods is developed and its application on breast cancer detection is presented. The k-means algorithm is applied to identify the patterns of the benign and malignant tumors which are used as features to build the dataset for SVM training. This approach achieves competitive performance results compared with other methods in cancer diagnosis. A multi-stage framework for sentiment analysis and opinion mining is proposed in [22]. This approach combines SVM and k-nearest neighbors methods, aiming to detect positive and/or negative opinion trends within weblogs containing knowledge written by baseline adopters. The authors in [23] introduced a SVM method for detection of American football head impacts using biomechanical features. A combined use of head impact sensors with video analysis was developed to the features extraction and to build training and validation datasets. A method of fault detection in wireless sensor networks based on SVM is presented in [24]. All data collected by the sensors of the network are redirected to the server that uses them to train an individual SVM with Gaussian kernel. Although this approach achieves good performance results, it requires additional communication overhead and a significant delay in data processing.

Although SVMs achieve excellent performance results, the computational time and memory requirements increase rapidly on complex and large datasets. For this reason, many research efforts have been conducted to design fast training algorithms of SVMs. The authors in [25] suggest a decomposed algorithm which divides the problem into smaller sub-problems that are solved iteratively. The method introduced in [26] proposes to reduce the size of the optimization problem by solving a sequence of sub-problems considering only a few features of the training dataset that are selected using a heuristic approach. Similarly to the aforementioned approaches, a decomposition method, called Sequential Minimal Optimization (SMO), is developed in [27]. The key idea behind the SMO method is to split the problem into the smallest possible sub-problems. Each sub-problem is solved analytically so the numerical optimization is avoided entirely, leading to a considerable reduction in computation time. More recent work [28] proposes a novel approach to select a representative subset from the training dataset using an algorithm based on convex hulls and extreme points.

2.2. Ensemble Learning

An ensemble of classifiers is a set of classifiers whose performance as a group improves the performance of individual classifiers. These individual classifiers are trained with subsets of the original training set and generate their own separating surfaces that will be later integrated in order to achieve more accurate and precise classification [29].

A nice theoretical property of ensembles is that the generalization error converges as the number of members of the ensemble increases. This property guarantees that overfitting will not become a problem [15]. Regarding accuracy, it can be demonstrated that an ensemble's accuracy depends on the strength of the individual classifiers and a measure of the dependence between them. To guarantee this property, the best members of the ensemble can be chosen during the training stage.

The widely used methods for constructing ensemble learning algorithms are boosting [30] and bagging [31]. Boosting is an algorithm that works by training base learners sequentially, so in each iteration the learner assigns higher weight to the observations of the dataset that have been misclassified by its predecessor. In bagging, different sample subsets are randomly drawn from the training dataset and each subset is used to train a basic learning model in a parallel manner. To obtain the global decision of the ensemble method, the outputs of the individual models are aggregated by voting.

Ensemble learning has been successfully used in diverse applications such as text classification [32], speech recognition [33,34], sentiment analysis [35], protein folding recognition [36], or streamflow forecasting [37]. Different learning algorithms have been used as base models to build ensemble methods such as neural networks, naive Bayesian, or decision trees, among others. An ensemble method based on neural network with random weights for online data stream regression is presented in [38]. The main idea of this method is to train various neural networks with subsets of the training dataset generated from combining bootstrap sampling with random feature selection. The results indicate an accuracy improvement and reduction in computational time compared to other available algorithms from literature. In [39], an ensemble of fine-tuned naive Bayesian classifiers for text classification is proposed. A bagging method is used for ensemble construction in combination with parameter modification over learning rate and number of iterations. In [40], a novel approach for constructing ensembles of decision trees is proposed, where each tree is trained with a subset containing all features of the training set, giving a different weight to every feature. All the nodes in a tree use the same vector of random weights, but different weights are used for each tree of the ensemble.

Finally, there is extensive research that has successfully applied SVMs as base models to build ensemble methods for solving machine learning problems, often leading to improved results compared with alternative techniques. An approach developed in [41] generates a new quality training dataset through the marginal density ratios transformation on the original features. The transformed data is used to train several SVM classifiers and feed their outputs to another SVM to train the final classification model. The results show that their method performs better than other ensemble approaches in terms of accuracy and training speed. The authors in [42] compared classification performance for breast cancer prediction of an individual SVM and various SVM ensemble methods. They used bagging and boosting methods for constructing the SVM ensembles combining different kernel functions. The experimental results showed that the radial basis function (RBF) kernel SVM ensemble based on the boosting method performed better than other classifiers.

2.3. Voronoi Diagrams

The Voronoi diagrams are an important method of computational geometry, designed primarily for evaluating nearest neighbor over two-dimensional spatial points [43]. A Voronoi diagram is characterized by regions of proximity, making the partitioning of a plane into disjoint convex polygons where the distance of points is defined by Euclidean distance so that all points in the same polygon have the same nearest neighbor, called the centroid. Thereby, from a given polygon, every point is closer to its centroid than to any other.

The Voronoi diagrams method has been used in a wide variety of applications [44] such as virus spread analysis among mobile devices [45], cluster analysis [46–48], continuous location-based services [49], or high-dimensional query evaluation [50].

In recent years, several works have been published presenting novel methods in diverse fields such as computer graphics, pattern recognition, or robotics. For instance, in [51], a method to achieve cost-effective 3-D printing of stiffened thin-shell objects is proposed. For that, they use the finite element analysis to determine the regions of the object with high stress and use a given number of seeds to create a Voronoi diagram to distribute these seeds in the areas with higher stress. These seeds are mapped from a 3-D mesh to a 2-D space with least squares conformal maps (LSCMs) [52]. The authors in [53] introduce the Voronoi diagrams for the analysis of the spatial organization in team sports, such as basketball, and define the behavioral team patterns during a positional attack. The approach in [54] proposes to reduce the computation time of the robots to make quick decisions before they collide with obstacles, using Voronoi diagrams for building a roadmap in the environment of the robot.

Finally, there are numerous studies using Voronoi diagrams to tackle imbalanced classification problems [55–57]. These kinds of problems arise when the distribution of examples among the classes is skewed. Real-world examples abound with problems of this type from fields such as visual computing, text classification, medicine, security, finance, among others. Furthermore, in the imbalanced classification problems, the class of interest is usually the minority class (e.g., credit card fraud detection, spam detection, disease risk detection) and traditional classifiers typically maximize an overall performance, which often results in the minority class being ignored. The synthetic minority oversampling technique (SMOTE) [58] is probably the most widely used method to mitigate this problem. It is based on the generation of synthetic samples for the minority class aiming to balance the dataset. An alternative approach is that of [55]. They proposed an over-sampling method based on Voronoi regions. The underlying idea of this method is to identify exclusive regions of the feature space where the generation of new instances by random resampling provides consistent data generalization. The results of this work suggest that, in certain cases where the complexity of the datasets is high, their proposed method leads to more accurate and better classification models than using SMOTE.

3. pSVM Algorithm

The key idea underpinning our novel method for pSVM is to build a guided ensemble of SVM classifiers. In this ensemble, each SVM can be trained separately and in a parallel environment. The ensemble is built using a clustering method over the training set that generates a Voronoi diagram, which splits the space into a specific number of regions defined by its center. Then, these regions are used to generate the ensembles in a guided manner.

3.1. Data Partitioning

Typically, in a binary classification problem, a training set consisting of n samples can be represented as:

$$D = \{(x_i, y_i)\}_{i=1}^n, \tag{1}$$

where $x_i \in \mathbb{R}^d$ denotes the training samples and $y_i \in C = \{-1, +1\}$ the associated labels.

In this phase, we split D into P training subsets $D_1, \ldots, D_P$, each consisting of n_p samples. Thus, the jth subset can be represented as:

$$D_j = \{(x_i, y_i)\}_{i=1}^{n_j}. \tag{2}$$

These subsets are created by ensuring that each D_j maintains a similar proportion of samples from each class as in the original dataset D. For this, a partitioning approach separately on each of the classes D^C is used. Thus, we can represent D as a collection of C classes:

$$D = \{D^i\}_{i=1}^C. \tag{3}$$

The next step is to generate a Voronoi diagram from the samples of each of the classes D^C. This can be achieved using a cluster algorithm such as k-means [59]. The idea of this method is to find k regions of the space, such that any point inside its region is closer to its region's center than to any other region's center. It is important to remark that we do not need to find a global minimum of the optimization problem involved in the k-means algorithm. For our purposes, it is enough with a single execution of a limited number of iterations of the k-means algorithm in order to obtain regions with a balanced number of data.

In order to determine the optimal number of clusters in a dataset, several methods have been proposed [60–63]. We have adapted the Sturges rule [64] to a multi-dimensional setting. Show, given a dataset of n samples, the number of clusters is estimated through the formula:

$$k = 1 + 3.332 \log n. \tag{4}$$

Therefore, as we mentioned above, we use k-means clustering on each of the classes separately leading to generate different Voronoi diagrams, one per class. Let us assume V_i^C denotes the ith Voronoi region of class C and c_i represents its associated centroid, the Voronoi diagrams of class -1 and class $+1$ could be represented as $V^- = \{(V_i^-, c_i^-)\}_{i=1}^{k^-}$ and $V^+ = \{(V_j^+, c_j^+)\}_{j=1}^{k^+}$, respectively, where k^- and k^+ are estimated by Equation (4).

Then, we generate the subset of all pairs, resulting as the combination of each Voronoi region of the class −1 with each of the regions of the class +1. Therefore, the new training subsets can be represented as:

$$D' = \{(V_i^-, V_j^+) \,|\, i = 1, \ldots, k^- \text{ and } j = 1, \ldots, k^+\}. \tag{5}$$

Figure 1 illustrates the steps carried out to perform data partitioning process.

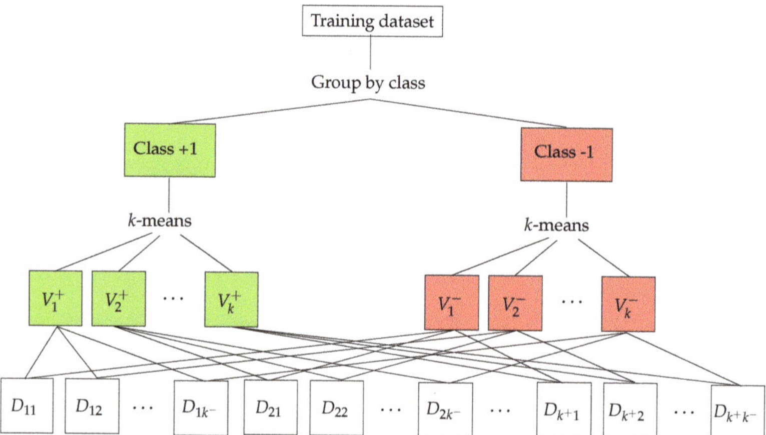

Figure 1. The flowchart of data partitioning method.

3.2. Training

Let $D_1, \ldots, D_P$ represent the P training subsets generated in the previous stage that contain data samples from both classes. Then, each of these subsets can be used to train a small SVM (sub-SVM) that can be trained independently using a standard SVM training algorithm. Each of the sub-SVM will generate a sub-model.

It is important to remark that the sub-models can be perfectly trained in a parallel manner, as the input data for the sub-SVM models are independent thanks to the Voronoi partitioning. Therefore, the training subsets are distributed among all available nodes. When the number of nodes is less than P, several sub-SVM are trained sequentially by each node. Otherwise, each node trains a sole sub-SVM. Formally, the parallelized system composed of N nodes, where H training subsets are allocated in each node, can be represented as:

$$SVM^{ensemble} = \{SVM_{lh} \,|\, l = 1 \ldots N, h = 1 \ldots H\}. \tag{6}$$

Additionally, in order to improve training times, the number of iterations required to converge toward the solution within each sub-SVM model could be limited. This is possible because the theory underlying ensembles guarantees that the accuracy of an ensemble depends on the strength of the individual classifiers [15].

Learning Strategy of Each Sub-SVM

Each sub-SVM follows the typical learning strategy based on regularization theory [20]. SVMs build a classification function through the solution of the following optimization problem:

$$\min_{f \in H_k} \frac{1}{n} \sum_{i=1}^{n} L(y_i, f(x_i)) + M\|f\|_k^2, \tag{7}$$

where (x_i, y_i), $i = 1, \ldots, n$, is a training dataset with $x_i \in \mathbb{R}^d$ and $y_i \in \{+1, -1\}$; H_K is a reproducing kernel Hilbert space (RKHS) with a kernel K; $\|f\|_K$ is the norm of f in the RKHS; $L(y_i, f(x_i))$ is a loss function; and the cost $M > 0$ is a constant that penalizes nonsmoothness of the possible solutions to optimization problem (7). The SVM loss function for classification purposes is:

$$L(y_i, f(x_i)) = \max(1 - y_i \times f(x_i), 0). \tag{8}$$

It can be shown that the solution of problem (7) using Equation (8) leads to a smooth function $f^* \in H_K$, such that:

$$f^*(x) = \sum_{i=1}^{n} \alpha_i K(x, x_i) + b^*, \tag{9}$$

where α_i and b^* are constants; $K(x, y) = \phi(x)^T \phi(y)$ is the kernel function that generates H_K; and $\phi : \mathbb{R}^n \to \mathbb{R}^p$ is a mapping defining K. ϕ maps the data from $\mathbb{R}^d$ (known as the "input space") into $\mathbb{R}^p$ (the so-called "feature space").

The main steps of the training algorithm are illustrated in Algorithm 1.

Algorithm 1: pSVM training algorithm.

Data: $D = \{(x_i, y_i)\}_{i=1}^{n}$, $x_i \in \mathbb{R}^d$, $y_i \in C = \{-1, +1\}$;
N: number of nodes;
Result: S (set of Voronoi regions pairs);
$SVM^{ensemble}$ (ensemble of SVM);
1   $\{D^c\}_{c=1}^{C} \leftarrow$ build a collection of C classes;
2   $k^c = 1 + 3.332 \log n^c$, where n^c is the number of samples in D^c and $c \in C$;
3   $V^c = \{(V_i^c, c_i^c)\}_{i=1}^{k^c} \leftarrow$ *k-means*(D^c, k^c), where c_i^c is the centroid of the Voronoi region V_i^c and $c \in C$;
4   $D' \leftarrow D'_{ij} = \{(V_i^-, V_j^+) | i = 1, \ldots, k^{-1} \text{ and } j = 1, \ldots, k^{+1}\}$;
5   $H \leftarrow \frac{length(D')}{N}$;
6   $S = \{S_{lh} | l = 1 \ldots N, h = 1 \ldots H\}$, where S is the distributed version of D' among the nodes ;
7   **for** $h \leftarrow 1$ *to* H **do**
8 | $SVM_{lh} = train\text{-}SVM(S_{lh}), l = 1, \ldots, N$;
9   $SVM^{ensemble} = \{SVM_{lh} | l = 1 \ldots N, h = 1 \ldots H\}$

3.3. Classification

Once the training phase is finished, an ensemble of sub-SVMs could be used to classify new data. Instead of using all sub-SVMs, the proposed algorithm selects a subset of them based on k nearest neighbor approach (k-NN) [65]. To achieve this, for each new individual, the Euclidean distance with the centroids of the Voronoi regions is computed and the γ closest ones of each class are selected. Let T^- and T^+ represent, respectively, the γ nearest Voronoi regions of class -1 and class $+1$ to the new individual. Then, a subset of the training subsets of Equation (5), $T \subset D'$, is generated as the Cartesian product of T^- and T^+:

$$T = \{(v^-, v^+), v^- \in T^- \text{ and } v^+ \in T^+\}. \tag{10}$$

Thereby, only the sub-SVM trained with the subsets on T are taken into account for prediction, discarding the remaining sub-SVM.

The pSVM uses a voting scheme similar to the one described in [66], where each new individual is evaluated by the selected sub-SVM, being the evaluation provided by each sub-SVM considered as a vote. Once all the votes are aggregated, the new individual is classified as a member of the most voted class. If there is an even number of sub-SVMs, ties during the voting of some individuals might take place. Those individuals are assigned at random, although more sophisticated schemes may classify those individuals as undetermined in order to evaluate their classification later by an expert. To be more specific, if t sub-SVMs are available, the class assigned to an individual z will be denoted as $class(z)$ and determined by Equation (11).

$$class(z) = \begin{cases} sgn\left(\sum_{i=1}^{t} prediction_i(z)\right) & \text{if } \sum_{i=1}^{t} prediction_i(z) \neq 0 \\ \pm 1\,(randomly) & \text{if } \sum_{i=1}^{t} prediction_i(z) = 0 \end{cases} \tag{11}$$

where $prediction_i(z)$ is the vote corresponding to sub-SVM i and sgn is a function defined as:

$$sgn(x) = \begin{cases} 1 & \text{if } x > 0 \\ 0 & \text{if } x = 0 \\ -1 & \text{if } x < 0 \end{cases} \tag{12}$$

The steps to perform the classification stage are summarized in Algorithm 2.

Algorithm 2: pSVM classification algorithm.

Data: $\{V^+, V^-\}$: the set of Voronoi regions of class $c \in \{-1, +1\}$ computed in data partitioning stage;
N: number of nodes ;
$\{S_j\}_{j=1}^{N}$, where $S_j = \{S_{jh}|h = 1 \ldots H\}$ is a set of pairs of Voronoi regions;
$\{SVM_j^{ensemble}\}_{j=1}^{N}$, where $SVM_j^{ensemble}$ corresponds to Voronoi pair S_j;
z: new point to classify;
γ: number of neighbors that we consider for voting;
Result: $class(z)$

1   Calculate $\{T^+, T^-\}$, where T^+ and T^- are, respectively, the γ closest regions to z in V^+ and V^-;
2   $ens \leftarrow 0$;
3   **for** $j \leftarrow 1$ *to length(S)* **do**
4     $l \leftarrow 1$;
5     $exit \leftarrow False$;
6   **while** $(l \leq \gamma)$ *AND (not exit)* **do**
7   **if** $(S_j$ *contains* $T_l^+)$ *OR* $(S_j$ *contains* $T_l^-)$ **then**
8         $ens = ens + 1$;
9         $p_{ens} \leftarrow prediction(p, SVM_j^{ensemble})$, with $p_{ens} \in \{+1, -1\}$;
10        $exit \leftarrow True$;
11       $l \leftarrow l + 1$
12   $class(z) = sgn(\sum_{j=1}^{ens} p_i)$;

3.4. Computational Complexity

The following theoretical result shows that the computational complexity of our proposal lowers the computational complexity of a single SVM.

Theorem 1. *For a bounded number of iterations of the k-means method, the worst case computational complexity of the pSVM training algorithm proposed in this work amounts to $O((\frac{n}{\log n})^3)$.*

Proof. The worst case computational complexity of using a single SVM is $O(n^3)$ [67]. Regarding the k-means algorithm, it is well known that the optimization problem involved in this method is NP-hard [68]. In practice, truncated versions of this algorithm are used, so that a rough worst case bound can be assumed to be $O(I * k * n)$, where I is the number of iterations and k is the number of Voronoi regions. In a typical truncated version of k-means method, the maximum number of iterations is fixed. Therefore, for a large k, this computational complexity can be considered to be lower than $O(n^2)$. By construction, each sub-SVM used in the pSVM algorithm has a computational complexity of:

$$O((\frac{n}{1 + 3.332 \log n})^3) = O((\frac{n}{\log n})^3).$$

Since each sub-SVM can be trained simultaneously to the rest, the overall computational complexity of our pSVM algorithm amounts to $O(n^2) + O((\frac{n}{\log n})^3)$, that is, $O((\frac{n}{\log n})^3)$. $\square$

4. Experimental Results

In this section, we provide empirical evidence of our analysis of guided pSVM based on Voronoi regions using two synthetic datasets and discuss the results. All experiments were conducted on a workstation running Linux with two Intel Xeon E5-2630 (6 cores per CPU, two threads per core), at 2.3 GHz and 64 GB of RAM memory. A prototype implementing the algorithms described in Section 3 was developed using the statistical software R v3.6.0 [69], RStudio v1.4.1106, and the following additional R packages. Data processing was carried out with packages `stats` v3.4.4 and `dplyr` v1.0.0. Visualization was undertaken using package `ggplot2` v3.3.3. We created a custom function based on package `e1071` v1.6-8 to build the SVM classifier, so that we can limit the number of iterations to achieve convergence. Finally, parallelization was carried out through package `doParallel` v1.0.16.

4.1. One Region with Two Partially Overlapping Classes

A first simple experiment consists of the classification of two partially overlapping classes, where all the data are located in the same space region. Figure 2 shows the situation for the two-dimensional case. We use two d-dimensional Gaussian distributions $(x, y) \sim \mathcal{N}_m(\mu_m, \sigma_m)$, $m \in \{1, 2\}$ to simulate each class, where $\mu_1 = (0, 0)$, $\mu_2 = (2, 2)$ and the covariance matrix is $([1, 0]; [0, 1])$ for both distributions. In particular, $\mu_1 = (0, \ldots, 0) \in \mathbb{R}^d$ and $\mu_2 = (2, \ldots, 2) \in \mathbb{R}^d$. The covariance matrices σ_m were randomly generated with diagonal $(1, \ldots, 1), \in \mathbb{R}^d$. The experiment is executed for $d = 2$. A balanced dataset is artificially generated by randomly sampling 500,000 training points and 50,000 testing points from each class.

As each class has a sample size of 500,000 points, from Equation (4), the number of clusters obtained is 45 ($k^{-1} = k^{+1} = 45$). Since the Voronoi diagrams corresponding to classes -1 and $+1$ are very similar, for conciseness, in Figure 3 we only show the diagram for class $+1$.

Results

Here, we evaluate the performance of pSVM versus a single SVM and a standard SVM ensemble [70]. We choose the well-known SVM implementation provided by the libSVM library [14]. We randomly split each dataset into a training and a testing group, where the training set is 10 times larger than the testing set, and run all methods using this setup. This procedure is repeated 10 times and we obtain the average value and standard deviation of the accuracy performance measure, that is, the fraction of individuals correctly classified, given by:

$$accuracy = \frac{t_p + t_n}{t_p + f_p + t_n + f_n},$$

where t_p (true positives) are defined as the set of individuals correctly classified in a certain class, t_n (true negatives) as the set of individuals correctly left out of a certain class, f_p (false positives) as individuals incorrectly classified in a certain class, and f_n (false negatives) as individuals that have been incorrectly left out of a certain class. Because we are using a balanced dataset, this measure will work correctly providing reliable information to assess the performance of these methods.

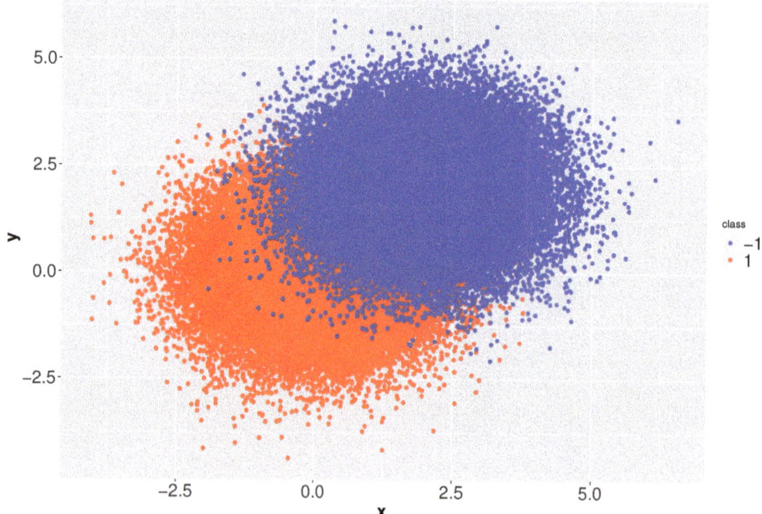

Figure 2. An example of the synthetic dataset in a 2-D feature space.

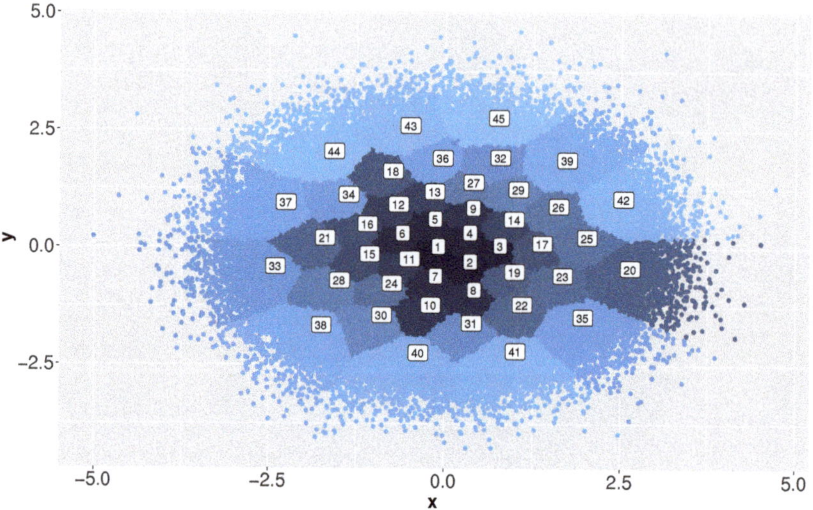

Figure 3. Voronoi diagram for class 1.

Each method is run with two different kernel functions (see [20] for different choices), namely a linear kernel and a radial basis function (RBF) kernel with parameters estimated by cross validation. Then, we compare the following approaches:

- Single SVM, ensemble, and pSVM with no limit of iterations;
- Single SVM, ensemble, and pSVM with a limit of 10 iterations;
- Single SVM, ensemble, and pSVM with a limit of 1 iteration.

As mentioned in Section 3.3, a k-NN approach based on Voronoi regions is used to select the sub-SVMs considered as classifiers. It seems obvious that different values of k lead to different performance results. To select the optimal value of this parameter empirically, we tested different choices for k: 1, 3, 5, 7, and 9. As Figure 4 shows, the accuracy improves while we increase k from 1 to 7, whereas it is relatively stable for k larger than 7. Therefore, $k = 7$ was chosen as the optimal number of Voronoi regions used in the classification scheme.

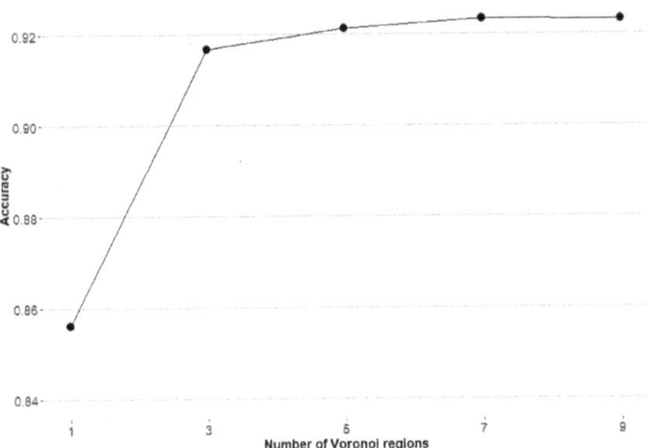

Figure 4. Number of Voronoi regions, for each class, selected for classification.

Table 1 shows the average classification accuracy and the standard deviation of the algorithms for ten runs on the synthetic dataset. As we can see, when the number of iterations required to converge to the solution is not limited, all approaches provide accuracy results over 91.0%. Best results are obtained by the ensemble with RBF kernel and the pSVM with RBF kernel being, respectively, 92.69% and 92.65%. However, when the number of iterations is limited, for energy saving reasons, the only methods providing consistent results are the two versions of the pSVM approach, which do not seem to be affected by the iteration limit. In these cases, the best accuracy results are 92.65% and 92.57%, for the two limited versions of the pSVM with RBF kernel. Furthermore, it is important to notice that the only method whose accuracy systematically remains over 91.0% is pSVM, for all versions.

Table 1. Average (standard deviation) for accuracy for each method. The method with the best accuracy is boldfaced.

Iterations	SVM (Linear Kernel)	SVM (RBF Kernel)	Ensemble (Linear Kernel)	Ensemble (RBF Kernel)	pSVM (Linear Kernel)	pSVM (RBF Kernel)
No limit	0.9223 (0.0030)	0.9246 (0.0125)	0.9237 (0.0011)	**0.9269** (0.0076)	0.9130 (0.0139)	0.9265 (0.0122)
10	0.6641 (0.1929)	0.4465 (0.1226)	0.8963 (0.0128)	0.4958 (0.0227)	0.9150 (0.0049)	**0.9265** (0.0120)
1	0.6543 (0.2790)	0.4241 (0.1266)	0.8887 (0.0167)	0.3107 (0.0189)	0.9107 (0.0078)	**0.9257** (0.0129)

4.2. Eight Multi-Dimensional Regions with Two Partially Overlapping Classes

This second experiment is based on a synthetic dataset that emulates a federated network of sensors. As mentioned above, such networks are characterized by providing data distributed in different regions in which it is necessary to categorize events in different classes. For this experiment, to simulate each class we generate sixteen d-dimensional Gaussian distributions $(x, y) \sim \mathcal{N}_m(\mu_m, \sigma_m)$, $m \in \{1, \ldots, 16\}$, paired two by two. For simplicity, 16,000 elements in a 10-dimensional space have been generated (1000 elements per class for each region), although similar results were obtained for larger dimensional settings and datasets, up to one million elements. Figure 5 depicts this dataset for $d = 2$.

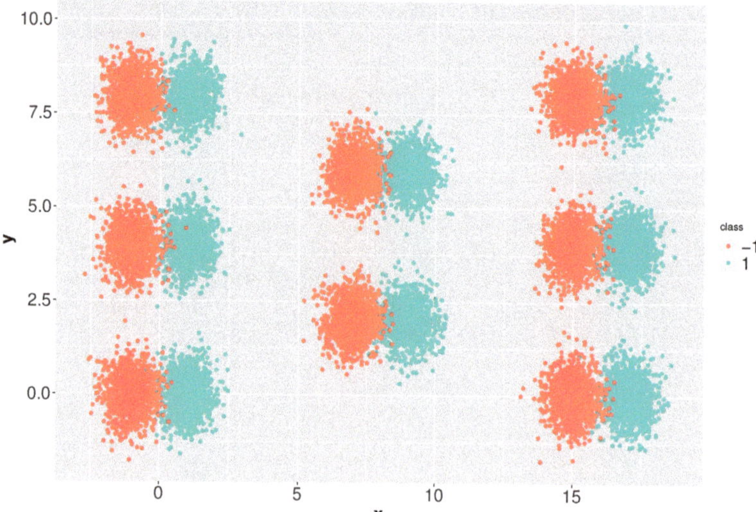

Figure 5. A two-dimensional example with two classes and eight regions.

Results

Again, on this dataset, we compare the performance of our pSVM approach to a single SVM and a standard SVM ensemble. For the three methods, two different versions are implemented: one using a linear kernel and another using an RBF kernel, with parameters estimated by cross validation. Moreover, for the ensemble and pSVM approaches, different classification schemes are used. In particular, for both methods, we implement the classification scheme described in Algorithm 2 for different values of the γ parameter, namely: $\gamma = 1$, $\gamma = 7$, and $\gamma = 15$. In the case of the single SVM, the classification scheme is made up of a single decisor and, in Table 2, the result appears in the row corresponding to $\gamma = 1$. In a similar manner to the previous example, we randomly split each dataset 10 times into a training and a testing set. Similarly, we run the methods and calculated the average value and standard deviation of the accuracy performance measure.

Table 2 presents the average classification accuracy and the standard deviation of the algorithms for ten runs on the multidimensional dataset. As we can observe, the best result for the linear kernel versions of the algorithms are always provided by the pSVM approach. This is because the method has been specifically designed for data whose structure is similar to that of a federated network of sensors. As expected, using a more complex kernel, the ensemble approach improved its results, especially for large values of γ. Unfortunately, this approach requires a cross-validation process to estimate the parameters of the kernel, whereas the linear kernel does not require this additional step. Finally, it is remarkable that, under a severe reduction in the number of training iterations up to a single one, the best overall accuracy result (94.95%) is obtained by the pSVM with an RBF kernel.

Table 2. Average (standard deviation) for accuracy for each method. The method with the best accuracy is boldfaced.

Iterations	γ	SVM (Linear Kernel)	SVM (RBF Kernel)	Ensemble (Linear Kernel)	Ensemble (RBF Kernel)	pSVM (Linear Kernel)	pSVM (RBF Kernel)
No limit	1	0.6183 (0.0580)	**0.9747** (0.0047)	0.4404 (0.1000)	0.9687 (0.0028)	0.9641 (0.0053)	0.9695 (0.0051)
	7	-	-	0.4033 (0.0100)	**0.9751** (0.0036)	0.8493 (0.0563)	0.8297 (0.0219)
	15	-	-	0.391 (0.0029)	**0.9763** (0.0042)	0.5437 (0.0799)	0.6566 (0.0931)
10	1	0.5730 (0.0025)	0.5506 (0.0100)	0.4779 (0.0097)	0.6513 (0.0183)	0.8970 (0.0249)	**0.9495** (0.0073)
	7	-	-	0.4289 (0.0083)	**0.8956** (0.0088)	0.8218 (0.0787)	0.7950 (0.0670)
	15	-	-	0.3910 (0.0029)	**0.9555** (0.0074)	0.5220 (0.0917)	0.6714 (0.1103)
1	1	0.5350 (0.0399)	0.5421 (0.0138)	0.4276 (0.0077)	0.5570 (0.0157)	0.7675 (0.0154)	**0.8331** (0.0111)
	7	-	-	0.4372 (0.0099)	0.6736 (0.0093)	**0.7637** (0.0449)	0.7616 (0.0288)
	15	-	-	0.4303 (0.0156)	**0.7600** (0.0151)	0.5329 (0.0400)	0.6639 (0.0153)

4.3. A Numerical Estimation of Training Time

Finally, for completeness, we provide a table with the execution time exhibited by the different methods on the 10-dimensional example in Section 4.2. It is important to notice that, although the smallest time results are obtained by the single SVM with a limited number of iterations, these implementations provide very poor classification results. Therefore, it would never be chosen in practice. Considering a tradeoff between accuracy and training times, the best implementations correspond to the pSVM approach with linear kernel. In particular, the pSVM version without a limit of iterations is, on average, up to 11.88 times faster than the single SVM with linear kernel. This magnitude is in accordance with the expected proportional reduction in the order of $log(n)$, shown in Section 3.4.

Table 3 summarizes the execution time (in seconds) for all versions of the methods implemented in this comparative.

Table 3. Average (standard deviation) for training time. The method with the shortest training time is boldfaced.

Iterations	SVM (Linear Kernel)	SVM (RBF Kernel)	Nodes	Ensemble (Linear Kernel)	Ensemble (RBF Kernel)	pSVM (Linear Kernel)	pSVM (RBF Kernel)
No limit	29.8763 (6.9040)	5.0406 (0.0184)	4	15.2576 (0.2009)	7.9840 (0.0817)	3.0220 (0.1934)	3.7600 (0.2626)
			9	9.1846 (0.4014)	5.3840 (0.1412)	**2.5140** (0.1260)	2.9566 (0.1526)
			16	8.6923 (0.1162)	4.4406 (0.0155)	2.6280 (0.2912)	2.6910 (0.0818)
10	0.1240 (0.0006)	**0.1202** (0.0015)	4	2.5753 (0.0307)	2.8840 (0.1424)	2.3566 (0.1353)	2.7400 (0.1582)
			9	1.9853 (0.0186)	2.2166 (0.0200)	1.8656 (0.0558)	2.0940 (0.1471)
			16	1.8790 (0.1065)	2.0703 (0.1079)	1.7970 (0.1799)	1.9126 (0.0489)
1	0.0933 (0.0011)	**0.0683** (0.0049)	4	2.4300 (0.1455)	2.5233 (0.1459)	2.4100 (0.2364)	2.5756 (0.0895)
			9	1.8416 (0.0256)	2.1013 (0.2426)	1.8203 (0.0592)	1.8686 (0.0499)
			16	1.8903 (0.0584)	1.9260 (0.0270)	1.8460 (0.1292)	1.8860 (0.0770)

5. Conclusions

In this paper, we present a novel method for accelerated training of parallel Support Vector Machines that is especially well-suited for problems involving a federated network of sensors where optimization of energy consumption is required. The proposed algorithm builds on a parallel training alternative of SVM ensembles (pSVMs), determined by Voronoi regions. Experimental results indicate that training time is reduced according to the analytical computational complexity analysis of the method. This method exhibits a stable performance when the convergence iterations within the training stage are limited. In particular, it is important to remark that the simplest version of this pSVM approach, that is, the one using a linear kernel, makes this method the most appropriate for a parallel implementation. In that case, the evaluation of the kernel function simply involves a scalar product without additional parameters, and thus cross validation is not needed.

Concerning further research, a more detailed complexity analysis including the effect of the dimension of the data may be interesting, especially for data coming from very high dimensional settings. Another interesting area of future research is the development of multiclass versions of the pSVM approach. As well, a drastic acceleration of the training stage could be achieved through a hardware implementation of this novel approach. To this aim, pSVM versions with a limited number of iterations are even more suitable.

Regarding possible shortcomings of our proposal, there is still room for improvement. Alternatives for constructing the Voronoi regions should be explored. Another limitation that requires future attention is that the Sturges formula was originally developed for one-dimensional data. Therefore, it would be advisable to develop a more sophisticated version, including in its closed-form the dimension d of the representation space. This is related to the necessary compromise between the number of Voronoi regions and the number of data elements comprised in each region, which may be crucial to improve the performance of this method.

Author Contributions: Conceptualization, J.M.M., J.C. and J.I.M.; methodology, C.A., J.G. and J.M.M.; software, C.A. and J.G.; validation C.A., J.G. and J.M.M.; writing—original draft preparation, C.A. and J.G.; writing—review and editing, C.A., J.G., J.M.M., J.C. and J.I.M.; supervision, J.M.M. and J.C.; funding acquisition, J.M.M. All authors have read and agreed to the published version of the manuscript.

Funding: This work was supported by research grant MODAS-IN of Spanish MICINN (ref. RTI2018-094269-B-I00).

Institutional Review Board Statement: Not applicable.

Informed Consent Statement: Not applicable.

Data Availability Statement: Not applicable.

Conflicts of Interest: The authors declare no conflict of interest.

References

1. LeCun, Y.; Boser, B.E.; Denker, J.S.; Henderson, D.; Howard, R.E.; Hubbard, W.E.; Jackel, L.D. Backpropagation Applied to Handwritten Zip Code Recognition. *Neural Comput.* **1989**, *1*, 541–551. [CrossRef]
2. LeCun, Y.; Bottou, L.; Bengio, Y.; Haffner, P. Gradient-based learning applied to document recognition. *Proc. IEEE* **1998**, *86*, 2278–2324. [CrossRef]
3. Schölkopf, B.; Smola, A.J. *Learning with Kernels: Support Vector Machines, Regularization, Optimization, and Beyond*; Adaptive Computation and Machine Learning Series; MIT Press: Cambridge, MA, USA, 2002.
4. Clarkson, K.L. Algorithms for Closest-Point Problems (Computational Geometry). Ph.D. Thesis, Stanford University, Stanford, CA, USA, 1985.
5. Graf, H.P.; Cosatto, E.; Bottou, L.; Durdanovic, I.; Vapnik, V. Parallel Support Vector Machines: The Cascade SVM. In Proceedings of the Advances in Neural Information Processing Systems 17 [Neural Information Processing Systems, NIPS 2004, Vancouver, BC, Canada, 13–18 December 2004; pp. 521–528.
6. Chang, E.Y. PSVM: Parallelizing Support Vector Machines on Distributed Computers. In *Foundations of Large-Scale Multimedia Information Management and Retrieval: Mathematics of Perception*; Springer: Berlin/Heidelberg, Germany, 2011; pp. 213–230. [CrossRef]
7. Caruana, G.; Li, M.; Qi, M. A MapReduce based parallel SVM for large scale spam filtering. In Proceedings of the 2011 Eighth International Conference on Fuzzy Systems and Knowledge Discovery (FSKD), Shanghai, China, 26–28 July 2011; Volume 4, pp. 2659–2662.

8. Arampatzis, T.; Lygeros, J.; Manesis, S. A Survey of Applications of Wireless Sensors and Wireless Sensor Networks. In Proceedings of the 2005 IEEE International Symposium on, Mediterrean Conference on Control and Automation Intelligent Control, Limassol, Cyprus, 27–29 June 2005; pp. 719–724. [CrossRef]

9. da Silva, A.P.R.; Martins, M.H.T.; Rocha, B.P.S.; Loureiro, A.A.F.; Ruiz, L.B.; Wong, H.C. Decentralized intrusion detection in wireless sensor networks. In Proceedings of the Q2SWinet'05—Proceedings of the First ACM Workshop on Q2S and Security for Wireless and Mobile Networks, Montreal, QC, Canada, 13 October 2005; Boukerche, A., de Araujo, R.B., Eds.; ACM: New York, NY, USA, 2005; pp. 16–23. [CrossRef]

10. Han, Z.; Gao, R.X.; Fan, Z. Occupancy and indoor environment quality sensing for smart buildings. In Proceedings of the 2012 IEEE International Instrumentation and Measurement Technology Conference Proceedings, Graz, Austria, 13–16 May 2012; pp. 882–887. [CrossRef]

11. Ko, J.; Lim, J.H.; Chen, Y.; Musvaloiu-E, R.; Terzis, A.; Masson, G.M.; Gao, T.; Destler, W.; Selavo, L.; Dutton, R.P. MEDiSN: Medical emergency detection in sensor networks. *ACM Trans. Embed. Comput. Syst.* **2010**, *10*, 11:1–11:29. [CrossRef]

12. Wan, S.; Zhao, Y.; Wang, T.; Gu, Z.; Abbasi, Q.H.; Choo, K.R. Multi-dimensional data indexing and range query processing via Voronoi diagram for internet of things. *Future Gener. Comput. Syst.* **2019**, *91*, 382–391. [CrossRef]

13. Voronoi, G. Nouvelles applications des paramètres continus à la théorie des formes quadratiques. Deuxième mémoire. Recherches sur les parallélloèdres primitifs. *J. Reine Angew. Math. (Crelles J.)* **1908**, *1908*, 198–287. [CrossRef]

14. Chang, C.C.; Lin, C.J. LIBSVM: A library for Support Vector Machines. *ACM Trans. Intell. Syst. Technol. (TIST)* **2011**, *2*, 1–27. [CrossRef]

15. Breiman, L. Some Infinity Theory for Predictor Ensembles. Available online: http://citeseerx.ist.psu.edu/viewdoc/download?doi=10.1.1.87.5037&rep=rep1&type=pdf (accessed on 29 November 2021).

16. Hu, Z.; Cai, Y.; Li, Y.; Xu, X. Support vector machine based ensemble classifier. In Proceedings of the 2005 American Control Conference, Portland, OR, USA, 8–10 June 2005; pp. 745–749.

17. Donta, P.K.; Amgoth, T.; Annavarapu, C.S.R. Machine learning algorithms for wireless sensor networks: A survey. *Inf. Fusion* **2019**, *49*, 1–25. [CrossRef]

18. Cortes, C.; Vapnik, V. Support-vector networks. *Mach. Learn.* **1995**, *20*, 273–297. [CrossRef]

19. Fischetti, M. Fast training of Support Vector Machines with Gaussian kernel. *Discret. Optim.* **2016**, *22*, 183–194. [CrossRef]

20. Moguerza, J.M.; Muñoz, A. Support Vector Machines with applications. *Stat. Sci.* **2006**, *21*, 322–336. [CrossRef]

21. Zheng, B.; Yoon, S.W.; Lam, S.S. Breast cancer diagnosis based on feature extraction using a hybrid of K-means and support vector machine algorithms. *Expert Syst. Appl.* **2014**, *41*, 1476–1482. [CrossRef]

22. Alfaro, C.; Cano-Montero, J.; Gómez, J.; Moguerza, J.M.; Ortega, F. A multi-stage method for content classification and opinion mining on weblog comments. *Ann. Oper. Res.* **2016**, *236*, 197–213. [CrossRef]

23. Wu, L.C.; Kuo, C.; Loza, J.; Kurt, M.; Laksari, K.; Yanez, L.Z.; Senif, D.; Anderson, S.C.; Miller, L.E.; Urban, J.E.; et al. Detection of American football head impacts using biomechanical features and support vector machine classification. *Sci. Rep.* **2017**, *8*, 1–14. [CrossRef]

24. Zidi, S.; Moulahi, T.; Alaya, B. Fault detection in wireless sensor networks through SVM classifier. *IEEE Sensors J.* **2017**, *18*, 340–347. [CrossRef]

25. Osuna, E.; Freund, R.; Girosi, F. An improved training algorithm for Support Vector Machines. In Proceedings of the Neural networks for signal processing VII. Proceedings of the 1997 IEEE Signal Processing Society Workshop, Amelia Island, FL, USA, 24–26 September 1997; pp. 276–285.

26. Joachims, T. Making Large-Scale SVM Learning Practical. Available online: https://www.cs.cornell.edu/people/tj/publications/joachims_99a.pdf (accessed on 29 November 2021).

27. Platt, J.C. Using Analytic QP and Sparseness to Speed Training of Support Vector Machines. In *Advances in Neural Information Processing Systems 11, NIPS Conference, Denver, CO, USA, 30 November–5 December 1998*; Kearns, M.J., Solla, S.A., Cohn, D.A., Eds.; The MIT Press: Cambridge, MA, USA, 1998; pp. 557–563.

28. Nandan, M.; Khargonekar, P.P.; Talathi, S.S. Fast SVM training using approximate extreme points. *J. Mach. Learn. Res.* **2014**, *15*, 59–98.

29. Kuncheva, L.I. Combining Pattern Classifiers: Methods and Algorithms, 2nd ed. Available online: https://www.wiley.com/en-in/Combining+Pattern+Classifiers%3A+Methods+and+Algorithms%2C+2nd+Edition-p-9781118315231 (accessed on 29 November 2021).

30. Schapire, R.E. The strength of weak learnability. *Mach. Learn.* **1990**, *5*, 197–227. [CrossRef]

31. Breiman, L. Bagging predictors. *Mach. Learn.* **1996**, *24*, 123–140. [CrossRef]

32. Kang, M.; Ahn, J.; Lee, K. Opinion mining using ensemble text hidden Markov models for text classification. *Expert Syst. Appl.* **2018**, *94*, 218–227. [CrossRef]

33. Deng, L.; Platt, J. Ensemble deep learning for speech recognition. Available online: https://www.microsoft.com/en-us/research/wp-content/uploads/2016/02/EnsembleDL_submitted.pdf (accessed on 29 November 2021).

34. Zvarevashe, K.; Olugbara, O. Ensemble learning of hybrid acoustic features for speech emotion recognition. *Algorithms* **2020**, *13*, 70. [CrossRef]

35. Araque, O.; Corcuera-Platas, I.; Sánchez-Rada, J.F.; Iglesias, C.A. Enhancing deep learning sentiment analysis with ensemble techniques in social applications. *Expert Syst. Appl.* **2017**, *77*, 236–246. [CrossRef]

36. Liu, B.; Li, C.C.; Yan, K. DeepSVM-fold: Protein fold recognition by combining Support Vector Machines and pairwise sequence similarity scores generated by deep learning networks. *Briefings Bioinform.* **2020**, *21*, 1733–1741. [CrossRef] [PubMed]

37. Tyralis, H.; Papacharalampous, G.; Langousis, A. Super ensemble learning for daily streamflow forecasting: Large-scale demonstration and comparison with multiple machine learning algorithms. *Neural Comput. Appl.* **2021**, *33*, 3053–3068. [CrossRef]

38. de Almeida, R.; Goh, Y.M.; Monfared, R.; Steiner, M.T.A.; West, A. An ensemble based on neural networks with random weights for online data stream regression. *Soft Comput.* **2020**, *24*, 9835–9855. [CrossRef]

39. El Hindi, K.; AlSalman, H.; Qasem, S.; Al Ahmadi, S. Building an ensemble of fine-tuned naive Bayesian classifiers for text classification. *Entropy* **2018**, *20*, 857. [CrossRef]

40. Maudes, J.; Rodríguez, J.J.; García-Osorio, C.; García-Pedrajas, N. Random feature weights for decision tree ensemble construction. *Inf. Fusion* **2012**, *13*, 20–30. [CrossRef]

41. Gu, J.; Wang, L.; Wang, H.; Wang, S. A novel approach to intrusion detection using SVM ensemble with feature augmentation. *Comput. Secur.* **2019**, *86*, 53–62. [CrossRef]

42. Huang, M.W.; Chen, C.W.; Lin, W.C.; Ke, S.W.; Tsai, C.F. SVM and SVM ensembles in breast cancer prediction. *PLoS ONE* **2017**, *12*, e0161501. [CrossRef]

43. Boots, B.; Okabe, A.; Sugihara, K. Spatial tessellations. *Geogr. Inf. Syst.* **1999**, *1*, 503–526.

44. Du, Q.; Faber, V.; Gunzburger, M. Centroidal Voronoi tessellations: Applications and algorithms. *SIAM Rev.* **1999**, *41*, 637–676. [CrossRef]

45. Wang, P.; González, M.C.; Menezes, R.; Barabási, A.L. Understanding the spread of malicious mobile-phone programs and their damage potential. *Int. J. Inf. Secur.* **2013**, *12*, 383–392. [CrossRef]

46. Hartigan, J.A. *Clustering Algorithms*; John Wiley & Sons, Inc.: Chichester, UK, 1975.

47. Jain, A.K.; Dubes, R.C. *Algorithms for Clustering Data*; Prentice Hall; Pearson Education, Inc.: Upper Saddle River, NJ, USA, 1988.

48. Preparata, F.P.; Shamos, M.I. *Computational Geometry: An Introduction*; Springer Science & Business Media: Dordrecht, The Netherlands, 2012.

49. Albers, G.; Guibas, L.J.; Mitchell, J.S.; Roos, T. Voronoi diagrams of moving points. *Int. J. Comput. Geom. Appl.* **1998**, *8*, 365–379. [CrossRef]

50. Berchtold, S.; Ertl, B.; Keim, D.A.; Kriegel, H.P.; Seidl, T. Fast nearest neighbor search in high-dimensional space. In Proceedings 14th International Conference on Data Engineering, Orlando, FL, USA, 23–27 February 1998; pp. 209–218.

51. Zheng, A.; Bian, S.; Chaudhry, E.; Chang, J.; Haron, H.; You, L.; Zhang, J.J. Voronoi diagram and Monte-Carlo simulation based finite element optimization for cost-effective 3D printing. *J. Comput. Sci.* **2021**, *50*, 101301. [CrossRef]

52. Haker, S.; Angenent, S.; Tannenbaum, A.; Kikinis, R.; Sapiro, G.; Halle, M. Conformal surface parameterization for texture mapping. *IEEE Trans. Vis. Comput. Graph.* **2000**, *6*, 181–189. [CrossRef]

53. Lopes, A.; Fonseca, S.; Lese, R.; Baca, A. Using voronoi diagrams to describe tactical behaviour in invasive team sports: An application in basketball. *Cuad. Psicol. Deporte* **2015**, *15*, 123–130. [CrossRef]

54. Ayawli, B.B.K.; Mei, X.; Shen, M.; Appiah, A.Y.; Kyeremeh, F. Mobile robot path planning in dynamic environment using Voronoi diagram and computation geometry technique. *IEEE Access* **2019**, *7*, 86026–86040. [CrossRef]

55. Young, W.A.; Nykl, S.L.; Weckman, G.R.; Chelberg, D.M. Using Voronoi diagrams to improve classification performances when modeling imbalanced datasets. *Neural Comput. Appl.* **2015**, *26*, 1041–1054. [CrossRef]

56. Silva, E.J.; Zanchettin, C. A voronoi diagram based classifier for multiclass imbalanced data sets. In Proceedings of the 2016 5th Brazilian Conference on Intelligent Systems (BRACIS), Recife, Brazil, 9–12 October 2016; pp. 109–114.

57. de Carvalho, A.M.; Prati, R.C. DTO-SMOTE: Delaunay Tessellation Oversampling for Imbalanced Data Sets. *Information* **2020**, *11*, 557. [CrossRef]

58. Chawla, N.V.; Bowyer, K.W.; Hall, L.O.; Kegelmeyer, W.P. SMOTE: Synthetic minority over-sampling technique. *J. Artif. Intell. Res.* **2002**, *16*, 321–357. [CrossRef]

59. Lloyd, S. Least squares quantization in PCM. *IEEE Trans. Inf. Theory* **1982**, *28*, 129–137. [CrossRef]

60. Milligan, G.W.; Cooper, M.C. An examination of procedures for determining the number of clusters in a data set. *Psychometrika* **1985**, *50*, 159–179. [CrossRef]

61. Tibshirani, R.; Walther, G.; Hastie, T. Estimating the number of clusters in a data set via the gap statistic. *J. R. Stat. Soc. Ser. B (Stat. Methodol.)* **2001**, *63*, 411–423. [CrossRef]

62. Sugar, C.A.; James, G.M. Finding the number of clusters in a dataset: An information-theoretic approach. *J. Am. Stat. Assoc.* **2003**, *98*, 750–763. [CrossRef]

63. Masud, M.A.; Huang, J.Z.; Wei, C.; Wang, J.; Khan, I.; Zhong, M. I-nice: A new approach for identifying the number of clusters and initial cluster centres. *Inf. Sci.* **2018**, *466*, 129–151. doi: 10.1016/j.ins.2018.07.034. [CrossRef]

64. Sturges, H.A. The choice of a class interval. *J. Am. Stat. Assoc.* **1926**, *21*, 65–66. [CrossRef]

65. Cover, T.; Hart, P. Nearest neighbor pattern classification. *IEEE Trans. Inf. Theory* **1967**, *13*, 21–27. [CrossRef]

66. Krebel, U.G. Pairwise classification and Support Vector Machines. Available online: https://dl.acm.org/doi/10.5555/299094.299108 (accessed on 29 November 2021).

67. Bordes, A.; Ertekin, S.; Weston, J.; Botton, L.; Cristianini, N. Fast kernel classifiers with online and active learning. *J. Mach. Learn. Res.* **2005**, *6*, 1579–1619.

68. Aloise, D.; Deshpande, A.; Hansen, P.; Popat, P. NP-hardness of Euclidean sum-of-squares clustering. *Mach. Learn.* **2009**, *75*, 245–248. [CrossRef]
69. R Core Team. *R: A Language and Environment for Statistical Computing*; R Foundation for Statistical Computing: Vienna, Austria, 2021.
70. Kim, H.C.; Pang, S.; Je, H.M.; Kim, D.; Bang, S.Y. Constructing support vector machine ensemble. *Pattern Recognit.* **2003**, *36*, 2757–2767. [CrossRef]

Article

An Information Gain-Based Model and an Attention-Based RNN for Wearable Human Activity Recognition

Leyuan Liu [1], Jian He [1,2,*], Keyan Ren [1,2,*], Jonathan Lungu [1], Yibin Hou [1,2] and Ruihai Dong [3]

[1]   Faculty of Information Technology, Beijing University of Technology, Beijing 100124, China; chinaleyuan@emails.bjut.edu.cn (L.L.); JonathanLungu@emails.bjut.edu.cn (J.L.); ybhou@bjut.edu.cn (Y.H.)

[2]   Beijing Engineering Research Center for IOT Software and Systems, Beijing University of Technology, Beijing 100124, China

[3]   School of Computer Science, University College Dublin, D04 V1W8 Dublin 4, Ireland; ruihai.dong@ucd.ie

*   Correspondence: jianhee@bjut.edu.cn (J.H.); keyanren@bjut.edu.cn (K.R.)

Abstract: Wearable sensor-based HAR (human activity recognition) is a popular human activity perception method. However, due to the lack of a unified human activity model, the number and positions of sensors in the existing wearable HAR systems are not the same, which affects the promotion and application. In this paper, an information gain-based human activity model is established, and an attention-based recurrent neural network (namely Attention-RNN) for human activity recognition is designed. Besides, the attention-RNN, which combines bidirectional long short-term memory (BiLSTM) with attention mechanism, was tested on the UCI opportunity challenge dataset. Experiments prove that the proposed human activity model provides guidance for the deployment location of sensors and provides a basis for the selection of the number of sensors, which can reduce the number of sensors used to achieve the same classification effect. In addition, experiments show that the proposed Attention-RNN achieves F1 scores of 0.898 and 0.911 in the ML (Modes of Locomotion) task and GR (Gesture Recognition) task, respectively.

Keywords: human activity recognition; information gain; attention mechanism; Attention-RNN

Citation: Liu, L.; He, J.; Ren, K.; Lungu, J.; Hou, Y.; Dong, R. An Information Gain-Based Model and an Attention-Based RNN for Wearable Human Activity Recognition. *Entropy* **2021**, *23*, 1635. https://doi.org/10.3390/e23121635

Academic Editors: Felipe Ortega and Emilio López Cano

Received: 28 October 2021
Accepted: 3 December 2021
Published: 6 December 2021

Publisher's Note: MDPI stays neutral with regard to jurisdictional claims in published maps and institutional affiliations.

1. Introduction

Human activity recognition (HAR) technology [1] has been widely used in various areas, such as security monitoring [2], human-machine interaction [3], sports analysis [4], medical treatment [5], and health care [6], etc. According to the types of sensors used, HAR systems can be mainly divided into environmental sensor-based HAR, video-based HAR, and wearable sensor-based HAR [7]. However, environmental sensor-based HAR requires placing sensors in a fixed environment, which may cause certain limitations [8,9]. Although video-based HAR systems have made great progress, as the nature of this kind of system requires using cameras to collect human activities and record as videos for data analysis, this would raise several issues, such as susceptibility to light and occlusion, vulnerability of privacy protection, and large data processing volume [10]. Wearable sensor-based HAR systems integrate sensors, e.g., accelerometers, magnetometers, and gyroscopes, into wearable devices such as smartphones, bracelets, smart glasses, helmets, etc., and human body data is collected through these devices [11]. Wearable sensor-based HAR has become popular due to its convenience of application and ability to protect user privacy. Researchers have developed a variety of wearable sensor-based HAR solutions. For example, Fu et al. integrated multiple heterogeneous sensors into a wireless wearable sensor node for HAR and proved that the multi-modal data could achieve a better accuracy [12]. Iqbal et al. used smartphones to collect the data and transferred these collected data to a data server for processing and analysis [13].

The wearable sensor-based HAR can be divided into three stages: data perception, feature extraction, and activity classification. In the data perception stage, since wearable sensor-based HAR systems lack unified protocols and specifications, the types, numbers, and deployment locations of sensors in each system are different. For example, Köping et al. deployed eight inertial sensors into an HAR system, which consisted of a mobile phone, a glass, and a watch [14]. Hegde et al. combined insole-based and wrist-worn wearable sensors for HAR [15]. Davidson et al. integrated accelerometers, gyroscopes, compasses, barometers, and a GPS receiver into a device on the back of the body for analysis of running mechanics [16]. Due to the different types and deployment locations of wearable HAR sensors, it is difficult to popularize and apply the HAR algorithms. In the past, there have been a few studies on the number and location of sensors for wearable sensor-based HAR. Sztyler et al. used a classifier for location selection and analyzed the impact of 7 different sensor locations on the HAR results [17]. However, this method relied heavily on the accuracy of the classifier and only obtained the position of one sensor. Atallah et al. measured the importance of each location by calculating the overall weight of 13 artificial characteristics [18]. This method relied too much on the selection of features by manual experience. In recent years, some researchers have applied some methods based on information theory in their perception systems. For example, Jin et al. used causal entropy to select high causal measures as input data, but did not study the location of sensor deployment [19]. Lee et al. estimated the posture stability of the elderly through permutation entropy, but only used a sensor fixed on the back [20].

In the feature extraction stage and activity classification stage of wearable sensor-based HAR, technology development has gone through the traditional machine learning period and the current deep learning period. Traditional machine learning relies on artificial features, while deep learning can automatically extract features. Artificial features refer to the features artificially constructed by experts through in-depth analysis and enlightening thinking of the original data with the help of domain knowledge, which requires a lot of human resources. Traditionally, various classical machine learning algorithms [21], such as random forest [22], Bayesian network [23], Markov model [24], and support vector machine (SVM) [25], were used for analyzing wearable HAR data. In a strictly controlled environment, the traditional machine learning algorithms discussed can obtain excellent results. However, they need professional domain knowledge for manual feature extraction and complex preprocessing steps [26]. In recent years, deep learning algorithms have been applied to HAR and achieved outstanding performances. For instance, Ignatov used a CNN to automatically extract features from human activity data and combined them with artificial features to achieve relatively excellent results on the WISDM dataset and UCI-HAR dataset [27]. The limitation of Ignatov's work is that artificial features were still necessary, i.e., its data processing was inefficient, as it still required professional domain knowledge. Ronao and Cho used mobile phone accelerometer data and gyroscope data to classify six human activities and achieved an overall accuracy of 95.75% [28]. Since only one mobile phone device was used, the range of perception was limited and only a few simple human activities could be recognized. Aiming to mine temporal and spatial characteristics of human activities, Ordóñez et al. proposed a deep neural network (namely DeepConvLSTM), which benefits from both LSTM and CNN architectures [29]. Its weighted F1 scores of the daily activity recognition task and the 18-class gesture recognition task on the UCI Opportunity Challenge dataset [30] reached 0.895 and 0.915, which was significantly higher than the pure CNN. Vaswani et al. used the attention mechanism for machine translation task and achieved excellent results [31]. Then the attention mechanism can also be applied in HAR. Although the deep learning algorithms work well in HAR, their complex structures require high computing and storage resources, and require special processor support, such as GPU, to meet the needs of real-time HAR.

Aiming at the problems of lacking unified standards for sensor placement and the over-complexity of deep learning classification algorithms in the current wearable sensor-based HAR, this paper proposes a new HAR method. First, an information gain-based human activity model is established according to the characteristics of the human skeleton structure. It serves as a standard for the placement location and number of sensors in the perception stage. Second, a deep neural network (namely Attention-RNN) combined with the attention mechanism and bidirectional LSTM (BiLSTM) is designed to extract the features of human activity data and classify the data. Finally, on the public UCI Opportunity Challenge dataset, the balance effect of Attention-RNN in F1 score and running speed is verified, and the effect of the information gain-based human activity model is verified. The follow-up content of this paper is organized as follows: in Section 2, the information gain-based human activity model is presented. Section 3 elaborates on the architecture and principles of Attention-RNN. Section 4 introduces the UCI Opportunity Challenge dataset, the Attention-RNN training, the performance metrics, the experiments on Attention-RNN, and the experiments on information gain-based human activity model. Because experiments on the information gain-based human activity model need to use the Attention-RNN for effect evaluation, experiments are performed first to verify the effectiveness of Attention-RNN. Section 5 summarizes the entire text and prospects for the follow-up research directions.

2. Information Gain-Based Human Activity Model

In the process of human activities, different parts of the human body can exhibit different movement characteristics. The location and number of sensors are key issues in wearable sensor-based HAR. A large number of studies have discovered the positions to place sensors on the human body: head, ears, neck, torso, chest, abdomen, back, waist, pelvis, buttocks, hands, wrists, arms, feet, ankles, calves, thighs, knees, and so on. Yu et al. summarized these positions into the following categories: head, upper limbs, chest, waist back hip, lower limbs, and feet [32]. In 2010, Microsoft released Kinect, a device that can collect color images and depth images. The skeleton API in the Kinect for Windows SDK could provide position information of up to two people in front of Kinect, including detailed postures and 3D coordinate information of bone points. In addition, Kinect for Windows SDK could support up to 20 bone points. The data object type was provided as skeleton frames, and each frame could save up to 20 points [33]. Based on the past research and the human skeleton model proposed by Microsoft, this paper proposes the information gain-based human activity model.

According to the relationship between bones and joints, bones can be regarded as rigid bodies, and joints can be regarded as connecting mechanisms [34]. Therefore, in the modeling of the articulated skeleton, the human body can be considered as a motion mechanism composed of multiple linkages and multiple joints. Figure 1 shows an example of the proposed human activity model. The skeleton of the model is composed of 15 linkages and 17 joints. Among them, 13 linkages are suitable for placing sensors, and the two linkages of the span are not suitable for placing sensors, which have been shown by dotted lines, as shown in Figure 1a. The deployable sensor nodes set of model is $P = \{K0, K1, K2, \ldots, K14\}$, as shown in Figure 1b, where $K0$ is the head perception node, $K1$ and $K2$ are shoulder perception nodes, $K3$, $K4$, $K5$ and $K6$ are upper limb perception nodes, $K7$ and $K8$ are hand perception nodes, $K9$, $K10$, $K11$ and $K12$ are lower limb perception nodes, and $K13$ and $K14$ are foot perception nodes.

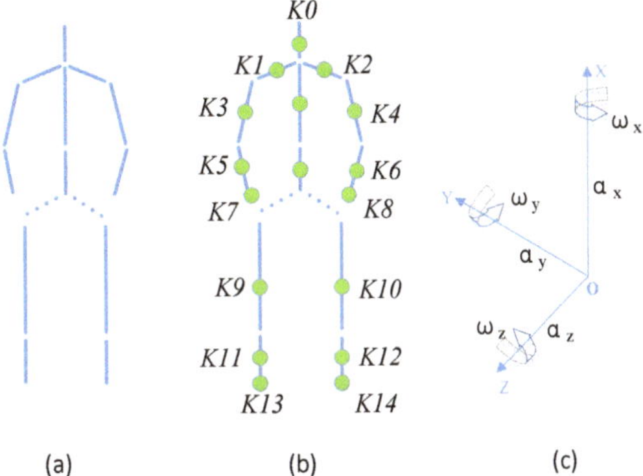

Figure 1. Information gain-based human activity model. (**a**) Human skeleton. (**b**) Positions of sensors can be fixed. (**c**) Cartesian coordinate system.

The joints of the proposed model all have three degrees of freedom, namely around the X-axis, around the Y-axis, and around the Z-axis. In order to standardize the expression of human activity, this paper adopts the spatial Cartesian rectangular coordinate system [35] to establish a unified human activity model. In Figure 1c, a_x, a_y, and a_z represent the acceleration component data collected by the 3-axis accelerometer along the X-axis, Y-axis, and Z-axis in the coordinate system during human activities; ω_x, ω_y, and ω_z represent the angular velocity component data of the human body sensed by gyroscope along X, Y and Z axes. Only the components of acceleration and angular velocity on each axis are shown in Figure 1c. In fact, each axis may contain other components, such as magnetic force. Suppose A is the human activity, F_{ext} is the feature extraction function, and F_{cls} is the human activity classification function, then the human activity can be expressed by Equation (1).

$$A = F_{cls} \left(F_{ext} \left(K0, K1, K2, \ldots , K14 \right) \right) \tag{1}$$

note that:

$$Ki = \left(a_x^i, a_y^i, a_z^i, \omega_x^i, \omega_y^i, \omega_z^i \ldots \right) \tag{2}$$

The contribution of Ki to HAR is an important basis for sensor deployment. The human activity model uses information gain [36] to measure the degree of contribution. Information gain is an evaluation method based on entropy. It measures the contribution of feature F to the classification model. It is generally defined as the difference between the information entropy of all category A before and after the feature F appears, as shown in Formulas (3)–(5).

$$\mathrm{InfoGain}(F, A) = H(A) - H(A|F) \tag{3}$$

$$H(A) = - \sum_{j=1}^{m} P\left(A_j\right) \log P\left(A_j\right) \tag{4}$$

$$H(A|F) = - \sum_{j} \sum_{v \in F} P\left(A_j | F = v\right) \log P\left(A_j | F = v\right) \tag{5}$$

where $H(A|F)$ and $H(A)$ are respectively the information entropy when the feature F appears or not. The v in Equation (5) belongs to the set F, that is, $v \in F$. In addition, $P(A_j)$ is the prior distribution of category probabilities and $P(A_j | F = v)$ is the posterior probabilities.

The information gain of Ki is the sum of the information gain of all its channels, as shown in Formula (6). Among them, $\mathrm{InfoGain}(K_{il})$ represents the information gain of Ki's lth channel, and C_i represents the total number of Ki's sensor channels.

$$\mathrm{InfoGain}(K_i) = \sum_{l=1}^{C_i} \mathrm{InfoGain}(K_{il}) \tag{6}$$

Then sort all sensor nodes according to the information gain value, and adopt the greedy strategy to select the optimal sensor combination with the top contribution. Human activity can finally be expressed by Equation (7). K_{top_i} represents the sensor whose information gain ranks i.

$$A = F_{cls} \left(F_{ext} \left(K_{top\_1}, K_{top\_2}, \ldots , K_{top\_i}, \ldots \right) \right) \tag{7}$$

3. Attention-RNN for Wearable HAR

A deep learning network based on an attention mechanism, named Attention-RNN, is designed to realize wearable HAR. The architecture of Attention-RNN is shown in Figure 2, including 1 input layer, 1 batch normalization (BN) layer, 2 BiLSTM layers, 1 attention layer, 1 dense layer, and 1 output layer.

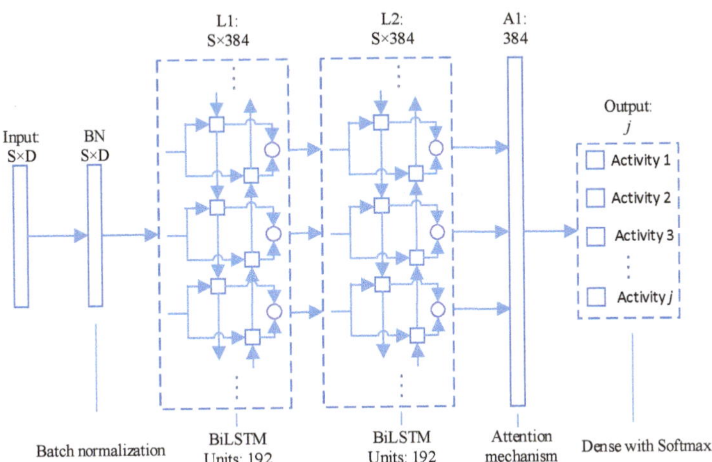

Figure 2. Network architecture of the Attention-RNN.

The first layer of Attention-RNN is the input layer. The input data $(X_1, X_2, X_3 \ldots X_t \ldots X_n)$ is a matrix of $n \times S \times D$, where D is the number of sensor channels, and S is the number of temporal data for each sensor channel.

The second layer is a batch normalization (BN) layer. Ioffe and Szegedy's research proved batch normalization method [37] could reduce the number of training steps required for model convergence, and could use a larger learning rate without paying too much attention to the initialization parameters and dropout. Therefore, a batch normalization layer is used here to simplify and speed up the training of the network.

The third layer (L1) and the fourth layer (L2) are both BiLSTM layers, and each layer has 192 units. The L1 layer outputs the sequence, which serves as the input of L2. Karpathy et al. proved through experiments that over two recurrent layers are more effective in predicting temporal events [38], so two BiLSTM layers are added after the BN layer. The Tanh function is used as the activation function when generating candidate memories. Because the output of the Tanh function is −1 to 1, which is consistent with the feature distribution of most scenes centered on 0, and the Tanh function has a larger gradient than the Sigmoid function near the input of 0, which can speed up the model convergence. L2 outputs the hidden state values of all time steps as the input to the next layer (A1). BiLSTM consists of forward LSTM and reverse LSTM. Each LSTM memory block is composed of a forget gate, an input gate, and a memory cell. The calculation process of BiLSTM is shown in Equations (8)–(16). In Equations (8)–(13), x_t is the input information at the current moment, f_t is the forgetting factor of the forgetting gate, i_t is the output of the input gate, $\widetilde{C}_t$ is the candidate value of the cell, C_t is the cell state, o_t is the output of the output gate, and h_t is the output of the LSTM memory block. In Equations (14)–(16), h_f and h_r represent the output of forward LSTM and reverse LSTM, respectively. The output of BiLSTM is H_t. In addition, w and b in the equations are the corresponding weight coefficient matrix and bias term.

$$f_t = \sigma\left(W_f\left[h_{(t-1)}, x_t\right] + b_f\right) \tag{8}$$

$$i_t = \sigma\left(W_i\left[h_{(t-1)}, x_t\right] + b_i\right) \tag{9}$$

$$\widetilde{C}_t = tanh\left(w_c * \left[h_{(t-1)}, x_t\right] + b_c\right) \tag{10}$$

$$C_t = f_t C_{(t-1)} + i_t * \widetilde{C}_t \tag{11}$$

$$o_t = \sigma\left(w_o * \left[h_{(t-1)}, x_t\right] + b_o\right) \tag{12}$$

$$h_t = o_t * tanh(C_t) \tag{13}$$

$$h_f = f\left(w_{f1}x_t + w_{f2}h_{t-1}\right) \tag{14}$$

$$h_r = f(w_{r1}x_t + w_{r2}h_{t+1}) \tag{15}$$

$$H_t = g\left(w_{o1} * h_f + w_{o2} * h_r\right) \tag{16}$$

The A1 layer is an attention mechanism layer. The attention mechanism is designed according to the importance of the temporal characteristics of human activities at different moments, as shown in Equations (17)–(19). Among them, u_t is the hidden layer unit, a_t is the weight coefficient vector, H_t is the output of BiLSTM, v_t is the output vector of the attention mechanism, w_w is the weight coefficient matrix from L2 to A1, and b is the bias. The vector u_w, which is randomly initialized and learned during training, is introduced to capture temporal context. The similarity, which is used as a measure of importance, is obtained by dot product u_t and u_w. The normalized weight coefficient vector a_t is obtained through the Softmax function. The time attention mechanism assigns different weights to the characteristics of human activities at different moments so that the characteristics at important moments receive more attention to improve the accuracy of HAR.

$$u_t = tanh(w_w H_t + b) \tag{17}$$

$$a_t = \text{softmax}\left(u_t^T u_w\right) \tag{18}$$

$$v_t = \sum a_t H_t \tag{19}$$

The last layer is a dense layer, which is also an output layer. The units of this layer are set to the number of human activity categories to be classified, which should be consistent with the number of label categories of the human activity dataset. Softmax is used as the activation function, as shown in Equation (20), where v_t is the output vector of A1, w_j is the weight matrix from A to the output layer, b_j is the offset corresponding to w_j. Softmax maps the results of various classes to the probability between 0 and 1, and the class with the highest probability is the predicted class.

$$y_j = \text{softmax}\left(w_j v_t + b_j\right) \tag{20}$$

4. Experiments and Analysis

4.1. Dataset

The public UCI Opportunity Challenge dataset is used as the experimental dataset, which has 113 data channels (each sensor axis one channel). The dataset was recorded by 19 sensors fixed on the body of the subjects and the sampling frequency was 30 Hz. As shown in Figure 3, five yellow squares represent the RS485-networked XSense inertial measurement unit (IMU). Two purple triangles represent InertiaCube3 inertial sensors, and 12 green circles represent Bluetooth acceleration sensors. Each XSense IMU comprised a 3-axis accelerometer, a 3-axis gyroscope, and a 3-axis magnetometer. Each InertiaCube3 included a gyroscope, magnetometers, and accelerometer. The dataset recorded two types of activity data: Drill type, where subjects performed a set of pre-defined activities in sequence, and ADL (activity of daily life) type, where subjects performed high-level activities (getting up, grooming, preparing breakfast, cleaning). These high-level tasks included multiple atomic activities (for example, preparing breakfast includes preparing sandwiches, preparing coffee, drinking water, and other atomic activities), and there was no limit to the order in which atomic activities were performed. The dataset contains 1 Drill activity and 5 ADL activities of 4 subjects. In the Opportunity Challenge, task A and task B were to classify 5 Modes of Locomotion (ML) and recognize 18 gestures (GR) respectively. Since the data of subject 4 added noise in the challenge to perform other tasks, we only used the data of subjects 1, 2, and 3. The dataset was divided into the training set and testing set consistent with the Opportunity Challenge. The ADL4 and ADL5 of subjects 2 and 3 constituted the testing set. The remaining activities of subjects 1, 2, and 3 were used as the training set.

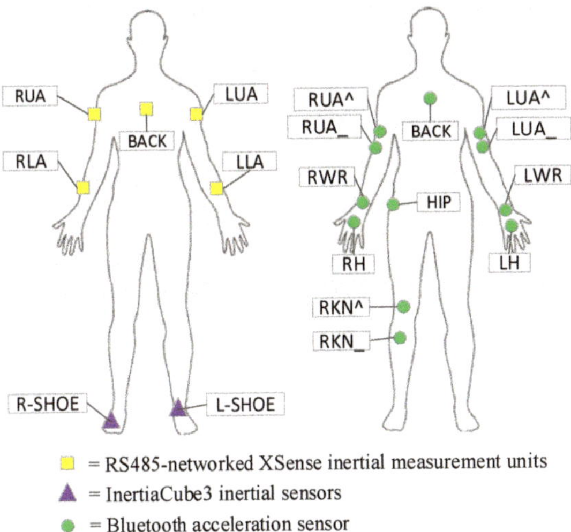

Figure 3. Sensors placement of the dataset.

The linear interpolation method was used to fill the missing values of the dataset in the temporal direction. Since the records of the dataset were continuous, a sliding window with a length of 24 and a sliding step of 12 was used to segment the continuous records. The label of the last data in the sliding window was used as the label of the intercepted sample. The final intercepted dataset is shown in Table 1. The Null class in the table represents data that is not of interest.

Table 1. Composition of the dataset intercepted by the sliding window.

Task	Activity Name	# of Training Instances	# of Testing Instances
	Open_Door1	864	58
	Open_Door2	887	95
	Close_Door1	806	60
	Close_Door2	846	83
	Open_Fridge	921	228
	Close_Fridge	850	160
	Open_Dishwasher	666	100
	Close_Dishwasher	628	77
	Open_Drawer1	490	39
GR	Close_Drawer1	413	42
	Open_Drawer2	457	40
	Close_Drawer2	416	26
	Open_Drawer3	566	67
	Close_Drawer3	564	61
	Clean_Table	904	99
	Drink_Cup	3246	317
	Toggle_Switch	623	105
	Null	32348	8237
	Stand	19321	3101
	Walk	10875	2272
ML	Sit	7410	2016
	Lie	1209	463
	Null	7680	2042

4.2. Attention-RNN Training

All experiments were carried out on a server with the Ubuntu system. The GPU of the server was TITAN Xp 12G, and the CPU was Intel Xeon E5-2620 v4. The RAM size of the server was 62 G. The experiments program was coded in Python 3.7. Pandas [39] and Numpy [40] were used for data processing, and Keras [41] was used to realize the Attention-RNN network. The CuDNNLSTM in Keras was used to construct the network to improve the speed of the network.

During training, a random 5% of the training data was used to verify the loss and F1 at the end of each epoch. The Adadelta method [42] with adaptive learning rate was used as the network parameters optimizer. The initial learning rate of 1.0 and the batch size of 16 were used for network training. The early stopping mechanism was used to stop the training automatically. If the training loss did not decrease after 50 epochs, the training would be stopped, otherwise, the training would continue. The verification F1 was monitored, and only the model with the highest verification F1 rate was saved.

4.3. Performance Metrics

Due to the imbalance of the dataset in different classes, it is more reasonable to use the F1 score as the performance metric. The F1 score combines the effects of precision rate and recall rate, as shown in Equation (21):

$$F_1 = \sum F_j = \sum \frac{N_j}{N} \cdot \frac{2P_j \cdot R_j}{P_j + R_j} \tag{21}$$

where j is the class index, and N_j is the number of samples of class j. N is the total number of samples. P_j and R_j are the precision rate and recall rate of class j, respectively.

The confusion matrix is suitable for visualizing the classification results of each class. The vertical axis of the confusion matrix is the actual class, and the horizontal axis is the predicted class. The sum of each column is the number of samples predicted as each class, and the sum of each row is the number of each class in the dataset. The background of each grid of the confusion matrix is filled with color according to the numerical value (the larger the numerical value, the darker the color).

4.4. Results and Discussion

4.4.1. Experiments on Attention-RNN

Table 2 shows the F1 comparison between the proposed Attention-RNN and the classification techniques published in the past. In the ML task, the F1 score of the proposed Attention-RNN was 0.898, which was over 3% higher than Random Forest [43] and was 0.03 higher than the best DeepConvLSTM [29]. In the GR task, the F1 score of the proposed Attention-RNN was 0.911, which was higher than Random Forest and CNN [44], but slightly lower than DeepConvLSTM. The classification time of testing instances (namely testing time) by Random Forest, DeepConvLSTM and Attention-RNN was 29.62 s, 9.82 s and 3.75 s, respectively. The test speed of Attention-RNN was 7.8 times that of Random Forest and 2.6 times that of DeepConvLSTM. The proposed Attention-RNN was more efficient than Random Forest and DeepConvLSTM. Although the test speed of Attention-RNN was slightly slower than that of CNN, the classification F1 value was greater than that of CNN. The above comparison results prove the beneficial effect of the proposed Attention-RNN. The proposed Attention-RNN had the largest F1 score in the ML task, the second F1 score in the GR task, and the second running speed. It achieved the optimal balance between F1 score and running efficiency.

Table 2. F1 comparison of different classification algorithms.

Method	F1 (ML Task)	F1 (GR Task)	Testing Time (S)
Random Forest [43]	0.870	0.900	29.62
CNN [44]	-[1]	0.851	2.29
DeepConvLSTM [29]	0.895	0.915	9.82
Attention-RNN (ours)	0.898	0.911	3.75

[1] "-" means there is no relevant data in the original paper.

The confusion matrix in Figure 4 shows the test results of Attention-RNN in the ML task. It can be seen from the figure that many Walk samples were misidentified as Stand and Null, and many Stand samples were misidentified as Walk and Null. Since the Walk samples were collected during daily indoor activities, the motion range was small. Therefore, Walk, Stand, and Null had certain similarities, and it was easy to identify them incorrectly.

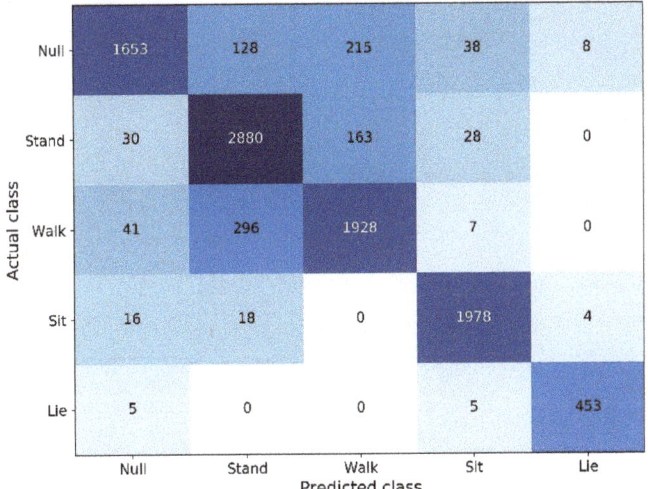

Figure 4. Confusion matrix of ML task.

The confusion matrix in Figure 5 shows the test results of Attention-RNN in the GR task. Most of the errors were related to the Null class. The main reason is that the classes of the dataset are extremely unbalanced, with Null classes accounting for 83.25% of the total samples.

The ablation experiments in Table 3 show the F1 score changes resulting from adding or removing different components of the Attention-RNN. The models of this set of experiments were all changed based on Attention-RNN. Model "A" removed the attention layer. Its F1 (ML) was 0.004 lower than Attention-RNN, and F1 (GR) was 0.008 lower than Attention-RNN. Model "B" removed the BN layer. Its F1 (ML) was 0.007 lower than Attention-RNN, and F1 (GR) was 0.005 lower than Attention-RNN. Models "D" and "E" changed the position of the BN layer, and their F1 scores were lower than the Attention-RNN. Models "I" and "J" changed the position of the Attention layer, and their F1 scores were not as good as Attention-RNN. Since Attention-RNN was only 0.01 orders higher than the F1 scores of the above models and the estimated F1 scores had uncertainty, it was unclear if it indicated an improvement. Models "C", "F", "G", and "H" changed the number of BiLSTM layers. The Attention-RNN model with 2 BiLSTM layers had a larger F1 score than other models. Model "K" and "L" had two attention layers, and model "M" had three attention layers. The F1 scores of models "K", "L" and "M" were all lower than Attention-RNN. The above results showed that increasing the number of attention layers or BiLSTM layers based on Attention-RNN did not improve the classification performance. In general, this set of experiments provided guidance for the establishment of the Attention-RNN.

Actual class \ Predicted class	Null	Open_Door1	Open_Door2	Close_Door1	Close_Door2	Open_Fridge	Close_Fridge	Open_Dishwasher	Close_Dishwasher	Open_Drawer1	Close_Drawer1	Open_Drawer2	Close_Drawer2	Open_Drawer3	Close_Drawer3	Clean_Table	Drink_Cup	Toggle_Switch
Null	7929	14	3	12	4	27	18	18	23	5	11	1	1	19	25	12	104	11
Open_Door1	13	39	1	5	0	0	0	0	0	0	0	0	0	0	0	0	0	0
Open_Door2	7	0	84	0	4	0	0	0	0	0	0	0	0	0	0	0	0	0
Close_Door1	4	10	0	46	0	0	0	0	0	0	0	0	0	0	0	0	0	0
Close_Door2	4	0	5	0	74	0	0	0	0	0	0	0	0	0	0	0	0	0
Open_Fridge	68	0	0	0	0	143	7	4	0	2	3	1	0	0	0	0	0	0
Close_Fridge	27	0	0	0	0	8	120	0	1	0	4	0	0	0	0	0	0	0
Open_Dishwasher	29	0	0	0	0	2	0	66	0	0	0	0	0	3	0	0	0	0
Close_Dishwasher	21	0	0	0	0	2	1	6	42	1	0	0	0	0	4	0	0	0
Open_Drawer1	10	0	0	0	0	1	0	1	2	12	4	5	1	0	3	0	0	0
Close_Drawer1	17	0	0	0	0	0	0	0	2	3	15	1	4	0	0	0	0	0
Open_Drawer2	10	0	0	0	0	2	0	2	2	9	0	13	0	2	0	0	0	0
Close_Drawer2	6	0	0	0	0	0	0	0	2	2	0	2	10	1	3	0	0	0
Open_Drawer3	13	0	0	0	0	0	0	0	1	0	0	0	0	50	3	0	0	0
Close_Drawer3	8	0	0	0	0	0	0	0	0	0	0	0	0	9	44	0	0	0
Clean_Table	41	0	0	0	0	2	0	3	0	0	0	4	0	0	0	49	0	0
Drink_Cup	81	0	0	0	0	0	0	2	0	0	0	0	0	0	0	0	234	0
Toggle_Switch	47	0	0	0	0	0	0	0	0	0	0	0	0	0	0	0	0	58

Figure 5. Confusion matrix of GR task.

Table 3. Experiments on different model structures.

Model	Structure	F1 (ML Task)	F1 (GR Task)
A	BN + 2BiLSTM + Dense	0.894	0.903
B	2BiLSTM + Attention + Dense	0.891	0.886
C	BN + 1BiLSTM + Attention + Dense	0.891	0.899
D	2BiLSTM + BN + Attention + Dense	0.893	0.903
E	2BiLSTM + Attention + BN + Dense	0.894	0.903
F	BN + 3BiLSTM + Attention + Dense	0.891	0.904
G	BN + 4BiLSTM + Attention + Dense	0.891	0.901
H	BN + 5BiLSTM + Attention + Dense	0.891	0.906
I	BN + Attention + 2BiLSTM + Dense	0.878	0.898
J	BN + BiLSTM + Attention + BiLSTM + Dense	0.892	0.891
K	BN + BiLSTM + Attention + BiLSTM + Attention + Dense	0.890	0.901
L	BN + Attention + BiLSTM + Attention + BiLSTM + Dense	0.881	0.899
M	BN + Attention + BiLSTM + Attention + BiLSTM + Attention + Dense	0.857	0.898
Attention-RNN	BN + 2BiLSTM + Attention + Dense	0.898	0.911

A set of cross-validation experiments was implemented to verify the stability of Attention-RNN. First, the training set in Section 4.1 was randomly divided into two sub-training sets of the same size. Then, in the ML task, the two sub-training sets were used to train two models, M1 and M2, respectively. In the GR task, the two sub-training sets were used to train two models G1 and G2, respectively. Finally, the above four trained models were tested on the test set in Section 4.1. The test F1 scores of M1, M2, G1, and G2 were 0.886, 0.894, 0.894, and 0.895, respectively. The results show that even if half of the training set is used to train Attention-RNN, good classification results can be achieved. Besides, the difference between M1 and M2 and the difference between G1 and G2 were relatively small. Then, the stability of Attention-RNN had been verified.

4.4.2. Experiments on Information Gain-Based Human Activity Model

To verify the validity of the human activity model, another set of experiments was carried out as follows: First, the information gain of each sensor was calculated according to the Formulas (3)–(6). The training set (including the validation set) without sliding window processing was used to calculate the information gain. Each sensor channel was selected as a feature, so that the F in the equations referred to each sensor channel, and the v referred to the data of the sensor channel. Since there are multiple feature selection methods, it may lead to different feature selection criteria and feature rankings. This set of experiments can only verify the effect of the proposed feature selection method. For the ML task and GR task, the information gain of each sensor was shown in Table 4. Second, the top n (1, 2, 3, … 18, 19) information gain sensors' data were used for training and testing Attention-RNN in turn, and the results are shown in Figures 6 and 7.

Table 4. Information gain and ranking of each sensor.

Sensor Name	Channels	InfoGain(K_i) of ML Task (Ranking)	InfoGain(K_i) of GR Task (Ranking)
RKNˆ	1–3	1.797 (8)	0.558 (15)
HIP	4–6	0.840 (18)	0.471 (19)
LUAˆ	7–9	1.092 (13)	0.615 (12)
RUA_	10–12	0.927 (16)	0.600 (14)
LH	13–15	1.617 (9)	0.972 (9)
BACK (Acc)	16–18	0.861 (17)	0.618 (11)
RKN_	19–21	1.332 (10)	0.603 (13)
RWR	22–24	1.308 (11)	1.464 (8)
RUAˆ	25–27	0.822 (19)	0.474 (18)
LUA_	28–30	1.119 (12)	0.510 (16)
LWR	31–33	1.011 (14)	0.492 (17)
RH	34–36	0.963 (15)	0.741 (10)
BACK (IMU)	37–45	2.817 (3)	2.088 (3)
RUA	46–54	2.610 (6)	1.890 (6)
RLA	55–63	2.241 (7)	1.971 (4)
LUA	64–72	2.664 (5)	1.818 (7)
LLA	73–81	2.772 (4)	1.899 (5)
L-SHOE	82–97	4.832 (1)	2.400 (2)
R-SHOE	98–113	4.784 (2)	2.448 (1)

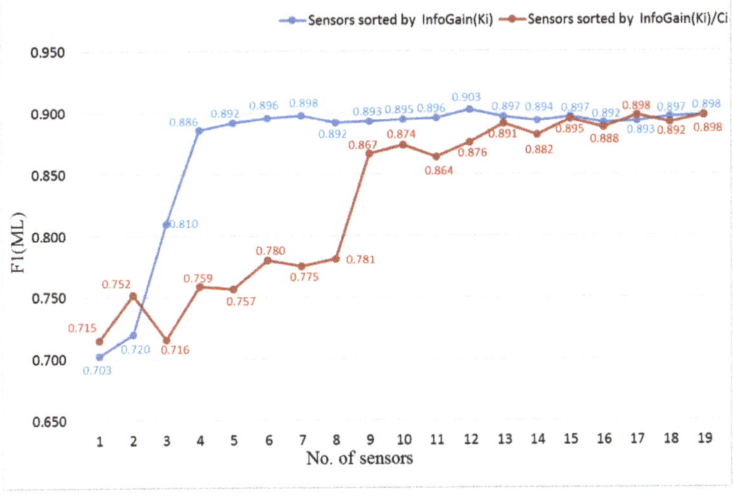

Figure 6. F1 scores for ML task with different numbers of sensors.

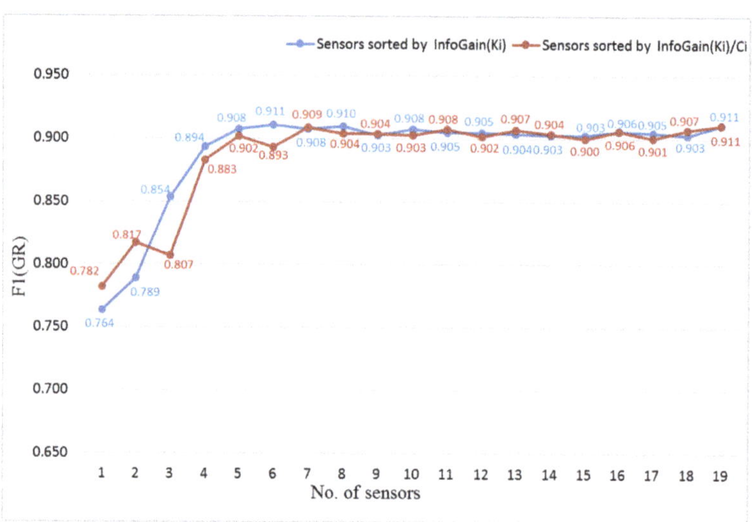

Figure 7. F1 scores for GR task with different numbers of sensors.

F1 scores for ML tasks with different numbers of sensors are shown in Figure 6. For example, when the number of sensors is 2, 2 refers to the sensors with the top 2 information gain, namely L-SHOE and R-SHOE. The blue line in Figure 6 represents that the sensors are sorted by the sensor information gain $\text{InfoGain}(K_i)$, which is the sum of the information gain over all channels of each sensor. The red line represents a set of comparative experimental results, and represents the sensors are sorted by $\text{InfoGain}(K_i)/C_i$, which is the average of information gain over all channels of each sensor. In the experiments represented by the blue line, the F1 value continued to increase as the number of sensors increased from 1 to 7. When the number of sensors was 7, the F1 score reaches the same maximum value as 19 sensors. When the number of sensors was 12, the F1 score was 0.903, which reached the maximum and exceeded 0.898 of 19 sensors. In the comparative experiments represented by the red line, the F1 score fluctuated and rose as the number of sensors increased from 1 to 17. When the number of sensors was 17, the F1 score reached the same maximum value of 0.898 as with all 19 sensors. The experiments represented by the blue line required fewer sensors than the experiments represented by the red line to achieve the high-level F1 score. Therefore, top 12 sensors sorted by the sensor information gain $\text{InfoGain}(K_i)$ can meet the requirements of ML task.

F1 scores for GR tasks with different numbers of sensors are shown in Figure 7. The blue and red lines in Figure 7 represent the experiments of two different sensor sorting methods, which are similar to Figure 6. In the experiments represented by the blue line, the F1 score steadily increased to the maximum value of 0.911 when the number of sensors gradually increased to 6. The F1 score of the experiment represented by the red line reached 9.09 when the number of sensors was 7, but it was smaller than that of the blue line with 6 sensors. Therefore, top 6 information gain sensors sorted by the sensor information gain $\text{InfoGain}(K_i)$ are enough to meet the requirements of GR task, and there is no need to continue increasing the number of sensors.

The red circle in Figure 8 marks the sensors with the top 6 information gain in the GR task, and the blue box marks the sensors with the top 12 information gain in the ML task. The sensors with top 6 information gain are mainly distributed on the arms and back, which are consistent with the characteristics of the upper limbs required to complete the GR task. Because completing the four activities in the ML task requires the cooperation of the upper and lower limbs, the top 12 information gain sensors that can achieve a good classification effect are distributed in the upper and lower limbs.

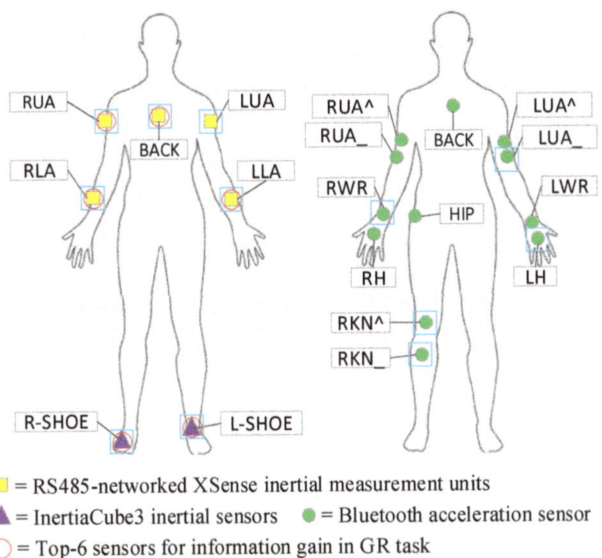

= RS485-networked XSense inertial measurement units

= InertiaCube3 inertial sensors   = Bluetooth acceleration sensor

= Top-6 sensors for information gain in GR task

= Top-12 sensors for information gain in ML task

Figure 8. Top 6 information gain sensors in GR task and top 12 information gain sensors in ML task.

5. Conclusions

This paper proposed an information gain-based human activity model and an Attention-RNN for wearable sensor-based HAR. The experimental results on the UCI Opportunity Challenge dataset show that the proposed Attention-RNN has high accuracy and operating efficiency. The F1 score of the proposed Attention-RNN was 0.03 higher than the Deep-ConvLSTM in the 5-class ML task and 0.04 lower in the 18-class GR task. The test speed of the proposed Attention-RNN was 2.6 times that of DeepConvLSTM. At the same time, experiments prove that the proposed information gain-based human activity model provides a quantitative basis for the deployment of the sensors and fills the research gap in this field. The same classification effect can be achieved by using fewer sensors with high information gain, which can reduce the amount of calculation.

In the future, classification algorithms will be studied to further improve the classification effect. In addition, methods to solve the problem of data imbalance will also be explored. Finally, the stability of the overall control will be proved and its complete theorem will be put forward.

Author Contributions: Conceptualization, L.L., J.H., K.R. and R.D.; methodology, L.L., J.H., K.R. and R.D.; software, L.L.; validation, L.L., K.R. and J.L.; formal analysis, J.H. and K.R.; investigation, L.L., K.R. and J.H.; resources, L.L. and K.R.; data curation, L.L. and J.H.; writing—original draft preparation, L.L.; writing—review and editing, L.L., Y.H., J.H., K.R., J.L. and R.D.; visualization, L.L.; supervision, Y.H., J.H. and K.R.; project administration, Y.H. and J.H.; funding acquisition, Y.H. All authors have read and agreed to the published version of the manuscript.

Funding: This research received no external funding.

Data Availability Statement: Publicly available dataset was analyzed in this study. This dataset can be found here: https://archive.ics.uci.edu/ml/datasets/OPPORTUNITY+Activity+Recognition, accessed on 2 December 2021.

Conflicts of Interest: The authors declare no conflict of interest.

References

1. Wang, J.; Chen, Y.; Hao, S.; Peng, X.; Hu, L. Deep learning for sensor-based activity recognition: A survey. *Pattern Recognit. Lett.* **2019**, *119*, 3–11. [CrossRef]
2. Ji, X.; Cheng, J.; Feng, W.; Tao, D. Skeleton embedded motion body partition for human action recognition using depth sequences. *Signal Process.* **2018**, *143*, 56–68. [CrossRef]
3. Anagnostis, A.; Benos, L.; Tsaopoulos, D.; Tagarakis, A.; Tsolakis, N.; Bochtis, D. Human Activity Recognition through Recurrent Neural Networks for Human–Robot Interaction in Agriculture. *Appl. Sci.* **2021**, *11*, 2188. [CrossRef]
4. Schuldhaus, D. *Human Activity Recognition in Daily Life and Sports Using Inertial Sensors*; FAU University Press: Erlangen, Germany, 2019.
5. Prati, A.; Shan, C.; Wang, K.I.-K. Sensors, vision and networks: From video surveillance to activity recognition and health monitoring. *J. Ambient Intell. Smart Environ.* **2019**, *11*, 5–22.
6. Aviles-Cruz, C.; Rodriguez-Martinez, E.; Villegas-Cortez, J.; Ferreyra-Ramirez, A. Granger-causality: An efficient single user movement recognition using a smartphone accelerometer sensor. *Pattern Recognit. Lett.* **2019**, *125*, 576–583. [CrossRef]
7. Cornacchia, M.; Ozcan, K.; Zheng, Y.; Velipasalar, S. A survey on activity detection and classification using wearable sensors. *IEEE Sens. J.* **2016**, *17*, 386–403. [CrossRef]
8. Taylor, W.; Shah, S.A.; Dashtipour, K.; Zahid, A.; Abbasi, Q.H.; Imran, M.A. An intelligent non-invasive real-time human activity recognition system for next-generation healthcare. *Sensors* **2020**, *20*, 2653. [CrossRef]
9. Gochoo, M.; Tan, T.-H.; Liu, S.-H.; Jean, F.-R.; Alnajjar, F.S.; Huang, S.-C. Unobtrusive activity recognition of elderly people living alone using anonymous binary sensors and DCNN. *IEEE J. Biomed. Health Inform.* **2018**, *23*, 693–702. [CrossRef]
10. Vijayaprabakaran, K.; Sathiyamurthy, K.; Ponniamma, M. Video-Based Human Activity Recognition for Elderly Using Convolutional Neural Network. *Int. J. Secur. Priv. Pervasive Comput.* **2020**, *12*, 36–48. [CrossRef]
11. Yao, R.; Lin, G.; Shi, Q.; Ranasinghe, D.C. Efficient dense labelling of human activity sequences from wearables using fully convolutional networks. *Pattern Recognit.* **2018**, *78*, 252–266. [CrossRef]
12. Fu, Z.; He, X.; Wang, E.; Huo, J.; Huang, J.; Wu, D. Personalized Human Activity Recognition Based on Integrated Wearable Sensor and Transfer Learning. *Sensors* **2021**, *21*, 885. [CrossRef]
13. Iqbal, A.; Ullah, F.; Anwar, H.; Ur Rehman, A.; Shah, K.; Baig, A.; Ali, S.; Yoo, S.; Kwak, K.S. Wearable Internet-of-Things platform for human activity recognition and health care. *Int. J. Distrib. Sens. Netw.* **2020**, *16*, 1550147720911561. [CrossRef]
14. Köping, L.; Shirahama, K.; Grzegorzek, M. A general framework for sensor-based human activity recognition. *Comput. Biol. Med.* **2018**, *95*, 248–260. [CrossRef]
15. Hegde, N.; Bries, M.; Swibas, T.; Melanson, E.; Sazonov, E. Automatic recognition of activities of daily living utilizing insole-based and wrist-worn wearable sensors. *IEEE J. Biomed. Health Inform.* **2017**, *22*, 979–988. [CrossRef]
16. Davidson, P.; Virekunnas, H.; Sharma, D.; Piché, R.; Cronin, N. Continuous analysis of running mechanics by means of an integrated INS/GPS device. *Sensors* **2019**, *19*, 1480. [CrossRef]
17. Sztyler, T.; Stuckenschmidt, H.; Petrich, W. Position-aware activity recognition with wearable devices. *Pervasive Mob. Comput.* **2017**, *38*, 281–295. [CrossRef]
18. Atallah, L.; Lo, B.; King, R.; Yang, G.-Z. Sensor positioning for activity recognition using wearable accelerometers. *IEEE Trans. Biomed. Circuits Syst.* **2011**, *5*, 320–329. [CrossRef]
19. Jin, X.-B.; Yu, X.-H.; Su, T.-L.; Yang, D.-N.; Bai, Y.-T.; Kong, J.-L.; Wang, L. Distributed deep fusion predictor for amulti-sensor system based on causality entropy. *Entropy* **2021**, *23*, 219. [CrossRef]
20. Lee, C.-H.; Chen, S.-H.; Jiang, B.C.; Sun, T.-L. Estimating postural stability using improved permutation entropy via TUG accelerometer data for community-dwelling elderly people. *Entropy* **2020**, *22*, 1097. [CrossRef]
21. Dang, L.M.; Min, K.; Wang, H.; Piran, M.J.; Lee, C.H.; Moon, H. Sensor-based and vision-based human activity recognition: A comprehensive survey. *Pattern Recognit.* **2020**, *108*, 107561. [CrossRef]
22. Rahman, A.; Nahid, N.; Hassan, I.; Ahad, M. Nurse care activity recognition: Using random forest to handle imbalanced class problem. In Proceedings of the Adjunct Proceedings of the 2020 ACM International Joint Conference on Pervasive and Ubiquitous Computing and Proceedings of the 2020 ACM International Symposium on Wearable Computers, Virtual Event, Mexico, 12–17 September 2020; pp. 419–424.
23. Liu, L.; Wang, S.; Su, G.; Huang, Z.-G.; Liu, M. Towards complex activity recognition using a Bayesian network-based probabilistic generative framework. *Pattern Recognit.* **2017**, *68*, 295–309. [CrossRef]
24. Asghari, P.; Soleimani, E.; Nazerfard, E. Online human activity recognition employing hierarchical hidden Markov models. *J. Ambient Intell. Humaniz. Comput.* **2020**, *11*, 1141–1152. [CrossRef]
25. Batool, M.; Jalal, A.; Kim, K. Sensors technologies for human activity analysis based on SVM optimized by PSO algorithm. In Proceedings of the IEEE 2019 International Conference on Applied and Engineering Mathematics (ICAEM), Taxila, Pakistan, 27–29 August 2019; pp. 145–150.
26. Portugal, I.; Alencar, P.; Cowan, D. The use of machine learning algorithms in recommender systems: A systematic review. *Expert Syst. Appl.* **2018**, *97*, 205–227. [CrossRef]
27. Ignatov, A. Real-time human activity recognition from accelerometer data using Convolutional Neural Networks. *Appl. Soft Comput.* **2018**, *62*, 915–922. [CrossRef]

28. Ronao, C.A.; Cho, S.-B. Human activity recognition with smartphone sensors using deep learning neural networks. *Expert Syst. Appl.* **2016**, *59*, 235–244. [CrossRef]

29. Ordóñez, F.J.; Roggen, D. Deep convolutional and lstm recurrent neural networks for multimodal wearable activity recognition. *Sensors* **2016**, *16*, 115. [CrossRef]

30. Chavarriaga, R.; Sagha, H.; Calatroni, A.; Digumarti, S.T.; Tröster, G.; Millán, J.D.R.; Roggen, D. The Opportunity challenge: A benchmark database for on-body sensor-based activity recognition. *Pattern Recognit. Lett.* **2013**, *34*, 2033–2042. [CrossRef]

31. Vaswani, A.; Shazeer, N.; Parmar, N.; Uszkoreit, J.; Jones, L.; Gomez, A.N.; Kaiser, Ł.; Polosukhin, I. Attention is all you need. In *Advances in Neural Information Processing Systems*; MIT Press: Cambridge, MA, USA, 2017; pp. 5998–6008.

32. Yu, H.; Cang, S.; Wang, Y. A review of sensor selection, sensor devices and sensor deployment for wearable sensor-based human activity recognition systems. In Proceedings of the IEEE 2016 10th International Conference on Software, Knowledge, Information Management & Applications (SKIMA), Chengdu, China, 15–17 December 2016; pp. 250–257.

33. Rahman, M. *Beginning Microsoft Kinect for Windows SDK 2.0: Motion and Depth Sensing for Natural User Interfaces*; Apress: New York, NY, USA, 2017.

34. Quoc, P.B.; Binh, N.T.; Tin, D.T.; Khare, A. Skeleton Formation From Human Silhouette Images Using Joint Points Estimation. In Proceedings of the IEEE 2018 Second International Conference on Advances in Computing, Control and Communication Technology (IAC3T), Allahabad, India, 21–23 September 2018; pp. 101–105.

35. Hosseini, M.; Hassanabadi, H.; Hassanabadi, S. Solutions of the Dirac-Weyl equation in graphene under magnetic fields in the Cartesian coordinate system. *Eur. Phys. J. Plus* **2019**, *134*, 1–6. [CrossRef]

36. Kullback, S.; Leibler, R.A. On information and sufficiency. *Ann. Math. Stat.* **1951**, *22*, 79–86. [CrossRef]

37. Ioffe, S.; Szegedy, C. Batch normalization: Accelerating deep network training by reducing internal covariate shift. In Proceedings of the International Conference on Machine Learning, Lille, France, 7–9 July 2015; pp. 448–456.

38. Krizhevsky, A.; Sutskever, I.; Hinton, G.E. ImageNet classification with deep convolutional neural networks. *Commun. ACM* **2017**, *60*, 84–90. [CrossRef]

39. McKinney, W. *Python for Data Analysis: Data Wrangling with Pandas, NumPy, and IPython*; O'Reilly Media, Inc.: Newton, MA, USA, 2012.

40. Van Der Walt, S.; Colbert, S.C.; Varoquaux, G. The NumPy array: A structure for efficient numerical computation. *Comput. Sci. Eng.* **2011**, *13*, 22–30. [CrossRef]

41. Gulli, A.; Pal, S. *Deep Learning with Keras*; Packt Publishing Ltd.: Birmingham, UK, 2017.

42. Zeiler, M.D. Adadelta: An adaptive learning rate method. *arXiv* **2012**, arXiv:1212.5701.

43. Schrader, L.; Vargas Toro, A.; Konietzny, S.; Rüping, S.; Schäpers, B.; Steinböck, M.; Krewer, C.; Müller, F.; Güttler, J.; Bock, T. Advanced sensing and human activity recognition in early intervention and rehabilitation of elderly people. *J. Popul. Ageing* **2020**, *13*, 139–165. [CrossRef]

44. Yang, J.; Nguyen, M.N.; San, P.P.; Li, X.L.; Krishnaswamy, S. Deep convolutional neural networks on multichannel time series for human activity recognition. In Proceedings of the Twenty-Fourth International Joint Conference on Artificial Intelligence, Buenos Aires, Argentina, 25–31 July 2015.

Article

Machine Learning Algorithm to Predict Acidemia Using Electronic Fetal Monitoring Recording Parameters

Javier Esteban-Escaño [1], Berta Castán [2], Sergio Castán [3,*], Marta Chóliz-Ezquerro [4], César Asensio [5], Antonio R. Laliena [5], Gerardo Sanz-Enguita [6], Gerardo Sanz [7], Luis Mariano Esteban [5,*] and Ricardo Savirón [8]

1   Department of Electronic Engineering and Communications, Escuela Universitaria Politécnica de La Almunia, Universidad de Zaragoza, Calle Mayor 5, 50100 La Almunia de Doña Godina, Spain; javeste@unizar.es
2   Department of Obstetrics and Gynecology, San Pedro Hospital, Calle Piqueras 98, 26006 Logroño, Spain; bcastan@riojasalud.es
3   Department of Obstetrics and Gynecology, Miguel Servet University Hospital, Paseo Isabel La Católica 3, 50009 Zaragoza, Spain
4   Department of Obstetrics, Dexeus University Hospital, Gran Via de Carles III 71-75, 08028 Barcelona, Spain; martacholiz@gmail.com
5   Department of Applied Mathematics, Escuela Universitaria Politécnica de La Almunia, Universidad de Zaragoza, Calle Mayor 5, 50100 La Almunia de Doña Godina, Spain; casencha@unizar.es (C.A.); arlalibi@unizar.es (A.R.L.)
6   Department of Applied Physics, Escuela Universitaria Politécnica de La Almunia, Universidad de Zaragoza, Calle Mayor 5, 50100 La Almunia de Doña Godina, Spain; cherraldin@unizar.es
7   Department of Statistical Methods and Institute for Biocomputation and Physics of Complex Systems-BIFI, University of Zaragoza, Calle Pedro Cerbuna 12, 50009 Zaragoza, Spain; gerardo@unizar.es
8   Department of Obstetrics and Gynecology, Hospital Clínico San Carlos and Instituto de Investigación Sanitaria San Carlos (IdISSC), Universidad Complutense, Calle del Prof Martín Lagos s/n, 28040 Madrid, Spain; rsaviron@gmail.com
*   Correspondence: scastan@salud.aragon.es (S.C.); lmeste@unizar.es (L.M.E.)

Citation: Esteban-Escaño, J.; Castán, B.; Castán, S.; Chóliz-Ezquerro, M.; Asensio, C.; Laliena, A.R.; Sanz-Enguita, G.; Sanz, G.; Esteban, L.M.; Savirón, R. Machine Learning Algorithm to Predict Acidemia Using Electronic Fetal Monitoring Recording Parameters. *Entropy* 2022, 24, 68. https://doi.org/10.3390/e24010068

Academic Editors: Felipe Ortega and Emilio López Cano

Received: 18 November 2021
Accepted: 27 December 2021
Published: 30 December 2021

Publisher's Note: MDPI stays neutral with regard to jurisdictional claims in published maps and institutional affiliations.

Abstract: Background: Electronic fetal monitoring (EFM) is the universal method for the surveillance of fetal well-being in intrapartum. Our objective was to predict acidemia from fetal heart signal features using machine learning algorithms. Methods: A case–control 1:2 study was carried out compromising 378 infants, born in the Miguel Servet University Hospital, Spain. Neonatal acidemia was defined as pH < 7.10. Using EFM recording logistic regression, random forest and neural networks models were built to predict acidemia. Validation of models was performed by means of discrimination, calibration, and clinical utility. Results: Best performance was attained using a random forest model built with 100 trees. The discrimination ability was good, with an area under the Receiver Operating Characteristic curve (AUC) of 0.865. The calibration showed a slight overestimation of acidemia occurrence for probabilities above 0.4. The clinical utility showed that for 33% cutoff point, missing 5% of acidotic cases, 46% of unnecessary cesarean sections could be prevented. Logistic regression and neural networks showed similar discrimination ability but with worse calibration and clinical utility. Conclusions: The combination of the variables extracted from EFM recording provided a predictive model of acidemia that showed good accuracy and provides a practical tool to prevent unnecessary cesarean sections.

Keywords: electronic fetal monitoring; fetal heart rate; sensors; acidemia; machine learning; random forest; clinical utility curve

1. Introduction

Currently, the universal method for the surveillance of intrapartum fetal well-being is the continuous monitoring of fetal heart rate (FHR) and maternal uterine contraction (UC) signals [1]. Electronic fetal monitoring (EFM) requires complex electronic devices developed to acquire, process, and display the signal. In the intrapartum period, an ultrasound transducer is used for the external FHR monitoring. This transducer contains piezoelectric effect crystals that convert electrical energy into ultrasound waves and uses the Doppler effect to detect movements of the cardiac structures [2,3]. In this context, several systems have been developed for central monitoring of fetal signals to provide simultaneous display of multiple tracings on several locations, allowing easier monitoring of signals [4]. The rate and pattern of the fetal heart are displayed on the computer screen and printed onto special graph paper.

Shannon defined entropy as a measure of the average information provided by a set of events and informs on its uncertainty [5]. The information theory is a mathematical theory of communication to quantify information. Information theory has been successfully used to evaluate biological biochemical signal networks [6] or in evolutionary biology [7]. Metrics such as mutual information have been used in the information theory in order to quantify the sharing of information in the presence of anomalies in electrocardiographic heart signals [8]. Fetal heart rate is altered in the presence of adverse fetal problems, the level of chaoticity in the signal may be measured using entropy. Higher entropy represents higher uncertainty and a more irregular behavior of the signal. Entropy can even explain how linked complex systems interact and exchange information.

The prediction of acidemia understood as fetal asphyxia was mainly based on the visualization of morphological aspects of fetal heart recording (FHR) with limited accuracy [9]. The quantification of the magnitude of this information becomes a goal in the study of FHR signals. Guidelines, such as the American College of Obstetricians and Gynecologists (ACOG) [10,11], proposed the categorization of FHR parameters to predict acidemia, but most categorization systems show lack of accuracy [12]. In addition, the interobserver agreement between experts shows the need to make the prediction of acidosis through the modeling of the EFM characteristics rather than the visual interpretation of the signal [13].

Two main objectives focused the effort on the improvement of the diagnosis of acidemia in recent years, the proposal of new predictors derived from the fetal cardiotocography (CTG) and their combination with previous features [14,15]. Automated systems can extract data on the FHR [16] or patterns can be obtained using signal processing as fractal analysis [17,18], but regarding combination of EFM variables, the artificial intelligence and machine learning algorithms have opened a range of possible applications with multiple development [19–22].

Machine learning algorithms had helped to improve prediction in different problems in medicine [23], although the nature of the used models is very diverse. Decision trees [24], support vector machines [24–26], adaptative boosting [24], convolutional neural networks [27,28], neuro fuzzy inference systems [29], neural networks [25,29], deep stacked sparse auto-encoders [29], or deep-ANFIS models [29] are machine learning techniques used for acidemia prediction. Machine learning algorithms are based on the minimization of a loss function. The cross-entropy is a generalized loss function that can be interpreted as an information measure [30], best models correspond to the minimum discrimination information [31]. Abnormalities in the FHR tend to increase the cross-entropy function, showing it as a candidate for quantifying the variety of physiological signals.

The success of machine learning models was distributed in a wide range, and can be classified in two groups, models that were built from the FHR signal and others built with the variables extracted from the signal. The most frequent parameters used to validate these previous models were the area under the receiver operating characteristic (ROC) curve [32], or the sensitivity and specificity that corresponds to a threshold probability of acidemia. To our knowledge, none, or very few of the developed machine learning models analyzed the clinical utility of these models although this is one the most important properties for the applicability of a prediction model [33].

Complementing the prediction of acidosis [34–36], recent publications have analyzed the importance of deceleration physiology and use parameters such as the deceleration area, that reports accumulated hypoxemia [14,37]. In addition, it is essential to know about the fetal time available to recover between deceleration and fetal ability to repeatedly activate the chemoreflex, fetal resilience [38,39]. Moreover, combining these parameters can provide better-adjusted predictions, the fetal reserve index is a promising classification system that proposed the improvement of EFM by adding three clinical variables: maternal, obstetrical, and fetal risk-related information in a scoring system to assess fetal perfusion and resilience rather than "hypoxia" [40].

In a previous study, we analyze a new parameter, the total reperfusion time (fetal resilience) to predict fetal acidemia [15]. In this study, we build a predictive model of acidemia using the FHR variables extracted from the EFM recording, including the reperfusion time, in a case–control study. For the combination of variables, we used the multivariate logistic regression, random forest, and neural networks models, performing a complete validation based on the analysis of the discrimination, calibration, and clinical utility of models.

2. Materials and Methods

2.1. Study Design and Patients Recruitment

The study was designed as a retrospective case–control analysis that involves pregnancy data recruited between June 2017 and October 2018 at the Miguel Servet University Hospital, in Zaragoza, Spain. The inclusion criteria were singleton term gestation between 37 and 42 weeks, cephalic presentation, and no fetal anomalies. In addition, we selected electrocardiographic recordings showing presence of a deceleration pattern in the EFM defined as two or more decelerations in the last 30 min. As exclusion criteria, we defined having experienced a sentinel event (uterine rupture, cord prolapse, or shoulder dystocia), EFM with less than 30 min registered period, or anomalies that do not enable the analysis of EFM. In the case of a monitoring that had not started active labor, the EFM register was discarded.

The outcome of the study was neonatal acidemia defined as pH < 7.10, measured by arterial cord blood at birth, these are the cases of the analysis. From the 5694 women in the initial cohort, 192 (3.4%) infants were acidotic. In Figure 1 we show the flowchart of the study, 72 acidemic fetuses were excluded from the analysis for lack of criteria. The remaining 120 infants with arterial acidemia were included as cases, together with 258 in the control group. The controls were selected using a non-randomized 1:2 consecutive type method; each selected control is chronologically consecutive to a case, selecting two controls for each case.

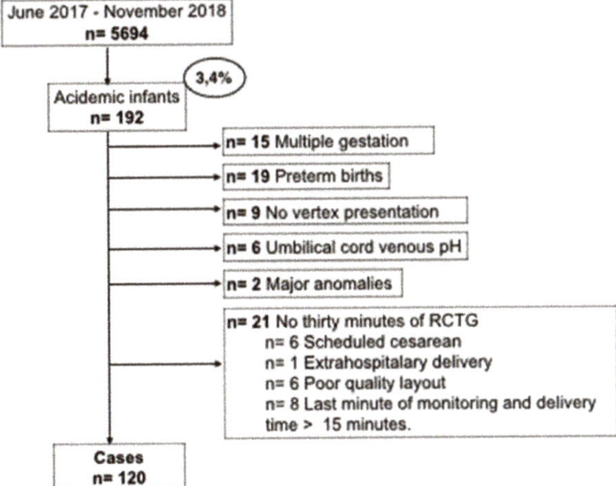

Figure 1. Flow chart of patient recruitment.

We additionally recruited maternal and pregnancy information on parity, maternal age, maternal pathologies, gestational age at birth, birthweight, estimated percentile weight, and fetal gender.

2.2. Electronic Fetal Monitoring

For the monitoring of fetal well-being, as can be seen in Figure 2, a fetal activity supervisor Corometrix 256CX was used. Two sensors were employed for this task: an ultrasonic transducer to capture the electrocardiographic (ECG) fetal activity and a TOCO (Tocotonometer) transducer to capture the uterine activity. Both were attached to the mother with binding bands and the coming signals were analyzed continuously by obstetricians during the final process of pregnancy prior to delivery.

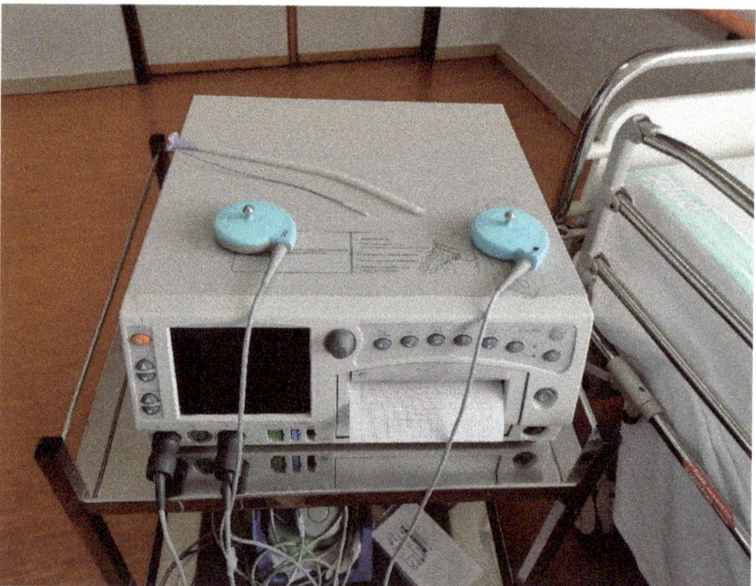

Figure 2. Electronic fetal monitoring.

The ultrasound transducer is placed on the maternal abdomen by means of one belt and transmits the ultrasonic signal of the fetal heart. It operates with a pulse repetition frequency of 4 kHz, a pulse duration of 92 uS, and a transmission frequency of 1151 MHz. It is capable of measuring heart rate from 50 to 210 bpm and its precision is 1 bpm.

The TOCO transducer is also placed on the maternal abdomen by means of one belt and it detects the forward displacement of the maternal abdominal muscles during a contraction. The TOCO transducer is composed of several strain gauges configurated to transduce pressure measurements into displacement. This device can measure pressures from 0 to 13.3 KPa with a resolution of 0.13 kPa and a bandwidth from 0 to 0.5 Hz.

In our study, the last 30 min of EFM prior to delivery were retrospectively analyzed and interpreted between two obstetricians attached to the delivery section, blind to the neonatal outcome, using the criteria and the patterns described in the Category system of the Eunice Kennedy Shriver National Institute of Child Health and Human Development (NICHD) [41]. Five elements of the EFM recording were extracted using the definitions from the NICHD criteria and then used to categorize the EFM recordings into one of the three accepted categories: Category I, Category II, or Category III to describe EFM data.

Additionally, as our purpose was to use machine learning algorithms in order to predict acidemia, we recruited information about the non-NICHD parameters These parameters were obtained from the EFM recordings as it is described in Figure 3. In the graph it can be seen the electrocardiographic fetus signal measured in beats per minute (above) and the mother's uterine contractions measured as mm Hg (below).

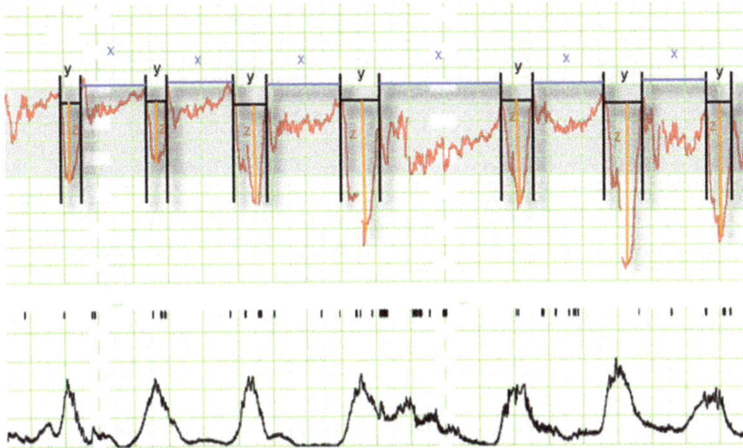

Figure 3. Intrapartum electronic fetal monitoring analysis (1 cm/min). The panel contains the fetus signal (**above**), where the following parameters can be observed: decelerations (y), time of reperfusion (x), and depth of deceleration (z); and the mother's uterine contractions measured in mm Hg (**below**) not used for the analysis.

We divided the EFM signal into deceleration (y) and interdeceleration (x) periods. The duration of reperfusions was defined as x (interdeceleration time), the duration of decelerations was defined as y (deceleration time), and the depth of decelerations as z.

From x, y, and z, we calculated the parameters:

- Total reperfusion time as the sum in minutes of the period that the fetus remains at baseline without deceleration during the last 30 min $\sum x$.
- Deceleration time as the sum in minutes of the period of time that the fetus is decelerating during the last 30 min $\sum y$.
- Total deceleration area as the sum of all areas of deceleration, being the deceleration area the product of the duration of deceleration in seconds and its maximum depth of fall from baseline expressed in beats per minute divided by two $\sum \frac{yz}{2}$.

Additionally, we considered for the multivariate model the following variables: number of decelerations, minimum beats per minute (bpm), number of decelerations greater than 60 s, number of decelerations greater than 60 beats per minute in depth, and the presence of decelerations in more than 50% of contractions, considered to be recurring, thus we defined the variables that describe the occurrence of recurrent decelerations greater than 60 s, and recurrent decelerations with depth > 60 bpm.

2.3. Statistical Analysis

We descriptively analyzed data comparing acidotic and non-acidotic infants. The continuous variables were summarized by median and interquartile range (IQ) and categorical variables by absolute and relative frequency of each category. Differences between acidotic and non-acidotic groups were analyzed using the Mann Whitney or Chi-square test for continuous or categorical data.

To predict acidotic infants in the last 30 min of labor, multivariate models were built using logistic regression models, random forest, and neural networks. For building and testing models the original database was randomly split into training (80%) and validation data (20%).

Validation of models was estimated by its discrimination measured by means of the area under the receiver characteristic curve (AUC), and its calibration through calibration curves and of the two informative parameters: 'intercept' (calibration-in-the-large) that measures the difference between average predictions and average outcome; and 'slope', which reflects the average effect of predictions on the outcome [42]. The AUC can be interpreted as the probability that the model assigns a greater probability of being acidotic for an acidotic case rather than a non-acidotic case, it ranges from 0 to 1, corresponding the 0.5 value to a random model, 0.7 to an acceptable model, 0.8 to a good model, 0.9 excellent model, and 1 perfect discrimination. The 95% confidence intervals for AUC were calculated using DeLong estimation [43]. The calibration curve analyzes graphically the concordance between predictions and the real occurrence of the outcome, a perfect calibration corresponds with the diagonal line. The predictive ability of the models summarized by their AUC was compared using the De Long test [43].

We also analyzed the clinical utility of the developed machine learning models. This property analyzes the practical use of a prediction model, that as a dichotomic classification model, using a cutoff point that classified individuals as positive (1) or negative (0), above or below the cutoff point. Several methods have been implemented for this purpose, probably the most used is the decision curve [44], that measures for different cutoff points the net benefit of the application of the model in comparison to classify all individual as 0 or 1, that also can be applied to compare models. Although this proposal provides a good guide to select the range of cutoff points with good net benefit, their interpretation is a weighted estimation and cannot be interpreted as a parameter with an easy clinical interpretation. Predictiveness curve also analyze the benefit of the application of a model, but with a less wide diffusion in this field [45].

Here, we used to analyze the clinical utility of the developed models the clinical utility curve [46] that we proposed previously in prostate cancer prediction with satisfactory results. In this curve, the X axis corresponds to the threshold probability to consider a neonate as acidotic, and on the Y axis we represent the percentage for two different measures. The first corresponds to the percentage of missing acidotic infants below the selected cut-off point, and the second one to the number of infants below the cut-off point. Using this curve for different cutoff points we can evaluate the percentage of acidotic fetuses with a wrong classification, and the fetuses with a very low risk of acidemia that are going to be saved from an unnecessary cesarean section for loss of fetal well-being, that are clinical practice parameters.

All analyses were performed using the R language programming v.4.0.3 (The R foundation for statistical computing, Vienna, Austria) with the addition of the rms, randomForestSRC, nnet, neuralnet, and NeuralNetTools libraries [47].

3. Results

3.1. Descriptive Analysis

Descriptive analysis of data is shown in Table 1. In the maternal–fetal variables of the study, we found statistically significant differences between acidotic and non-acidotic groups in the nulliparity, type of delivery, and SGA variables. Regarding EFM variables, the ACOG categories, № Decelerations > 60 sg, Recurrent decelerations > 60 sg, № Decelerations depth > 60 bpm, Recurrent decelerations depth > 60 bpm, Deceleration area, Minimum deep bpm, Maximum deep bpm, and Mean deep bpm showed differences between groups.

Table 1. Descriptive characteristics.

Variable	Acidotic (*n* = 120)	Non-Acidotic (*n* = 258)	*p*-Value
Maternal–fetal variables			
Maternal age	33 (29–37)	34 (30–36)	0.499
Hypertension disorders	5 (4.2%)	5 (1.9%)	0.362
Gestational diabetes	15 (12.5)	29 (11.2%)	0.855
Nulliparity	132 (51.2%)	80 (66.7%)	0.007
Gestational age	280 (274–285)	280 (273–286)	0.841
Male gender	64 (53.3%)	145 (56.2%)	0.681
Delivery			<0.001
Vaginal	60 (50.0%)	187 (72.5%)	
Operative vaginal	30 (25.0%)	52 (20.1%)	
Cesarean	30 (25.0%)	19 (7.4%)	
Birthweight	3238 (2918–3638)	3295 (2975–3620)	0.645
Percentile birthweight	43. 1 (20.0–74.5)	49.3 (24.1–77.4)	0.553
Small for gestational age	22 (18.3%)	28 (10.9%)	0.066
Large for gestational age	21 (17.5%)	32 (12.4%)	0.185
EFM variables			
ACOG categories			<0.001
Category 1	13 (10.8%)	123 (47.7%)	
Category 2	57 (47.5%)	110 (42.6%)	
Category 3	50 (41.7%)	25 (9.7%)	
Reperfusion time (min)	18.1 (14.8–20.8)	21.8 (18.2–25.2)	<0.001
Number of decelerations	8 (5–10)	7.5 (4–10)	0.509
№ Decelerations > 60 sg	2.5 (0–5)	0 (0–2)	<0.001
Recurrent decelerations > 60 sg	25 (20.8%)	20 (7.8%)	<0.001
№ Decelerations depth > 60 bpm	3 (1–5)	0 (0–3)	<0.001
Recurrent decelerations depth > 60 bpm	33 (27.5%)	43 (16.7%)	0.021
Deceleration area	16.5 (11.3–22.6)	9.6 (5.1–15.5)	<0.001
Minimum deep bpm	40 (30–54)	31 (24–40)	<0.001
Maximum deep bpm	79 (68–92)	60 (52–78)	<0.001
Mean deep bpm	58 (48–69)	48 (40–69)	<0.001

EFM: electro fetal monitoring; ACOG: American College of Obstetricians and Gynecologists; bpm: beats per minute.

3.2. Multivariable Prediction of Acidemia

3.2.1. Building Models

To predict acidemia we used a traditional approach in classification problems as the logistic regression model, and the machine learning algorithms: random forest and neural networks.

The logistic regression model was built using a backward stepwise selection process. In Table 2 we show the significant variables in the multivariate analysis.

Table 2. Multivariate logistic regression model.

Variable	Odds Ratio (95% C.I.)	*p*-Value
Nulliparity	0.413 (0.217–0.763)	0.006
Large for gestational age	4.562 (1.969–10.840)	<0.001
Reperfusion time (min)	0.809 (0.729–0.889)	<0.001
Number of decelerations	0.804 (0.694–0.919)	0.002
№ Decelerations > 60 sg	1.190 (1.037–1.369)	0.013
№ Decelerations depth > 60 bpm	1.328 (1.111–1.599)	0.002
Recurrent decelerations depth > 60 bpm	0.178 (0.056–0.530)	0.005
Minimum deep bpm	1.034 (1.010–1.060)	<0.001

The model showed good accuracy, with an AUC value in training data (80% data) of 0.826 (0.778–0.875) (95% confidence interval (C.I.)), and 0.840 (0.750–0.930 95% C.I.) in validation data (20% data).

Regarding the additive model of classification trees that is the random forest, it was training with different set of parameters, and the best model was attained with the configuration shown in Table 3.

Table 3. Random forest parameter configuration.

Parameter	Value
Number of trees	100
Forest terminal node size	5
Average number of terminal nodes	29.9
Resampling used to grow trees	SWOR
Resample size used to grow trees	191
Splitting rule	MSE
Number of random split points	10

SWOR: sampling without replacement; MSE: mean squared error.

The AUC value in training data was 0.991 (0.984–0.999 95% C.I.), and 0.865 (0.774–0.955 C.I.) in validation data. We found a slightly greater discrimination ability than that obtained with the logistic regression model in the validation data. Random forest is an additive model of classification trees where each model is built with different data and set of variables, to quantify the effect of the predictor variables to predict acidemia, we show in Figure 4 the variable importance (VIMP) plot. The VIMP measures the difference between prediction error under a perturbed predictor, where a permutation is designed to push a variable to a terminal node different than its original assignment, and the original predictor, these are calculated for each tree and averaged over the forest. This yields Breiman–Cutler VIMP [48]. The most influential variables in the prediction of acidemia were the number of decelerations with a deep greater than 60 beats per minute, the reperfusion time and the number of decelerations greater than 60 s.

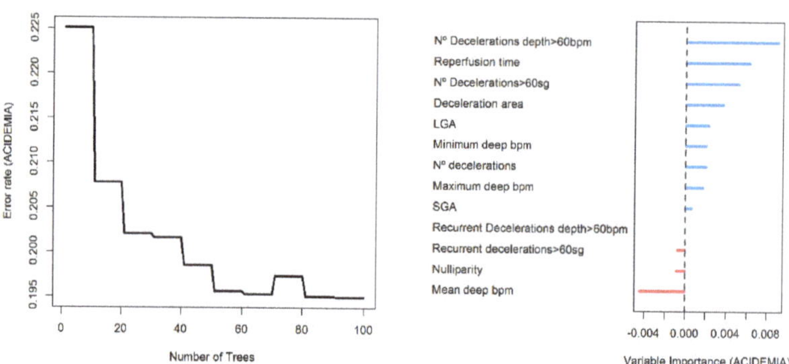

Figure 4. Error rate plot (**left** panel) and Breiman–Cutler variable importance plot (**right** panel) in random forest model.

Additionally, neural networks were trained with different architectures. We used the multilayer perceptron model with 1 or 2 hidden layers, different activation functions, initial weights, and training parameters. The best model on validation data was attained using the 13-10-1 architecture with 151 weights, and the activation function was logistic. The cross-entropy was used as the optimization function, this loss function measures the discrepancy between predictions and real occurrence of acidemia.

$$E = -\frac{1}{N} \sum_{i=1}^{N} y_i \cdot \log(p(y_i)) + (1 - y_i) \cdot \log(1 - p(y_i)) \tag{1}$$

being y_i the dichotomic outcome, acidotic ($y_i = 1$) or non-acidotic ($y_i = 0$), and $p(y_i)$ the predicted probability of being acidotic for observation i out of N observations.

The architecture of the network is plotted in Figure 5, positive weights between layers are plotted as black lines, and the negative weights as grey lines. Line thickness is in proportion to relative magnitude of each weight.

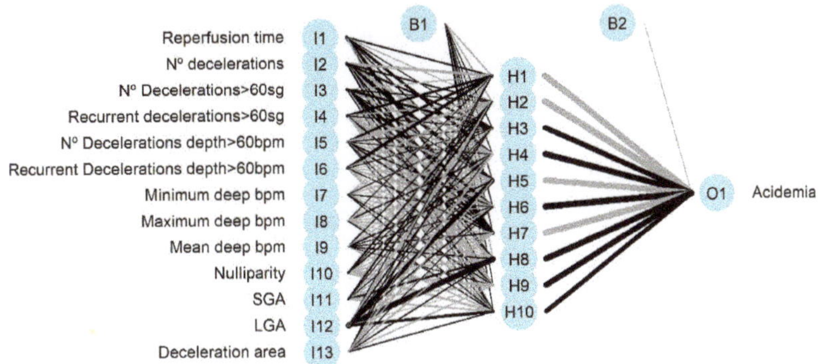

Figure 5. Neural network architecture with input (I), hidden (H), and output (O) layers. (B) is the result obtained after applying the activation function.

The neural networks had an AUC value of 0.995 (0.985–1) (95% C.I.) for training data and 0.857 (0.751–0.963 95% C.I.) for validation data, greater than that obtained using logistic regression model but lower than the AUC that corresponds to random forest. Additionally, we present the variable importance plot for the multilayer perceptron, shown in Figure 6, following the method described by Garson 1991 [49], where the relative importance of explanatory variables for a single response in a supervised neural network is estimated by deconstructing the model weights. The most influential variables were the number of decelerations, being large for gestational age fetus, and the number of decelerations greater than 60 s.

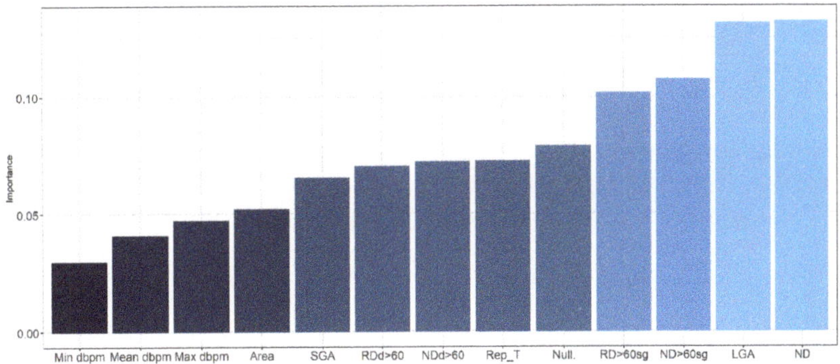

Figure 6. Variable importance in neural network. dbpm: deep in beats per minute; SGA: small for gestational age; RDd > 60: recurrent decelerations depth > 60 beats per minute; NDd > 60: number of decelerations depth > 60 beats per minute; Rep_T: reperfusion time; Null: nulliparity; RD > 60 sg: recurrent decelerations > 60 s; ND > 60 sg: number of decelerations > 60 s; LGA: large for gestational age; ND: number of decelerations.

3.2.2. Validation of Models

In this section, we present the validation of the models developed using the validation data. The agreement between predictions and real outcomes was analyzed by calibration curves in Figure 7. For the logistic regression model, we found an overestimation of real acidemia occurrence, this is even more clear for neural networks. In the X axis of the graph, we show the predicted probabilities provided by models, for a 60% probability of acidemia, the actual occurrence of acidosis (Y axis) was 40% for logistic regression model, and 30% for neural networks, therefore, both models overestimate the real occurrence of acidosis.

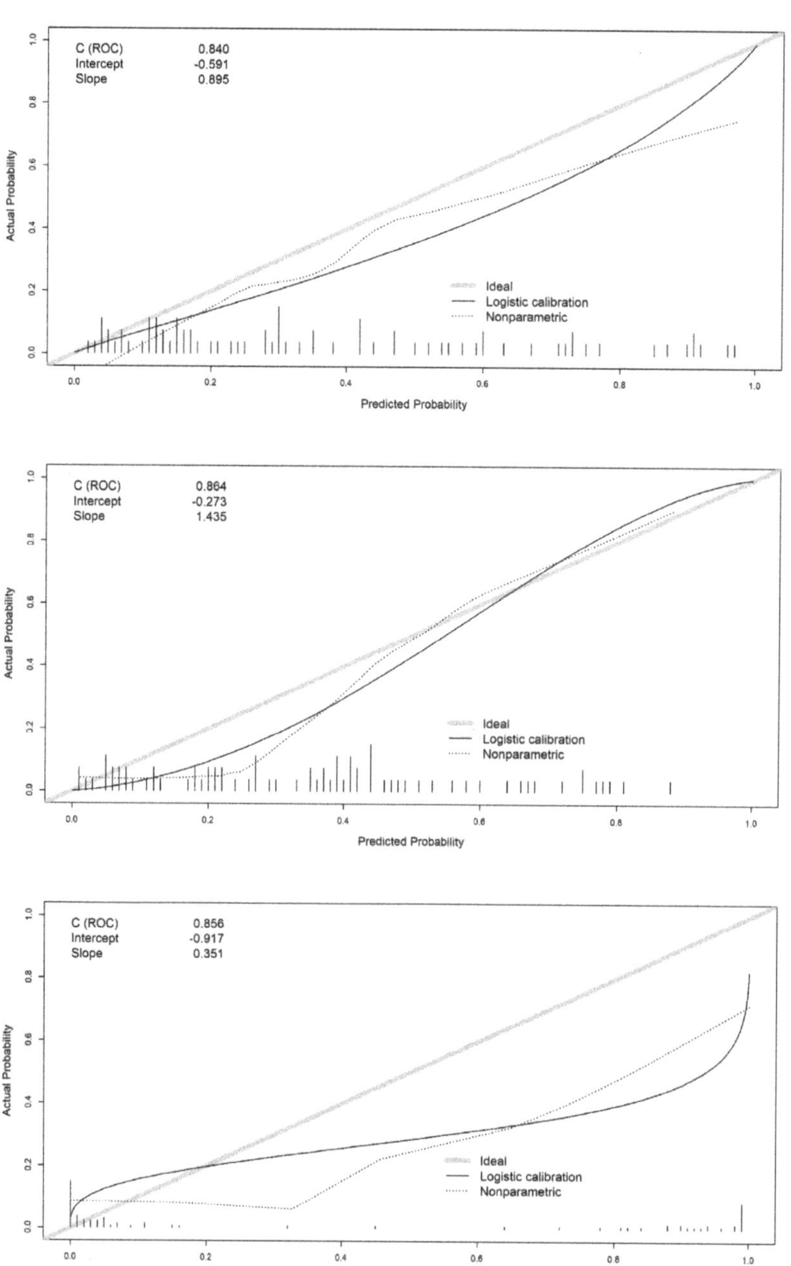

Figure 7. Calibration curves of logistic regression (**top** panel), random forest (**center** panel), and neural network (**bottom** panel) models.

For the random forest model, this overestimation was present only for probabilities below 0.4. The intercept showed also worse mean predictions for logistic regression (-0.591) and neural networks (-0.917) than random forest (-0.273) which is closer to 0. The slope was closer to 1 for logistic regression (0.895) with better concordance between predicted probabilities and real outcome.

The discrimination ability of models is shown by ROC curves in Figure 8. All models show a good discrimination capacity. To compare the AUC of the models, we used the Delong comparison test. Differences between areas were not significant in our study, logistic regression vs. random forest ($p = 0.561$), logistic regression vs. neural networks ($p = 0.736$), random forest vs. neural networks ($p = 0.888$).

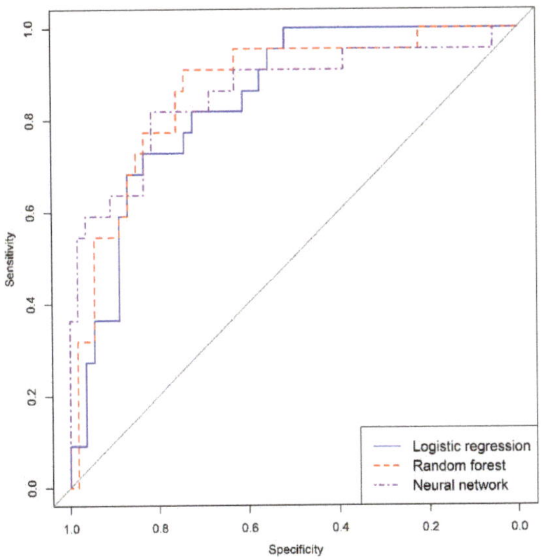

Figure 8. Receiver characteristic curves of logistic regression, random forest, and neural network models.

Finally, the clinical utility of models was analyzed. As our purpose was to predict acidemia, the most important issue was to analyze, for different threshold points, the false negative cases, that is, patients that by means of a cut-off point are going to be classified as non-acidotic below the cut-off point being acidotic. In the clinical utility curve, we analyzed this measure and the number of cases below a cut-off point, which in our study are candidates to a cesarean section that are going to avoid it.

Figure 9 presents the clinical utility curves. If we choose a maximum admissible level of 5% missing acidemia cases wrongly classified, in the curves we can analyze the threshold point that corresponds to this value. For the logistic regression model, this corresponds to a 23% cut-off point, and the number of deliveries saved with a minimum loss of acidotic cases was 40.8%. For the random forest model, it corresponds to a 33% acidotic probability threshold point, with 46.1% saved deliveries. Finally, for the neural network, it corresponds to a 1% cut-off point with 25% saved deliveries. Considering the clinical utility of the models, it is clear that the random forest proved superior.

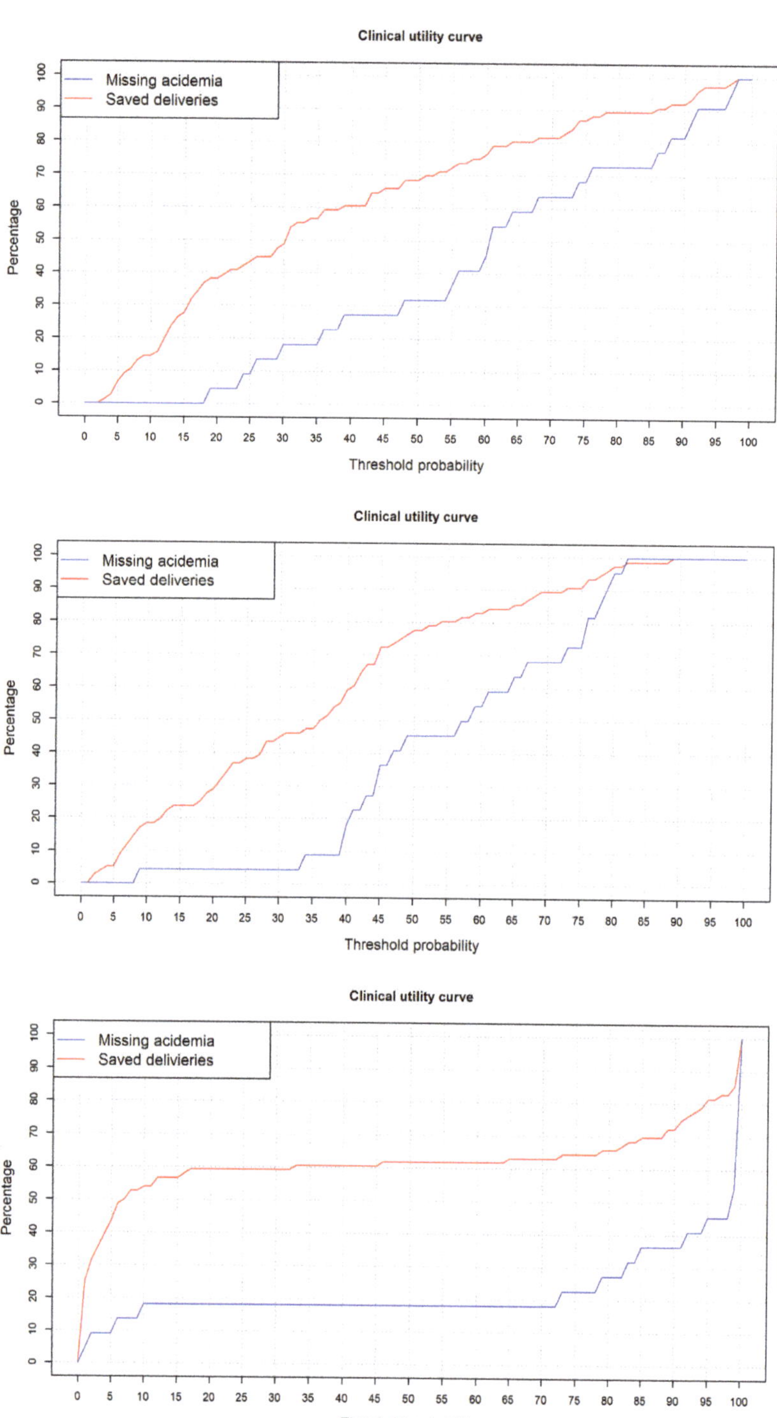

Figure 9. Clinical utility curve for logistic regression (**top** panel), random forest (**center** panel), and neural network (**bottom** panel) models.

4. Discussion

Here, we developed a comparison analysis of machine learning techniques to predict acidemia using FHR variables derived from the last 30 min of a continuous electronic fetal monitoring during intrapartum period. Built models showed a good and similar discrimination ability, but with clear differences in the calibration and clinical utility analysis, in which the random forest model showed the best performance.

The external monitoring of FHR is based mainly on the transmission of a transducer placed on the maternal abdomen, binding by an elastic band encircling the abdomen, localized at the fetal heart, although there is variability on CGT monitors [2]. Conductive gel placed between the abdomen and sensor favors the transmission of sound waves, but the signal can be affected by movement of maternal vessels or the fetus extremities, causing artefacts. This is a limitation for all systems that try to predict acidemia in real time, specially, in cases where the signal must be processed as in fractal analysis [17,18].

The development of devices to extract and monitor data should be followed by new software to analyze the FHR. The information theory is an essential issue to transmit, process, analyze data, and provide accurate information to the obstetrician in real time. In this context, there is a variety of applications of the theory of information in signal processing [50]. The digitalization of the signal provides the possibility of processing it by means of convolutional type networks or even more complex encoder–decoder deep learning structures in order to predict acidemia. Tang [27] designs a convolution neural network (CNN) model named MKNet with an AUC value of 0.95, they proposed their use by a real-time monitoring of fetal health on portable devices. Zhao [28] also uses CNN to provide predictions with an AUC above 0.95 in a 10-fold cross validation procedure. The accuracy of both models is very high but there is no analysis of calibration and clinical utility.

A different approach to the modeling of the complete signal is the extraction of variables from the signal that are combined in binary classification models of acidemia. In our analysis, we trained logistic regression, random forest, and neural networks using as predictor variables EFM features easily obtained from the EFM recording. Our best model was reached using random forest algorithm. These additive models provide robust models as their prediction is based on the sum of combination of trees building using different sets of data and variables. In our study, the best model was found using 100 trees, those trees are built using the 40% of predictor variables and 63% of the training data sample. The purpose of this selection is to guarantee that each tree explores the predictive ability of predictor variables in different data sample and over a different set of variables. In addition, the trees had a maximum number of cases at a terminal node of 5, preventing the overfitting that occurs in trees with too many branches.

The AUC obtained in validation data was 0.86, below results of the previous CNN models [27,28], but with good accuracy. Unfortunately, these studies lack a complete validation analysis, this would make them more comparable with ours. In our calibration analysis, we found that probabilities of acidemia provided by logistic regression and random forest model are well distributed in a wide range between 0 and 1. By contrast, in neural networks most probabilities are very close to 0 and 1, this is a clear sign of overfitting in the model. As a consequence, it is very difficult to choose a threshold probability point that separates acidotic and non-acidotic cases because probabilities are very concentrated in a narrow range. Logistic regression and random forest are more robust models, allowing the analysis of the advantages and disadvantages in terms of wrong classification of acidotic cases and avoided cesarean sections. In the case of the random forest model, to prevent 46% of unnecessary deliveries with a minimum loss of 5% of acidotic cases is a promising result.

Zhao [24] used an AdaBoost model with sensitivity of 92%, and specificity of 90%, similar to our results, showing the robustness of the additive tree models, although there is no information about how many cesarean sections could be saved with the 10% of academic cases wrongly classified. Iraji [29] used neural networks to reach a sensitivity of 99% and specificity of 97% which is near perfect classification. These values are extremely high and probably need an external validation to verify them. Balayla [20] in a metanalysis conclude that the use of AI and computer analysis for the interpretation of EFM during labor does not improve neonatal, but their conclusions are based only on risk ratio analysis. As we showed in our study, global measures of accuracy such as AUC can give the appearance that models are very similar, but their performance should be further explored using a complete validation process.

As a strength of our study, we found a classification model developed by means of a machine learning algorithm applied to EFM features that are easy to obtain from EFM recording. These predictor variables have proved as good predictors of acidemia in previous studies [14,15], but few studies have combined them in a predictive model using different machine learning algorithms. In addition, this model has shown good clinical utility to apply it in real clinical practice.

A limitation of the study is that it was a retrospective analysis with data sourced from a unique hospital without an external validation.

5. Conclusions

Using EFM recording, based on fetal resilience parameters, we developed a random forest model to predict acidemia that showed good accuracy, with AUC = 0.86 in validation data. This model can be applied in clinical practice using a cutoff point of 33% for the probability of acidemia, that showed 5% of missing acidemia but prevented 46% of unnecessary cesarean sections.

Author Contributions: Conceptualization, J.E.-E., B.C., R.S., L.M.E. and S.C.; methodology, J.E.-E., B.C. and L.M.E.; software, J.E.-E., C.A., A.R.L., G.S.-E. and L.M.E.; validation, B.C., R.S. and G.S.; formal analysis, L.M.E.; investigation, J.E.-E., B.C., R.S. and S.C.; resources, M.C.-E.; data curation, M.C.-E.; writing—original draft preparation, J.E.-E., B.C. and L.M.E.; writing—review and editing, all authors.; visualization, G.S. and S.C.; supervision, R.S. and S.C.; project administration, M.C.-E. and S.C.; funding acquisition, G.S. All authors have read and agreed to the published version of the manuscript.

Funding: This research was funded by Government of Aragon, grant number E46_20R and Ministerio de Ciencia e Innovación, MCIN/AEI/10.13039/501100011033, grant number PID2020-116873GB-I00.

Institutional Review Board Statement: The study was conducted in accordance with the Declaration of Helsinki, and approved by Clinical Research Ethics Committee of Aragon (CEICA, PI 20/137).

Informed Consent Statement: Due to the retrospective observational nature of this study, data were fully anonymized, and the need to obtain informed consent was waived.

Data Availability Statement: The data analyzed were retrieved from the Miguel Servet University Hospital database.

Acknowledgments: We gratefully thank all the staff of the Ultrasound and Prenatal Diagnosis Unit of the Miguel Servet University Hospital (Zaragoza, Spain) for their help in the acquisition and management of data.

Conflicts of Interest: The authors declare no conflict of interest. The funders had no role in the design of the study; in the collection, analyses, or interpretation of data; in the writing of the manuscript, or in the decision to publish the results.

References

1. Nunes, I.; Ayres-de-Campos, D. Computer analysis of foetal monitoring signals. *Best Pract. Res. Clin. Obstet. Gynaecol.* **2016**, *30*, 68–78. [CrossRef]
2. Ayres-de-Campos, D.; Nogueira-Reis, Z. Technical characteristics of current cardiotocographic monitors. *Best Pract. Res. Clin. Obstet. Gynaecol.* **2016**, *30*, 22–32. [CrossRef]
3. Docker, M. Doppler ultrasound monitoring technology. *BJOG Int. J. Obstet. Gynaecol.* **1993**, *100*, 18–20. [CrossRef]
4. Nunes, I.; Ayres-de-Campos, D.; Figueiredo, C.; Bernardes, J. An overview of central fetal monitoring systems in labour. *J. Perinat. Med.* **2013**, *41*, 93–99. [CrossRef]
5. Shannon, C.E. A mathematical theory of communication. *Bell Syst. Tech. J.* **1948**, *27*, 379–423. [CrossRef]
6. Cheong, R.; Rhee, A.; Wang, C.J.; Nemenman, I.; Levchenko, A. Information transduction capacity of noisy biochemical signaling networks. *Science* **2011**, *334*, 354–358. [CrossRef]

7.   Adami, C. The use of information theory in evolutionary biology. *Ann. N. Y. Acad. Sci.* **2012**, *1256*, 49–65. [CrossRef]
8.   Frénay, B. Uncertainty and Label Noise in Machine Learning. Ph.D. Dissertation, Catholic University of Louvain, Louvain-la-Neuve, Belgium, 2013.
9.   Clark, S.L.; Hamilton, E.F.; Garite, T.J.; Timmins, A.; Warrick, P.A.; Smith, S. The limits of electronic fetal heart rate monitoring in the prevention of neonatal metabolic acidemia. *Am. J. Obstet. Gynecol.* **2017**, *216*, 163.e1–163.e6. [CrossRef]
10. American College of Obstetricians and Gynecologists. Fetal heart rate monitoring: Guidelines. *ACOG Tech. Bull.* **1974**, *32*, 1–10.
11. American College of Obstetricians and Gynecologists. Practice bulletin no. 116: Management of intrapartum fetal heart rate tracings. *Obstet. Gynecol.* **2010**, *116*, 1232–1240. [CrossRef]
12. Zamora, C.; Chóliz, M.; Mejía, I.; Díaz de Terán, E.; Esteban, L.M.; Rivero, A.; Castán, B.; Andeyro, M.; Savirón, R. Diagnostic capacity and interobserver variability in FIGO, ACOG, NICE and Chandraharan cardiotocographic guidelines to predict neonatal acidemia. *J. Matern. Fetal Neonatal Med.* **2021**, *80*, 6479. [CrossRef]
13. Rei, M.; Tavares, S.; Pinto, P.; Machado, A.P.; Monteiro, S.; Costa, A.; Costa-Santos, C.; Bernardes, J.; Ayres-De-Campos, D. Interobserver agreement in CTG interpretation using the 2015 FIGO guidelines for intrapartum fetal monitoring. *Eur. J. Obstet. Gynecol. Reprod. Biol.* **2016**, *205*, 27–31. [CrossRef]
14. Cahill, A.G.; Tuuli, M.G.; Stout, M.J.; López, J.D.; Macones, G.A. A prospective cohort study of fetal heart rate monitoring: Deceleration area is predictive of fetal acidemia. *Am. J. Obstet. Gynecol.* **2018**, *218*, 523.e1–523.e12. [CrossRef]
15. Chóliz, M.; Savirón, R.; Esteban, L.M.; Zamora, C.; Espiau, A.; Castán, B.; Castán Mateo, S. Total intrapartum fetal reperfusion time (fetal resilience) and neonatal acidemia. *J. Matern. Fetal Neonatal Med.* **2021**, *91*, 5977. [CrossRef]
16. Sbrollini, A.; Agostinelli, A.; Marcantoni, I.; Morettini, M.; Burattini, L.; Di Nardo, F.; Fioretti, S.; Burattini, L. eCTG: An automatic procedure to extract digital cardiotocographic signals from digital images. *Comput. Methods Programs Biomed.* **2018**, *156*, 133–139. [CrossRef]
17. Doret, M.; Helgason, H.; Abry, P.; Goncalves, P.; Gharib, C.; Gaucherand, P. Multifractal analysis of fetal heart rate variability in fetuses with and without severe acidosis during labor. *Am. J. Perinatol.* **2011**, *28*, 259–266. [CrossRef]
18. Doret, M.; Spilka, J.; Chudáček, V.; Gonçalves, P.; Abry, P. Fractal analysis and hurst parameter for intrapartum fetal heart rate variability analysis: A versatile alternative to frequency bands and LF/HF ratio. *PLoS ONE* **2015**, *10*, e0136661. [CrossRef]
19. Desai, G.S. Artificial intelligence: The future of obstetrics and gynecology. *J. Obstet. Gynecol. India* **2018**, *68*, 326–327. [CrossRef]
20. Balayla, J.; Shrem, G. Use of artificial intelligence (AI) in the interpretation of intrapartum fetal heart rate (FHR) tracings: A systematic review and meta-analys.sis. *Arch. Gynecol. Obstet.* **2019**, *300*, 7–14. [CrossRef]
21. Iftikhar, P.; Kuijpers, M.V.; Khayyat, A.; Iftikhar, A.; De Sa, M.D. Artificial intelligence: A new paradigm in obstetrics and gynecology research and clinical practice. *Cureus* **2020**, *12*, e7124. [CrossRef]
22. Emin, E.I.; Emin, E.; Papalois, A.; Willmott, F.; Clarke, S.; Sideris, M. Artificial intelligence in obstetrics and gynaecology: Is this the way forward? *Vivo* **2019**, *33*, 1547–1551. [CrossRef]
23. Aznar-Gimeno, R.; Esteban, L.M.; Labata-Lezaun, G.; del-Hoyo-Alonso, R.; Abadia-Gallego, D.; Paño-Pardo, J.R.; Esquillor Rodrigo, M.J.; Lanas, A.; Serrano, M. A Clinical Decision Web to Predict ICU Admission or Death for Patients Hospitalised with COVID-19 Using Machine Learning Algorithms. *Int. J. Environ. Res. Public Health* **2021**, *18*, 8677. [CrossRef]
24. Zhao, Z.; Zhang, Y.; Deng, Y. A comprehensive feature analysis of the fetal heart rate signal for the intelligent assessment of fetal state. *J. Clin. Med.* **2018**, *7*, 223. [CrossRef]
25. Cömert, Z.; Kocamaz, A.F. Open-access software for analysis of fetal heart rate signals. *Biomed. Signal Process. Control* **2018**, *45*, 98–108. [CrossRef]
26. Cömert, Z.; Kocamaz, A.F.; Subha, V. Prognostic model based on image-based time-frequency features and genetic algorithm for fetal hypoxia assessment. *Comput. Biol. Med.* **2018**, *99*, 85–97. [CrossRef]
27. Tang, H.; Wang, T.; Li, M.; Yang, X. The design and implementation of cardiotocography signals classification algorithm based on neural network. *Comput. Math. Methods Med.* **2018**, *2018*, 8568617. [CrossRef]
28. Zhao, Z.; Deng, Y.; Zhang, Y.; Zhang, Y.; Zhang, X.; Shao, L. DeepFHR: Intelligent prediction of fetal Acidemia using fetal heart rate signals based on convolutional neural network. *BMC Med. Inform. Decis. Mak.* **2019**, *19*, 1–15. [CrossRef]
29. Iraji, M.S. Prediction of fetal state from the cardiotocogram recordings using neural network models. *Artif. Intell. Med.* **2019**, *96*, 33–44. [CrossRef]
30. Shore, J.; Johnson, R. Properties of cross-entropy minimization. *IEEE Trans. Inf. Theory* **1981**, *27*, 472–482. [CrossRef]
31. Aznar-Gimeno, R.; Labata-Lezaun, G.; Adell-Lamora, A.; Abadía-Gallego, D.; del-Hoyo-Alonso, R.; González-Muñoz, C. Deep Learning for Walking Behaviour Detection in Elderly People Using Smart Footwear. *Entropy* **2021**, *23*, 777. [CrossRef]
32. Hanley, J.A.; McNeil, B.J. The meaning and use of the area under a receiver operating characteristic (ROC) curve. *Radiology* **1982**, *143*, 29–36. [CrossRef]
33. Steyerberg, E.W. *Clinical Prediction Models*; Springer: Berlin/Heidelberg, Germany, 2019.
34. Malin, G.L.; Morris, R.K.; Khan, K.S. Strength of association between umbilical cord pH and perinatal and long term outcomes: Systematic review and meta-analysis. *BMJ* **2010**, *340*, c1471. [CrossRef]
35. Cahill, A.G.; Roehl, K.A.; Odibo, A.O.; Macones, G.A. Association and prediction of neonatal acidemia. *Am. J. Obs. Gynecol.* **2012**, *207*, 206.e1–206.e8. [CrossRef]
36. Ogunyemi, D.; Jovanovski, A.; Friedman, P.; Sweatman, B.; Madan, I. Temporal and quantitative associations of electronic fetal heart rate monitoring patterns and neonatal outcomes. *J. Matern. Fetal Neonatal Med.* **2019**, *32*, 3115–3124. [CrossRef]

37. Martí, S.; Lapresta, M.; Pascual, J.; Lapresta, C.; Castán, S. Deceleration area and fetal acidemia. *J. Matern. Fetal Neonatal Med.* **2017**, *30*, 2578–2584. [CrossRef]
38. Lear, C.A.; Galinsky, R.; Wassink, G.; Yamaguchi, K.; Davidson, J.O.; Westgate, J.A.; Bennet, L.; Gunn, A.J. The myths and physiology surrounding intrapartum decelerations: The critical role of the peripheral chemoreflex. *J. Physiol.* **2016**, *594*, 4711–4725. [CrossRef]
39. Lear, C.A.; Wassink, G.; Westgate, J.A.; Nijhuis, J.G.; Ugwumadu, A.; Galinsky, R.; Bennet, L.; Gunn, A.J. The peripheral chemoreflex: Indefatigable guardian of fetal physiological adaptation to labour. *J. Physiol.* **2018**, *596*, 5611–5623. [CrossRef]
40. Eden, R.D.; Evans, M.I.; Evans, S.M.; Schifrin, B.S. The "Fetal Reserve Index": Re-engineering the interpretation and responses to fetal heart rate patterns. *Fetal Diagn.* **2018**, *43*, 90–104. [CrossRef]
41. Eunice Kennedy Shriver National Institute of Child Health and Human Development (NICHD). Available online: https://www.nih.gov/about-nih/what-we-do/nih-almanac/eunice-kennedy-shriver-national-institute-child-health-human-development-nichd (accessed on 28 November 2021).
42. Steyerberg, E.W.; Van Calster, B.; Pencina, M.J. Performance measures for prediction models and markers: Evaluation of predictions and classifications. *Rev. Esp. Cardiol.* **2011**, *64*, 788–794. [CrossRef]
43. DeLong, E.R.; DeLong, D.M.; Clarke-Pearson, D.L. Comparing the areas under two or more correlated receiver operating characteristic curves: A nonparametric approach. *Biometrics* **1988**, *44*, 837–845. [CrossRef]
44. Vickers, A.J.; van Calster, B.; Steyerberg, E.W. A simple, step-by-step guide to interpreting decision curve analysis. *Diagn. Progn. Res.* **2019**, *3*, 18. [CrossRef]
45. Pepe, M.S.; Feng, Z.; Huang, Y.; Longton, G.; Prentice, R.; Thompson, I.M.; Zheng, Y. Integrating the predictiveness of a marker with its performance as a classifier. *Am. J. Epidemiol.* **2008**, *167*, 362–368. [CrossRef]
46. Borque-Fernando, A.; Esteban-Escaño, L.M.; Rubio-Briones, J.; Lou-Mercade, A.C.; Garcia-Ruiz, R.; Tejero-Sanchez, A.; Muñoz-Rivero, M.V.; Cabañuz-Plo, T.; Alfaro-Torres, J.; Marquina-Ibañex, I.M.; et al. A preliminary study of the ability of the 4Kscore test, the Prostate Cancer Prevention Trial-Risk Calculator and the European Research Screening Prostate-Risk Calculator for predicting high-grade prostate cancer. *Actas Urológicas Españolas* **2016**, *40*, 155–163. [CrossRef]
47. R Core Team. *R: A Language and Environment for Statistical Computing*; R Foundation for Statistical Computing: Vienna, Austria, 2020; Available online: https://www.R-project.org/ (accessed on 28 November 2021).
48. Ishwaran, H.; Lu, M. Standard errors and confidence intervals for variable importance in random forest regression, classification, and survival. *Stat. Med.* **2019**, *38*, 558–582. [CrossRef]
49. Garson, G.D. Interpreting neural network connection weights. *Artif. Intell. Expert* **1991**, *6*, 46–51.
50. Cruces Álvarez, S.A.; Martín Clemente, R.; Samek, W. Information Theory Applications in Signal Processing. *Entropy* **2019**, *21*, 653. [CrossRef]

Article

Analysis of Accelerometer and GPS Data for Cattle Behaviour Identification and Anomalous Events Detection

Javier Cabezas [1,*], Roberto Yubero [1], Beatriz Visitación [1], Jorge Navarro-García [1], María Jesús Algar [1], Emilio L. Cano [1,2] and Felipe Ortega [1]

[1] Data Science Laboratory, University Rey Juan Carlos, 28933 Móstoles, Spain; roberto.yubero@urjc.es (R.Y.); beatriz.visitacion@urjc.es (B.V.); j.navarro.2016@alumnos.urjc.es (J.N.-G.); mariajesus.algar@urjc.es (M.J.A.); emilio.lopez@urjc.es (E.L.C.); felipe.ortega@urjc.es (F.O.)

[2] Quantitative Methods and Socioeconomic Development Group, Institute for Regional Development, University of Castilla-La Mancha, 02071 Albacete, Spain

* Correspondence: javier.cabezas@urjc.es; Tel.: +34-615-926-812

Abstract: In this paper, a method to classify behavioural patterns of cattle on farms is presented. Animals were equipped with low-cost 3-D accelerometers and GPS sensors, embedded in a commercial device attached to the neck. Accelerometer signals were sampled at 10 Hz, and data from each axis was independently processed to extract 108 features in the time and frequency domains. A total of 238 activity patterns, corresponding to four different classes (*grazing, ruminating, laying* and *steady standing*), with duration ranging from few seconds to several minutes, were recorded on video and matched to accelerometer raw data to train a random forest machine learning classifier. GPS location was sampled every 5 min, to reduce battery consumption, and analysed via the k-medoids unsupervised machine learning algorithm to track location and spatial scatter of herds. Results indicate good accuracy for classification from accelerometer records, with best accuracy (0.93) for *grazing*. The complementary application of both methods to monitor activities of interest, such as sustainable pasture consumption in small and mid-size farms, and to detect anomalous events is also explored. Results encourage replicating the experiment in other farms, to consolidate the proposed strategy.

Keywords: animal behaviour; pattern recognition; anomaly detection; clustering; spectral analysis; accelerometer sensor; GPS sensor

Citation: Cabezas, J.; Yubero, R.; Visitación, B.; Navarro-García, J.; Algar, M.J.; Cano, E.L.; Ortega,F. Analysis of Accelerometer and GPS Data for Cattle Behaviour Identification and Anomalous Events Detection. *Entropy* **2022**, *24*, 336. https://doi.org/10.3390/e24030336

Academic Editor: Sotiris Kotsiantis

Received: 23 December 2021
Accepted: 23 February 2022
Published: 26 February 2022

Publisher's Note: MDPI stays neutral with regard to jurisdictional claims in published maps and institutional affiliations.

1. Introduction

Monitoring activity of animals in livestock farms can provide relevant indicators about their health and welfare level. In fact, ensuring animal well-being through objective evidence has become a major concern for both cattle producers and consumers [1]. For example, EU Directive 98/58/EC regarding the protection of animals kept in farms [2] introduces general rules for protecting all animals species for production of food, wool, skin, fur or other farming purposes. Later on, EU legislation has been progressively extended to increase the well-being of farmed animals. Additionally, current EU regulation regarding organic farming rules encourages high standards for animal welfare, requiring farmers to meet specific behavioural needs of animals [3].

The development of systems to gather and analyse animal behaviour data can certainly help cattle producers to meet these high quality standards. In recent years, wireless sensor networks (WSN) and Internet of Things (IoT) technologies have paved the way for implementing monitoring systems on farms [4–6]. Various methods have been proposed for automated recording and identification of animal activity in this context. Sensors embedded in electronic devices attached on animals legs or using neckbands can record activity information, with great detail. Then, activity patterns of interest can be revealed through the analysis of these behavioural records.

A frequent case is the use of accelerometers to create motion logs by tracking movement in a 3-D coordinate system. These devices have been used to register movements of human users [7]. Moreover, it is possible to estimate the vertical component and magnitude of the horizontal component of the user's motion, even in absence of precise information about the position and orientation of the device with respect to the body [8,9]. Many previous studies of livestock behavioural activity on farms use accelerometers to gather data describing animal movements [10–15]. Furthermore, certain studies also use low-cost GPS devices to register the location of animals, augmenting the information obtained from accelerometers [16].

Machine learning (ML) classification algorithms can be used to classify cattle activity patterns automatically, based on registers from accelerometer and GPS sensors [17,18]. Previous studies have documented accurate identification of cattle standing and walking behaviour through accelerometer data [11,19], along with precise estimation of the duration of standing behavioural patterns [20].

A comprehensive survey [21], comparing previous studies on ruminant behaviour prediction, indicates that most of them are focused on identifying a predefined set of activities. For instance, Smith et al. [22] differentiate among five possible classes (*grazing, walking, ruminating, resting* and *other activities*), whereas Riaboff et al. [23] distinguish up to 13 different behavioural patterns. Usually, the initial problem is broken down into a set of "one-vs-all" binary classification tasks. Hence, individual outputs from each classifier must be integrated, which leads to some practical challenges. For instance, it is desirable to follow a robust methodology for data acquisition and feature engineering that can be shared among different classifiers [21,24]. In the same way, another limitation of previous studies is the relatively narrow focus on specific cases of outlier detection techniques, such as lameness [25], oestrus periods [14] or parturition events [18,26].

Likewise, current research works exhibit a noticeable scarcity with respect to the early detection of specific cattle social interactions at group level, especially when they lead to anomalous situations involving potential economic impact on livestock farm operations.

Wolf and other predator attacks constitute a prominent example of such anomalies. When there exists a potential threat of attack, herds change their behavioural patterns to put on an alert. They could also stop grazing and ruminating, or even move away to a different location. The growing and perceptible concern among farm producers on this matter, and the pressing need to find sustainable trade-off solutions, that preserves both protected species, such as wolves in the northwest of Spain and other countries, and farmers' rights to continue their normal operations and guarantee their animals well-being, constitute a challenging issue yet to be solved.

Disease transmission represents another good example. Early detection and subsequent application of proper corrective actions bring in an opportunity to avoid a severe impact on productivity. Unusual resting behaviours, abnormal stance and gaits, the absence of vertical or horizontal neck movements or the observation of too slow displacements can provide key signals of possible ongoing diseases.

Similarly, despite not representing an anomalous activity itself, the detection of an unbalanced use of pasture land can also help farmers to develop strategies aimed at a more rational consumption of natural resources, achieving better management and saving costs. In this regard, recurrent grazing habits and lack of displacement to alternative areas may render valuable information to farmers on pasture land usage. Along these lines, recent results shown in [27] suggest that the combination of movement records and GPS location data can improve detection of anomalous situations on farms.

In this paper, we present a method to classify cattle behaviour from accelerometer and GPS data, collected from collars attached to cows in two field experiments. Time and frequency-domain features are extracted from accelerometer data, to train a supervised ML classification model for cattle behaviour. GPS data is processed with an unsupervised clustering method to estimate the number of herds and their spatial scattering. This general method can be applied to a wide range of scenarios. Furthermore, new activities could be incorporated to the classifier, provided that customised training data describing the new patterns of interest are obtained, following the same preparation procedure. In addition, potential applications for tracking interesting or anomalous activities, such as unbalanced use of pasture land, disease transmission or predator attacks, are also explored.

The rest of the paper is organized as follows. Section 2 describes the equipment and experimental setup for this study, along with our proposed method to analyse animal behaviour records. Section 3 presents the main results from the two field studies to validate the suggested approach. In Section 4, we discuss the main implications that can be drawn from experimental results, as well as potential practical applications of the proposed method. Finally, Section 5 concludes and describes further research directions.

2. Materials and Methods

2.1. Farms and Animals

In our study, we focus on beef cattle located on two different commercial farms, located in the Spanish provinces of Avila and Segovia, respectively. Herds raised on these farms comprise widespread breeds, including *Fleckvieh* or *Salers*, among others, along with native Spanish breeds, such as the *Berrenda en Colorado* (brindle cow in red) or the *Avileña-negra Ibérica* (Iberian Avila's black). Most of time, animals were kept on pasture and moved freely within the farm limits. Cows were mainly fed with pasture, although they also received hay and concentrate supplements. A random sample of 30 cows in both livestock farms were equipped with accelerometer and GPS devices (see Section 2.2, below). Selected cows are representative of the most prevalent breeds in cattle from both farms, namely, *Fleckvieh*, *Salers*, *Berrenda en Colorado* and cross-bred dairy specimens.

2.2. Device and Data Loggers

We explore the use of two different procedures to monitor and analyse animal behaviour:

- *Tracking movement*: Detailed movement registries are recorded through triaxial accelerometers attached to the neck. In this way, we can identify more different behavioural patterns than when the accelerometer is installed on the leg.
- *Tracking location*: Animal location is registered with GPS sensors that periodically transmit this information to a centralized server in a cloud computing infrastructure.

Accelerometers and GPS sensors for this study were provided and installed by Digitanimal (https://digitanimal.com/?lang=en, accessed on 25 February 2022), a private company based in Madrid that develops innovative hardware and software animal monitoring solutions on farms. These sensors are integrated in an electronic device developed by this company, mounted inside a weatherproof plastic case and attached to the cow using a neckband. Figure 1 depicts a model for the collar case containing the device, and the three coordinate axes monitored by the accelerometer sensor. In turn, Figure 2 shows a cow of Fleckvieh breed wearing the neckband attaching the device to monitor movement and location.

Acceleration levels on cows necks are measured by using MEMS (Micro Electro Mechanical System) accelerometers. This type of accelerometer measures acceleration in 3 orthogonal directions (triaxial accelerometer, see Figure 1). The sensor captures DC (direct current or *offset*) acceleration (earth gravity), providing not only acceleration levels but also sensor orientation. It is a low-power consumption sensor, with a working temperature range of $-40°$ to $85°$ Celsius, suitable for the required application. Raw data are acquired at a 10 Hz sampling frequency, using a dynamic range of ±2 g. Data are retrieved continuously since the sensor is connected and directly stored in plain text format in a SD memory card.

With respect to GPS sensors, they must send information at more widely spaced intervals than in the case of accelerometers in order to optimize battery consumption and, therefore, avoid premature battery draining. We must take into account that the monitor device is a commercial hardware solution, conceived to be affordable and remain operative over relatively long time periods (usually, 2–3 months). The GPS device is configured to use a maximum DOP (Dilution of Precision) threshold of 1, and to seek signal reception from a minimum of 7 different satellites. With this configuration, the estimated average measurement error is 1.7 m, and 90% of measurements present an error lower than 5.2 m.

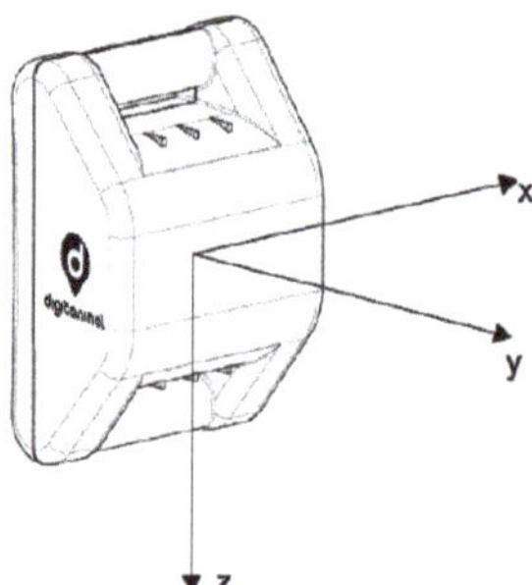

Figure 1. Monitoring device with 3-D accelerometer and GPS sensors. Coordinate axes represent movement directions tracked by the accelerometer.

Figure 2. A Fleckvieh breed cow wearing the monitoring device, attached with a neckband.

A primary goal for this product is to avoid the need of frequent maintenance tasks (such as replacing the battery or the SD card), that would interfere with normal farm routines. In consequence, the sampling rate of GPS data is set to 5 min, that is, a single message is sent at the end of each 5-min interval. In spite of this initial specification, it is also possible that the GPS signal is lost in certain shadow regions on the farm, or that transmitted data do not successfully arrive at the server, due to propagation issues, network problems or other causes. For this reason, the system must be prepared to deal with missing data in location records.

Next, we describe the procedure for data acquisition, the proposed method for processing accelerometer and GPS data, as well as the approach for identification of animal behaviours using ML algorithms.

2.3. Accelerometer and GPS Data Collection

Figure 3 shows an example of 3 raw signals produced by one of the accelerometers for 220 s. Each individual signal is the result of monitoring acceleration changes along a single axis. The blue signal corresponds to the X-axis, the orange signal represents oscillations along the Y-axis and the green signal stands for acceleration changes along the Z-axis. Raw signals recorded by each accelerometer are inputs for the feature extraction step in data analysis (see Section 2.6 for further details).

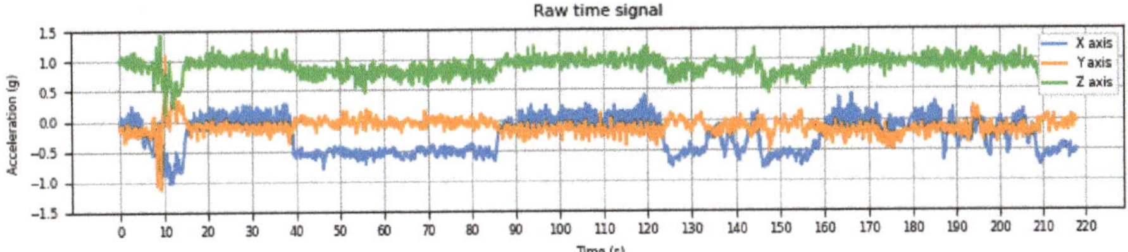

Figure 3. Raw signals recorded by the 3-D accelerometer for each coordinate axis.

While triaxial accelerometers in collars store signal data locally, the GPS sensor monitors the location of the animal and periodically transfer these data to a central server each 5-min interval, containing the following attributes:

- `id`: Unique identifier for the monitored cow.
- `timestamp`: A timestamp value in the format YYYY-MM-DD HH:MM:SS.
- `longitude`: Longitude coordinate for the current animal position.
- `latitude`: Latitude coordinate for the current animal position.

2.4. Behavioural Observations

One of the main limitations to validate the automated detection of behavioural patterns in ruminants is the lack of a validation database, providing examples of specific behaviours and their associated patterns captured by the accelerometer. In our experiments, this is addressed by taking video recordings of a sample of animals on pasture fields, wearing the monitoring devices described above. Hence, the main goal of these recordings is to match each logged signal with its corresponding recorded behaviour. Furthermore, video recordings also allow double-checking the correct alignment between signals and video timestamps, a problem addressed via a specific methodology described in Section 2.5.

A team of 10 scientists were trained to supervise the recorded scene, annotating the timestamp and observed behaviours. Each scientist was responsible for tracking a single animal, annotating behavioural patterns over a 5 h session. Operators encoded activities on log files using a shared predefined nomenclature, described in the ethogram shown in Table 1. Annotated behaviours include: *grazing, ruminating, steady standing, laying* and *others*. The last category encompasses less frequent behaviours, such as running, scratching, drinking, calf nursing, etc.). The duration of individual behaviours was quite variable, ranging from few seconds (e.g., for *scratching*) to several minutes, in some cases (e.g., ~16 min for a single instance of *ruminating* or ~11.5 min for one instance of *laying*). Some behaviours required immediate reactions from human observers, therefore reducing the length of video recordings to identify them (e.g., in *running* operators must relocate to follow the animals).

Table 1. Behavioural ethogram describing frequent activities observed by operators in the experiment, ordered by total duration of recorded video evidence.

Behaviour	Code	Total Durat. (sec.)	Description
Grazing	GRA	12,056	Regularly lowering and raising its head to eat pasture, while standing or walking slowly
Ruminating	RUM	4429	Ruminating previously eaten food, while standing or laying
Laying	LAY	1940	Laying on the ground without performing any other relevant activity
Steady standing	STA	1011	Standing almost still without performing any other relevant activity
Walking	OTH#WALK	509	Walking at normal pace with calm steps
Licking	OTH#LICK	414	Noticeably turning its neck to lick itself
Scratching	OTH#SCRA	159	Raising one leg to scratch its head or body (also specified if scratching against a tree)
Running	OTH#RUN	94	Moving at high pace with quick steps
Drinking	OTH#DRI	93	Lowering its head to drink water
Calf nursing	OTH#NUR	30	Steady while nursing a calf

Video recordings and observational log files derived from them were subsequently reviewed by independent supervisors, to ensure consistency of activity labelling between files. After an initial screening, 3 animals were selected for this analysis, as their activity logs provided the most accurate registries. A total of 238 unique behavioural patterns were identified from these recordings. These patterns, together with their associated signals recorded by accelerometers, constitute the gold standard for this study.

Figure 4 reports the percentage of samples corresponding to each individual activity, over the total number of logged behaviours. Despite other activities were also annotated by scientists in observational logs, only the most frequent ones are considered in this classification analysis.

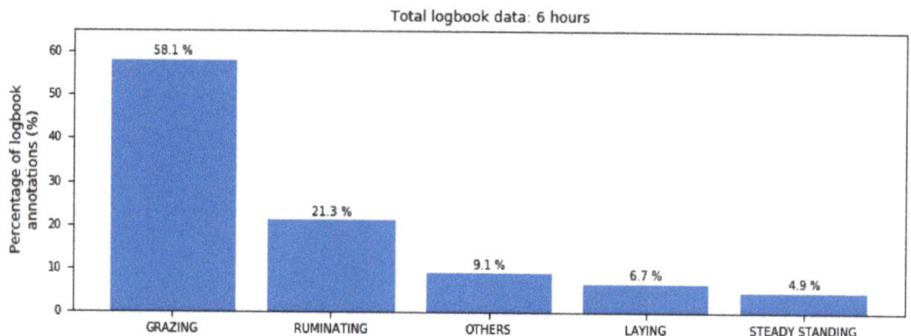

Figure 4. Proportion of observed behaviours of cows on the field, annotated by scientists.

2.5. Alignment between Accelerometer Data and Observations

To facilitate the matching of signal and video records, a special procedure was followed to create a distinctive signature that clearly marked the start and the end of the experimental scope:

- Before the sensor collar is installed, the operator swings the collar for 1 min, so that a unique oscillation pattern is produced by the accelerometer on the 3 axes.
- When the experiment is finished and just after the collar is taken off from the animal, the operator swings again the collar for 1 min to reproduce the same unique pattern as in the starting point.

This pattern marking the start and the end of the experimental observation time cannot be reproduced naturally by cows while wearing the collar. Hence, this signature signal can be employed to fine tune the alignment between internal clocks in video cameras and the accelerometer clock.

2.6. Processing Accelerometer Data

The procedure for accelerometer raw data processing consists of different steps, which are depicted in Figure 5.

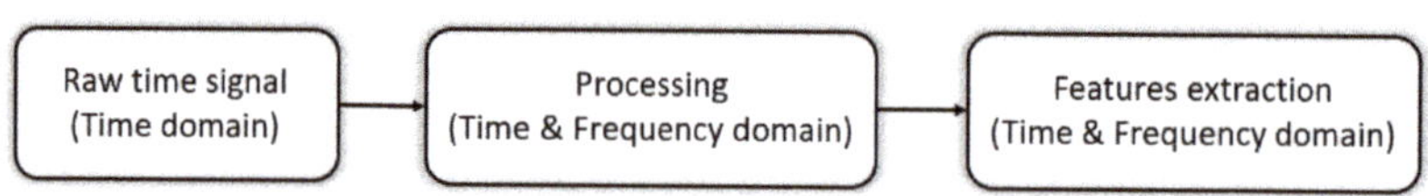

Figure 5. Overview of the proposed procedure for accelerometer data processing.

In the first step, time signals are divided in 10 s consecutive, non-overlapping *time intervals* or *time windows*. As a result, each interval contains 100 consecutive samples, since 10 samples per second are generated using a 10 Hz sampling rate. From now on, we refer to the time intervals obtained from this process as $x_i(t)$, where i denotes the interval index. Figure 6a illustrates the result from this step, dividing the original signal (in this case, for the X-axis) in 4 different time windows, spanning adjacent intervals of 10 s. It must be remarked that, unlike many previous studies of this kind of data (see [21] for a comprehensive survey), we process the signals from each of the 3 accelerometer axes (X, Y, Z) separately. Previous studies analysing animal behaviour in wild habitats [28–30] suggest that this alternative data processing method can provide advantages for accurate detection of behavioural patterns, especially dynamic ones.

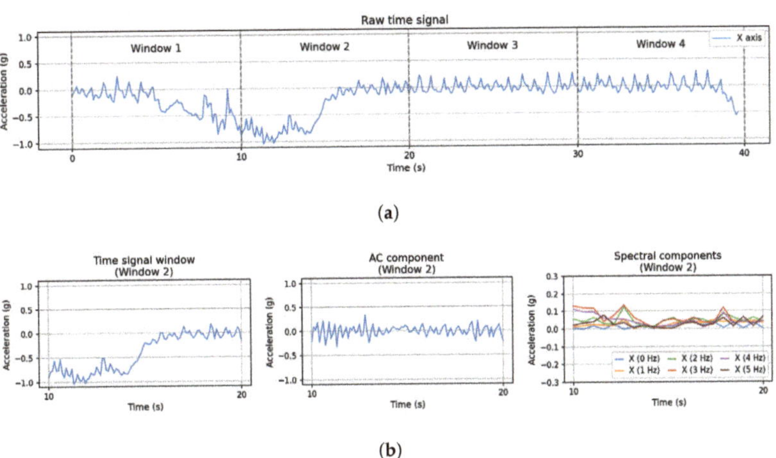

(a)

(b)

Figure 6. Time windows extracted from original signal generated by the accelerometer and their corresponding components. (**a**) Raw time signal divided in 4 windows. (**b**) Time-domain signal and components extracted from *Window 2* in Figure 6a.

The second step in Figure 5 involves processing the time interval to obtain their AC (alternating current) component, along with its representation in the frequency domain, which are illustrated in Figure 6b. In this study, the AC component is extracted using a method that differs from several previous research works, where digital filters are applied to remove high-frequency noise and eliminate the DC component [31]. Hämäläinen et al. [9] show that problems may arise when the orientation of the sensor changes (e.g., due to sudden shakes) during data acquisition. Thus, they propose a simple alternative method to avoid these problems, calculating instead the "jerk" (acceleration change) between two consecutive samples. This approach renders orientation-independent features, avoiding the need to estimate the actual acceleration accurately.

As a result, the AC component, identified as $x_{i-AC}(t)$, is computed as the regular difference between two consecutive time windows. Finally, the frequency-domain representation of the AC component is calculated in the last stage of this pipeline, as the basis for subsequent spectral processing. To achieve this, we compute the Fast Fourier Transform (FFT) [32,33] of the AC component, using a 1-s window size, a Hanning window type [34] and 50% overlapping between consecutive time windows. Figure 7 summarizes the pipeline for processing the raw signal from the accelerometer in the time and frequency domains.

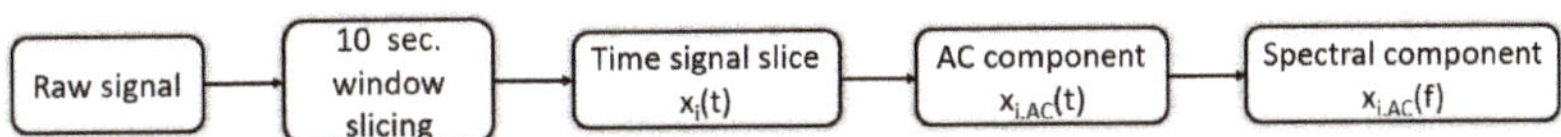

Figure 7. Pipeline performed within the accelerometer signal processing stage.

Calculating the FFT of the AC component renders a spectrogram in a frequency range from 0 Hz to 5 Hz (according to Shannon's theorem), with a 1 Hz resolution. For this purpose, the Python SciPy signal processing toolbox (https://docs.scipy.org/doc/, accessed on 25 February 2022) is used. This will be denoted as the *spectral component*, identified as $X_{i-AC}(f)$. The complete spectral component comprises 6 individual frequency components, corresponding to each of the 1-Hz resolution bands. The panels in the lower part of Figure 6 represent the three elements calculated in this second step. The lower-left panel represents `slice` 2 from the original signal in the top panel. The AC component extracted from the original signal in time `interval` 2 is shown in the lower-centre panel. Finally, the time-domain representation of the spectral component for each frequency band is depicted in the lower-right panel.

The three elements obtained from the preprocessing step (time window, AC component and spectral component) are inputs for the extraction of different features, performed in the third step of Figure 5. Finally, these features are used for automated behaviour detection with ML algorithms. When the data inputs are the time window and the AC component, features obtained in this way correspond to the time domain. In turn, when the input is the spectral component, resulting features correspond to the spectral domain. Table 2 describes the list of features extracted from each data input. Details about their computation are explained below.

Table 2. Features extracted from each input generated after preprocessing the accelerometer signal.

Data Input	Feature	Description
Raw accelerometer axis (X,Y,Z)	Mean	Average value of signal
	Max	Maximum value of signal
	Min	Minimum value of signal
	Q5	5th percentile of signal values
	Q95	95th percentile of signal values
AC component (time domain) $x_{i-AC}(t)$	Mean	Average value
	STD	Standard deviation of values distribution
	Kurtosis	Kurtosis of values distribution
	Skewness	Skewness of values distribution
	Max	Maximum value
	Q5	5th percentile of values
	Q95	95th percentile of values
AC component (freq. domain) $x_{i-AC}(f)$	RMS	Root mean square spectral density
	STD	Standard deviation spectral density
	Min	Minimum value spectral density
	Max	Maximum value spectral density

2.6.1. Time Domain Features

Table 2 shows the features obtained for each type of data input from accelerometer signals. Using the raw accelerometer axis (X, Y, Z) input, the following features are obtained (see first row in Table 2): mean, maximum, minimum, 5th percentile and 95th percentile. A total of 15 time features are extracted, 5 features per each accelerometer axis (X, Y, Z). Likewise, using the AC component representation in the time domain as a data input (see second row in Table 2) the mean, maximum, standard deviation (STD), skewness, kurtosis, 5th percentile and 95th percentile features are computed. A total of 21 features are extracted using this data source, that is, 7 features per each accelerometer axis.

2.6.2. Frequency Domain Features

The spectrogram represents how acceleration levels progress for each frequency and time instant. According to the processing parameters previously defined, the spectrogram of the AC component comprises 6 different frequency bands, at 0 Hz, 1 Hz, 2 Hz, 3 Hz, 4 Hz and 5 Hz. Since the spectrogram represents information in 3 dimensions simultaneously (time, frequency and amplitude), it cannot be used directly for feature extraction. To achieve this, the spectrogram is decomposed of frequency, obtaining 6 spectral series that progress along time. Figure 6b shows the resulting spectral series for a certain spectrogram.

Once these spectral series are computed for the AC component, they can be used for feature extraction. The third row of Table 2 presents the features that are extracted for each spectral series: root mean square (RMS) value, standard deviation (STD), minimum and maximum value. A total of 72 spectral features are obtained, that is, 4 features for 6 spectral series, resulting in 24 features per axis.

2.7. Processing GPS Data

Location records transferred by GPS sensors to the server are stored in CSV files. There is one file for each tracking collar attached to a cow. To start off, data in all CSV files are coalesced into a single file and records are ordered according to their timestamp value. Duplicate entries that might have been incorrectly recorded or transferred are also elided in this first step. After this, we must clean the dataset filtering incorrect location entries. This may be caused, for instance, due to inaccurate location detection by the sensor in areas of the farm where GPS coverage is insufficient. To attain this, the daily average values for latitude and longitude coordinates are calculated for each monitored farm. Then, any location registry further than 1 km from the average position is eliminated. This figure is well above the average value of the overall area of both livestock farms under analysis (about 50 hectares, in both cases), to filter out clear data registration errors. Once GPS data are completely prepared, the relevant attributes (id, timestamp, longitude and latitude) can be used.

2.8. Machine Learning Algorithms

Features extracted from accelerometer signals are used to train a supervised ML algorithm for behavioural pattern classification, whereas GPS location data is analysed through an unsupervised machine learning method, to detect anomalous activity patterns. Details on these analyses using machine learning models are provided below.

2.8.1. Behaviour Classification Based on Accelerometer Data

Classification of behavioural data from field experiments is performed using the random forests (RF) algorithm [35]. This tool has been selected due to the high number of descriptive features available and the capacity of RF to automatically identify important features to detect each individual behaviour.

The complete set of 238 behavioural samples, including the 108 features extracted from accelerometer signal processing, is split into 5 different folds, following a stratified random sampling approach [36]. Then, a multi-class RF classification algorithm is trained for every fold using 75% of data and the remaining 25% for testing. Video recordings are combined with these input data to produce a validation database. The target categories for the classification task are *grazing*, *ruminating*, *laying* and *steady standing*, while patterns included in category *others* are filtered out, since not enough samples for each individual behaviour in this group are available to identify them accurately. The hyperparameters selected for RF are the following: we use information gain (*entropy*) to measure the quality of splits; the minimum number of samples required to split an internal node is set to 20; we select using out-of-bag samples to estimate the generalization score and we build 100 trees for each forest. Then, the importance of each feature to identify individual activities is obtained in every trained model. Finally, all feature importance values per activity are averaged over the 5 folds to report the final results.

2.8.2. Automated Detection of Herd Scattering Using GPS

The main objective in our analysis of GPS location data is to automatically identify groups of monitored cows within the limits of the farm and sudden changes in the scattering of a given group. Rapid modifications in animal dispersion within a certain group may indicate the occurrence of anomalous events that must be reported to farm operators and managers.

In the first place, data for each livestock farm is identified and analysed separately. Given a location dataset describing the situation in a farm, a centre location for every group of animals must be identified. Then, the dispersion of animals around their corresponding group centre must be estimated and tracked, to account for abrupt alterations. We use the *Euclidean* distance (L_2 norm) [37] to measure the separation between any two cows, and generate the distance matrix for all animals in the farm.

Identifying the groups and their representative location leads to an unsupervised learning task. Among the different alternative algorithms that can be applied, partitioning clustering algorithms [38,39] provide a convenient solution, as the total number of location points in each farm is not large. Although the k-means algorithm [37] is a popular solution for this kind of problems, we found that, in many cases, it does not provide representative locations for each group of animals in this application. The main cause behind this problem is the frequent presence of outliers in animal groups, that is, cows that are well-separated from the rest of members of the same cluster, thus pulling the location of the k-means centre for that group.

Due to this, a more robust clustering algorithm, insensitive to the presence of outliers in a cluster, must be employed. The k-medoids algorithm [38,39] forces the selection of one of the actual location points in a certain cluster to act as the centre for that group. We found that cluster identification following this approach is much more reliable and better matches extant information from farm workers and managers about the number and location of herds. The appropriate number of clusters for each farm is selected by calculating the within-clusters sum of squares (WCSS) for different values of k, evaluating the cohesion of clusters in each case. Then, a scree plot of WCSS against k is generated and we choose the value for k using the elbow method [37,40]. Alternatively, farm managers could override this choice of k by entering extant information about the estimated number of herds.

Once the number of herds and a reference location for each group are found, we turn to the problem of estimating the scattering of animals in a given group from their reference point. In this case, we opt for choosing the farthest animal assigned to the cluster as the delimiter of the maximum scattering range for that group, as shown in Figure 8. Since we use the Euclidean distance to measure proximity between cows, we effectively establish a circular region of radius r equal to the distance from the reference location in the group to the farthest member of that herd.

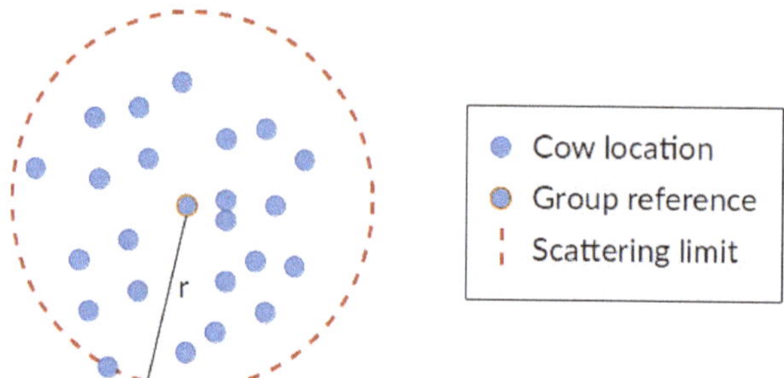

Figure 8. Detection of the scattering limits for a herd. The farthest animal assigned to that group determines the scattering radius r.

This procedure is periodically repeated for every new sample of locations sent by GPS sensors from the farm. For each new sample, the total number of groups, the reference location and the estimated value of r for each group are computed and stored.

3. Results

In this section, we summarize the results from the field experiments to identify animal behavioural patterns using the features extracted from accelerometer signals represented in the time domain and the frequency domain, and GPS location data.

3.1. Relevant Classification Features

Table 3 shows the rank and feature importance values [35,41] (mean accumulation of impurity decrease within each tree, known as Mean Decrease in Impurity or MDI), averaged from the five RF models trained with time domain and frequency domain features obtained from accelerometer signals. For the sake of conciseness, here we only report the top-five features identified for each activity. Graphs displaying the complete set of features for each activity and their associated importance values are presented in Appendix A.

Table 3. Identified animal behaviours, top-5 features used by trained RF models to classify them and their importance (MDI), averaged over the 5 RF models.

Behaviour	Rank	Feature	Avg. MDI
Grazing	1	Z_AC_Q5	0.06798
	2	Z_AC_STD	0.06274
	3	Z_2Hz_RMS	0.06189
	4	Z_AC_Q95	0.06115
	5	Z_1Hz_RMS	0.06007
Laying	1	Y_Q95	0.06129
	2	Z_AC_Q5	0.05319
	3	Y_MAX	0.04246
	4	Y_AC_Q95	0.03975
	5	Y_MEAN	0.03601
Ruminating	1	Z_AC_Q5	0.06787
	2	Z_AC_Q95	0.04863
	3	X_Q5	0.04849
	4	Z_AC_STD	0.03537
	5	Y_Q_95	0.03486
Steady standing	1	X_1Hz_MIN	0.06075
	2	X_5Hz_MIN	0.04870
	3	X_3Hz_MIN	0.04429
	4	X_2Hz_MIN	0.04194
	5	X_AC_KURT	0.04022

We can spot several interesting traits regarding the most important features used by the RF algorithm to identify each behaviour. In the case of activity *grazing*, the most important features to detect this pattern are related to movement along the Z-axis. This is consistent with the observed movements, involving vertical necks displacements as the cow lows down its head to eat pasture and raise it up to continue chewing. Moreover, we also notice that two out of the top five features come from the frequency domain representation of the AC component. This confirms the usefulness of the spectral analysis of accelerometer signals for animal behaviour recognition. Another salient example of the key role of spectral components in activity detection is the case of *steady standing*. four out of the top five features come from the AC component processing in the frequency domain.

3.2. Classification Performance Metrics

Table 4 presents several performance metrics computed for the RF classification model, namely, accuracy, recall and AUC [42,43]. In general, classification accuracy attained by this algorithm was good for all behavioural patterns, with the highest score for *grazing* and the lowest for *ruminating*. However, recall metrics drop for activities with fewer samples in the dataset, such as *laying* or *steady standing*. Since we are developing a general detection procedure, that targets a variety of activities, the algorithm still presents limitations detecting all instances from under-represented categories, with fewer samples in the dataset.

Table 4. Performance metrics for the RF classification model. All metrics are average values over the 5 folds.

Behaviour	Accuracy	Recall	AUC
Grazing	0.93	0.945	0.974
Laying	0.907	0.611	0.894
Ruminating	0.881	0.893	0.967
Steady standing	0.922	0.58	0.912

4. Discussion

Previous research has shown the high interest of animal behaviour identification on farms [4–6]. Therefore, this work aims to propose a general procedure to recognize multiple activities based on accelerometer and GPS data. On top of this, previous studies has been restricted, so far, to the use of one of these two types of data sources for tracking animal behaviour, with only recent exceptions [16,18,44]. In this work, we explore the potential of combining data from both types of sensors to achieve a more advanced activity pattern identification.

4.1. Classification Model from Accelerometer Data

As described in Section 2.6, a separate analysis of accelerometer signals over each axis (X, Y, Z) along with the use of jerk filters and spectrograms to compute relevant features is proposed. Previous studies have shown [30] that the combination of this data processing method with classification trees ML algorithms (like the RF ensemble learning method applied in this work) can render good results for identification of animal behavioural patterns.

According to the feature importance metrics reported by the assessment of the RF classification model, shown in Table 3, time-domain features play an important role in the classification of certain behaviours such as *laying* or *ruminating*, where animals tend to remain relatively still. In our data processing method, this is linked to the absence of sudden shakes ("jerk" or "AC component" in this study), which turn the AC signal quite stable over time. In turn, frequency-domain features are also relevant for detection of dynamic behaviours such as *grazing*, or patterns with sudden activity peaks in any axis such as *steady standing*, better captured by our definition of AC component.

In this regard, it is of key importance that the internal clock used by the accelerometer marks precise regular intervals between samples. Otherwise, digital signal processing techniques to obtain the spectrogram of the AC component for different frequency bands will not be applicable, in case that sampling intervals present irregularities. However, results from this field study with commercial, low-cost equipment are limited by the accuracy of captured signals (that can be subject to sensor failures, battery drain due to climatic conditions and other adverse situations) and the ability to precisely correlate behaviours observed by human operators and registered on video recordings with the corresponding patterns captured by sensor devices. For example, as shown in Section 3.2, *grazing* was the most frequent activity pattern detected, which is in line with results from previous studies [16]. Possibly due to this high number of available samples identification of most frequent behaviours is more accurate than for other less represented patterns, according to performance metrics in Table 4.

On top of this, 9.1% of behaviours included in the study were labelled as *other*. However, detailed annotations were taken by operators regarding actions jointly accounted for in the omnibus *other* category. These include, among others, cows feeding younger calves, running cows or animals licking themselves. Some of these behaviours were correctly logged by human operators but not enough signal samples were obtained to generalize their detection to other cases. As a result, this study confirms that the proposed methodology could be generalized to other behavioural patterns, as soon as new data becomes available. An important implication in this sense is the absence of publicly available online reference datasets, registering data captured by sensors and their related activity patterns.

Therefore, addressing this lack of validation databases could be a very useful contribution in further research works.

4.2. Potential of GPS Data for Activity Detection Based on Herd Scattering

Figure 9 presents the result of a preliminary algorithm for automated detection of herds and within-herd spread, based on GPS data, corresponding to the livestock farm in Avila. The red dots depict the location of animals tracked by the GPS sensors. The map shows two separate herds, represented by the algorithm via the identified k-medoids for each group (black point). Then, the algorithm calculates the scattering of animals around the cow selected as the representative centre for that group. The algorithm could also detect changes in the radii calculated for each herd, following a basic procedure based on change point detection [45].

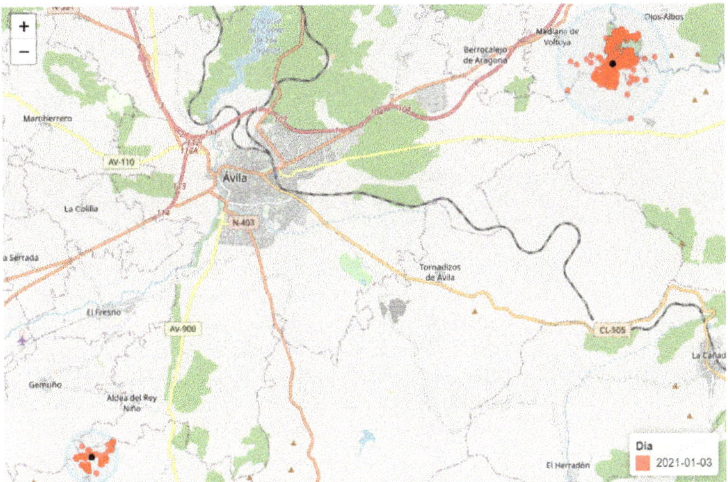

Figure 9. Example of automated detection of location and scattering of two different herds in one of the farms, near Ávila (Spain).

As shown in previous research [27], accelerometer and GPS data can be combined to detect anomalous events, such as unbalanced use of pasture land or disease transmission, among others. Table 5 describe some potential cases in which both components could be combined to eventually provide farmers with the proper tool for an early detection.

Table 5. Examples of anomalous/interesting activities and how analysis of accelerometer and GPS data can be applied to detect them.

Activity of Interest	Accelerometer Data	GPS Data
Predator attacks	Vertical axis with no movement	Quick displacement to alternative location; possible successive relocations
Pasture land use	Detection of grazing behaviour	Mapping of areas under use (presence longer than a certain time threshold)
Disease transmission	Detection of steady-standing or laying behaviours	Erratic movements; very slow transitions to alternative areas

In the case of predator attacks, cows are vigilant and in state of alert. This natural response to a feasible external threat translates into the detection of noticeable periods of time in which cows are not moving their heads (grazing and ruminating activities are stopped). Likewise, herds may move away to an alternative location quite rapidly to mitigate the detected risk [46,47].

As for the use of pasture land, accelerometer data obtained from monitored cows would inform about grazing activities. For its part, GPS data would provide clear indication of the areas on which such activity occurs. Despite not being an anomalous activity, this information may assist farmers to better manage resources and costs, or even reduce pasture land required which is considered as a top priority demand [16,48,49].

Finally, the lack of vertical or horizontal movements in cow necks, an abnormal stance and gait, an unusual resting behaviour or too slow (or even non-existent) displacements detected via GPS data could offer an complementary perspective to detect disease transmission, whose modelling process would also require health scoring for each monitored animal. Early disease detection could prevent severe cases and facilitate immediate application of treatment measures, reducing productivity loss [50,51].

This procedure is also compatible with the automated detection of the number of herds within the farm limits or with manual configuration of the number of herds to be tracked, introduced by human users. Additionally, the proposed method can also be integrated in existing tools for animal monitoring on farms. Operators can configure the appropriate parameters to raise notifications, based on their own management experience with animals. As additional data are tagged and become available, the tool can be linked to the detection of particular patterns of interest (predator attacks, parturition, etc.). Again, the absence of publicly available datasets that can serve as a benchmark for this type of automated tools in animal behaviour recognition calls for filling this gap in further research.

Tracking the evolution of these indicators over time, it would be possible to identify two types of interesting changes:

- As herds move around the terrain, the reference animal representing that herd will register such displacement. Therefore, at the end of the day farm operators and managers can review the trajectory followed by different herds, leading to a more precise estimation of pasture consumption.
- Changes in the scattering radius r calculated for each herd may indicate interesting behavioural patterns happening to that group of animals. In particular, a sudden increase in the value of r may indicate among other possibilities) the attack of potential predators or other threats.

Another interesting line for further research is exploring the formal combination of activity records from accelerometers and GPS, for instance, through information fusion techniques [44]. The validity of this approach has already been tested for the case of outlier detection. Moreover, the only previous work that combines GPS and accelerometer datasets [16] is tailored to detecting a single behaviour (grazing) and just employ the GPS coordinates to locate every behaviour interval. However, there is a clear potential in the simultaneous utilization of features extracted from both accelerometer and GPS location data analysis to improve the recognition of animal activity patterns on farms.

5. Conclusions

In this work, we present a new method for automated classification of animal behavioural patterns, through the analysis of activity data registered by a triaxial accelerometer and a GPS sensor. A unique aspect introduced in this approach is the application of techniques for spectral analysis of accelerometer signals in the frequency domain. Descriptive features derived from the spectrogram of these signals play an important role in detecting certain patterns of interest, such as grazing (the most frequent activity observed) or steady standing. Likewise, this method is not restricted to a particular behavioural pattern and it can be readily generalized to any behaviour of interest, provided that labelled activity data is available. Furthermore, the analysis of GPS data recording animals locations through unsupervised machine learning algorithms enables the detection of groups of animals and their dispersion, which can be regularly tracked and reported to users. Jointly, results from these two analyses can build a more complete picture of activity logs and facilitate decision-makers the necessary information to oversee pasture consumption, develop actions in response to anomalous events and improve animal welfare in their farms.

Author Contributions: Conceptualization, R.Y., B.V. and F.O.; methodology, J.C., R.Y., B.V. and F.O.; software, R.Y. and B.V.; validation, M.J.A., E.L.C. and F.O.; formal analysis, F.O.; investigation, J.C., R.Y., B.V. and F.O.; resources, J.N.-G. and F.O.; data curation, J.C., R.Y., B.V. and J.N.-G.; writing—original draft preparation, J.C. and F.O.; writing—review and editing, M.J.A. and E.L.C.; visualization, R.Y. and B.V.; supervision, F.O.; project administration, F.O.; funding acquisition, F.O. All authors have read and agreed to the published version of the manuscript.

Funding: This research was supported by the Spanish Ministry of Agriculture, Fisheries and Food under project GELOB (ref. 20190020007471), the Spanish Ministry of Economy, Industry and Competitivity under project Advances in Risks Management for Security (ref. AEI/MTM2017-86875-C3-1-R) and the Spanish Ministry of Science and Innovation under grant MODAS-IN (ref. RTI2018-094269-B-I00).

Institutional Review Board Statement: Ethical review and approval were waived for this study by the Spanish Ministry of Agriculture, Fisheries and Food, under project GELOB (ref. 20190020007471).

Informed Consent Statement: Not applicable.

Acknowledgments: We thank Rubén Blanco, Diego Varona and Ignacio Gómez Maqueda from Digitanimal, for providing the sensors used in this study and facilitating access to servers in which GPS location data are stored.

Conflicts of Interest: The authors declare no conflict of interest.

Abbreviations

The following abbreviations are used in this manuscript:

AC	Alternating Current
AUC	Area Under the Curve
CSV	Comma Separated Values
DC	Direct Current
FFT	Fast Fourier Transform
GPS	Global Positioning System
IoT	Internet of Things
MDI	Mean Decrease in Impurity
MEMS	Micro Electro Mechanical System
ML	Machine Learning
PCA	Principal Components Analysis
RF	Random Forests
RMS	Root Mean Square
ROC	Receiver Operating Characteristic
SD	Secure Digital
STD	Standard Deviation
WCSS	Within-Cluster Sum of Squares

Appendix A

Next, we provide several graphs presenting the importance values of each feature extracted from the analysis of signals captured by triaxial accelerometers, for identification of the main behavioural patterns considered in this study, as reported by the RF algorithm.

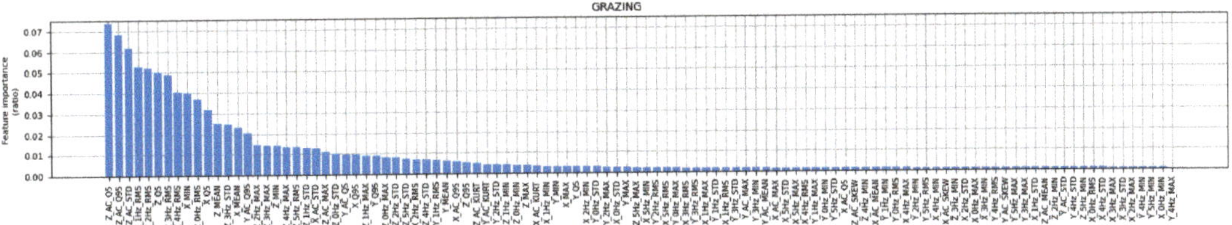

Figure A1. Feature importance in detection of grazing behaviour.

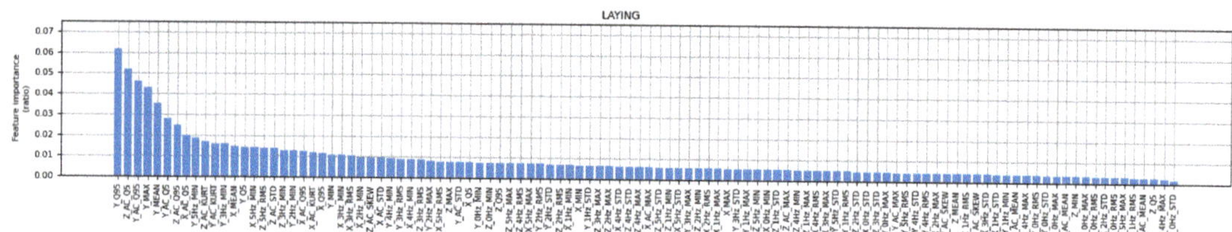

Figure A2. Feature importance in detection of lying behaviour.

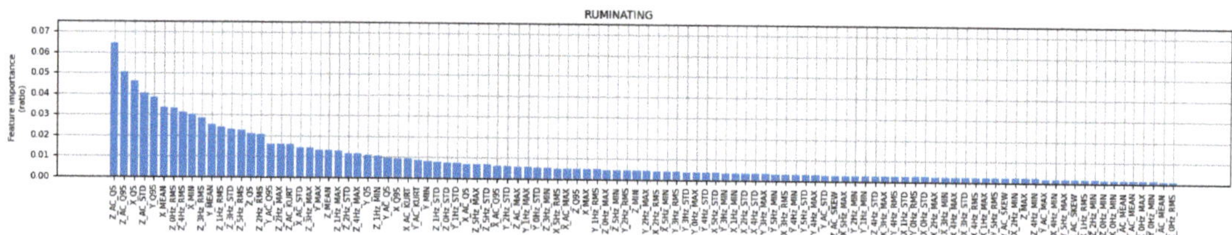

Figure A3. Feature importance in detection of ruminating behaviour.

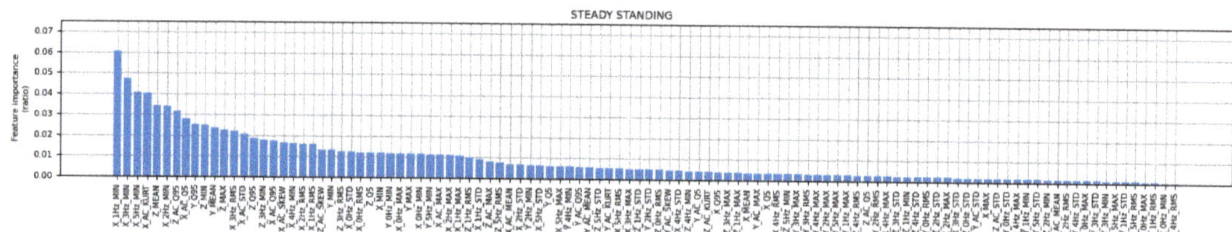

Figure A4. Feature importance in detection of steady standing behaviour.

References

1. Wolf, C.; Tonsor, G.; McKendree, M.; Thomson, D.; Swanson, J. Public and farmer perceptions of dairy cattle welfare in the United States. *J. Dairy Sci.* **2016**, *99*, 5892–5903. [CrossRef] [PubMed]
2. European Council. Directive (EC) 98/58/EC of the European Council of 20 July 1998, concerning the protection of animals kept for farming purposes. *Off. J. L221* **1998**, *41*, 23–27. Available online: http://data.europa.eu/eli/dir/1998/58/oj (accessed on 25 February 2022).
3. European Commission. Commission Regulation (EC) No 889/2008 of 5 September 2008 laying down detailed rules for the implementation of Council Regulation (EC) No 834/2007, on organic production and labelling of organic products with regard to organic production, labelling and control. *Off. J. L250* **2008**, *51*, 1–84. Available online: http://data.europa.eu/eli/reg/2008/889/oj (accessed on 25 February 2022).
4. Kwong, K.H.; Wu, T.T.; Goh, H.G.; Sasloglou, K.; Stephen, B.; Glover, I.; Shen, C.; Du, W.; Michie, C.; Andonovic, I. Practical considerations for wireless sensor networks in cattle monitoring applications. *Comput. Electron. Agric.* **2012**, *81*, 33–44. [CrossRef]
5. Barriuso, A.L.; Villarrubia González, G.; De Paz, J.F.; Lozano, Á.; Bajo, J. Combination of Multi-Agent Systems and Wireless Sensor Networks for the Monitoring of Cattle. *Sensors* **2018**, *18*, 108. [CrossRef] [PubMed]
6. Michie, C.; Andonovic, I.; Davison, C.; Hamilton, A.; Tachtatzis, C.; Jonsson, N.; Duthie, C.A.; Bowen, J.; Gilroy, M. The Internet of Things enhancing animal welfare and farm operational efficiency. *J. Dairy Res.* **2020**, *87*, 20–27. [CrossRef]
7. Lee, C.H.; Chen, S.H.; Jiang, B.C.; Sun, T.L. Estimating Postural Stability Using Improved Permutation Entropy via TUG Accelerometer Data for Community-Dwelling Elderly People. *Entropy* **2020**, *22*, 1097. [CrossRef] [PubMed]
8. Mizell, D. Using gravity to estimate accelerometer orientation. In Proceedings of the Seventh IEEE International Symposium on Wearable Computers, White Plains, NY, USA, 21–23 October 2003; p. 252. [CrossRef]
9. Hamäläinen, W.; Järvinen, M.; Martiskainen, P.; Mononen, J. Jerk-based feature extraction for robust activity recognition from acceleration data. In Proceedings of the 2011 11th International Conference on Intelligent Systems Design and Applications, Cordoba, Spain, 22–24 November 2011; pp. 831–836. [CrossRef]
10. Robert, B.; White, B.; Renter, D.; Larson, R. Evaluation of three-dimensional accelerometers to monitor and classify behavior patterns in cattle. *Comput. Electron. Agric.* **2009**, *67*, 80–84. [CrossRef]
11. Nielsen, L.R.; Pedersen, A.R.; Herskin, M.S.; Munksgaard, L. Quantifying walking and standing behaviour of dairy cows using a moving average based on output from an accelerometer. *Appl. Anim. Behav. Sci.* **2010**, *127*, 12–19. [CrossRef]

12. Vázquez Diosdado, J.A.; Barker, Z.E.; Hodges, H.R.; Amory, J.R.; Croft, D.P.; Bell, N.J.; Codling, E.A. Classification of behaviour in housed dairy cows using an accelerometer-based activity monitoring system. *Anim. Biotelemetry* **2015**, *3*, 1–14. [CrossRef]

13. Mattachini, G.; Riva, E.; Perazzolo, F.; Naldi, E.; Provolo, G. Monitoring feeding behaviour of dairy cows using accelerometers. *J. Agric. Eng.* **2016**, *47*, 54–58. [CrossRef]

14. Shahriar, M.S.; Smith, D.; Rahman, A.; Freeman, M.; Hills, J.; Rawnsley, R.; Henry, D.; Bishop-Hurley, G. Detecting heat events in dairy cows using accelerometers and unsupervised learning. *Comput. Electron. Agric.* **2016**, *128*, 20–26. [CrossRef]

15. Arablouei, R.; Currie, L.; Kusy, B.; Ingham, A.; Greenwood, P.L.; Bishop-Hurley, G. In-situ classification of cattle behavior using accelerometry data. *Comput. Electron. Agric.* **2021**, *183*, 106045. [CrossRef]

16. Brennan, J.; Johnson, P.; Olson, K. Classifying season long livestock grazing behavior with the use of a low-cost GPS and accelerometer. *Comput. Electron. Agric.* **2021**, *181*, 105957. [CrossRef]

17. Dutta, R.; Smith, D.; Rawnsley, R.; Bishop-Hurley, G.; Hills, J.; Timms, G.; Henry, D. Dynamic cattle behavioural classification using supervised ensemble classifiers. *Comput. Electron. Agric.* **2015**, *111*, 18–28. [CrossRef]

18. Fogarty, E.S.; Swain, D.L.; Cronin, G.M.; Moraes, L.E.; Bailey, D.W.; Trotter, M. Developing a Simulated Online Model That Integrates GNSS, Accelerometer and Weather Data to Detect Parturition Events in Grazing Sheep: A Machine Learning Approach. *Animals* **2021**, *11*, 303. [CrossRef]

19. Arcidiacono, C.; Porto, S.; Mancino, M.; Cascone, G. Development of a threshold-based classifier for real-time recognition of cow feeding and standing behavioural activities from accelerometer data. *Comput. Electron. Agric.* **2017**, *134*, 124–134. [CrossRef]

20. Busch, P.; Ewald, H.; Stüpmann, F. Determination of standing-time of dairy cows using 3D-accelerometer data from collars. In Proceedings of the 2017 Eleventh International Conference on Sensing Technology (ICST), Sydney, NSW, Australia, 4–6 December 2017; pp. 1–4. [CrossRef]

21. Riaboff, L.; Shalloo, L.; Smeaton, A.; Couvreur, S.; Madouasse, A.; Keane, M. Predicting livestock behaviour using accelerometers: A systematic review of processing techniques for ruminant behaviour prediction from raw accelerometer data. *Comput. Electron. Agric.* **2022**, *192*, 106610. [CrossRef]

22. Smith, D.; Rahman, A.; Bishop-Hurley, G.J.; Hills, J.; Shahriar, S.; Henry, D.; Rawnsley, R. Behavior classification of cows fitted with motion collars: Decomposing multi-class classification into a set of binary problems. *Comput. Electron. Agric.* **2016**, *131*, 40–50. [CrossRef]

23. Riaboff, L.; Poggi, S.; Madouasse, A.; Couvreur, S.; Aubin, S.; Bédère, N.; Goumand, E.; Chauvin, A.; Plantier, G. Development of a methodological framework for a robust prediction of the main behaviours of dairy cows using a combination of machine learning algorithms on accelerometer data. *Comput. Electron. Agric.* **2020**, *169*, 105179. [CrossRef]

24. Kamminga, J.W.; Le, D.V.; Meijers, J.P.; Bisby, H.; Meratnia, N.; Havinga, P.J. Robust sensor-orientation-independent feature selection for animal activity recognition on collar tags. In *Proceedings of the ACM on Interactive, Mobile, Wearable and Ubiquitous Technologies*; ACM: New York, NY, USA, 2018; Volume 2, pp. 1–27. [CrossRef]

25. Haladjian, J.; Haug, J.; Nüske, S.; Bruegge, B. A wearable sensor system for lameness detection in dairy cattle. *Multimodal Technol. Interact.* **2018**, *2*, 27. [CrossRef]

26. Fogarty, E.; Swain, D.; Cronin, G.; Moraes, L.; Trotter, M. Can accelerometer ear tags identify behavioural changes in sheep associated with parturition? *Anim. Reprod. Sci.* **2020**, *216*, 106345. [CrossRef]

27. Navarro, J.; Martín de Diego, I.; Carballo Pérez, P.; Ortega, F. Outlier detection in animal multivariate trajectories. *Comput. Electron. Agric.* **2021**, *190*, 106401. [CrossRef]

28. Wilson, R.P.; White, C.R.; Quintana, F.; Halsey, L.G.; Liebsch, N.; Martin, G.R.; Butler, P.J. Moving towards acceleration for estimates of activity-specific metabolic rate in free-living animals: the case of the cormorant. *J. Anim. Ecol.* **2006**, *75*, 1081–1090. [CrossRef]

29. Shepard, E.L.; Wilson, R.P.; Quintana, F.; Laich, A.G.; Liebsch, N.; Albareda, D.A.; Halsey, L.G.; Gleiss, A.; Morgan, D.T.; Myers, A.E.; et al. Identification of animal movement patterns using tri-axial accelerometry. *Endanger. Species Res.* **2008**, *10*, 47–60. [CrossRef]

30. Lush, L.; Ellwood, S.; Markham, A.; Ward, A.; Wheeler, P. Use of tri-axial accelerometers to assess terrestrial mammal behaviour in the wild. *J. Zool.* **2016**, *298*, 257–265. [CrossRef]

31. Riaboff, L.; Aubin, S.; Bédère, N.; Couvreur, S.; Madouasse, A.; Goumand, E.; Chauvin, A.; Plantier, G. Evaluation of pre-processing methods for the prediction of cattle behaviour from accelerometer data. *Comput. Electron. Agric.* **2019**, *165*, 104961. [CrossRef]

32. Cooley, J.W.; Tukey, J.W. An Algorithm for the Machine Calculation of Complex Fourier Series. *Math. Comput.* **1965**, *19*, 297–301. [CrossRef]

33. Cochran, W.; Cooley, J.; Favin, D.; Helms, H.; Kaenel, R.; Lang, W.; Maling, G.; Nelson, D.; Rader, C.; Welch, P. What is the fast Fourier transform? *Proc. IEEE* **1967**, *55*, 1664–1674. [CrossRef]

34. Proakis, J.G.; Manolakis, D.G. *Digital Signal Processing: Principles, Algorithms and Applications*; Pearson: London, UK, 2006.

35. Breiman, L. Random Forests. *Mach. Learn.* **2001**, *45*, 5–32. [CrossRef]

36. Lohr, S.L. *Sampling: Design and Analysis*, 3rd ed.; Chapman and Hall/CRC: Boca Raton, FL, USA, 2021.

37. Duda, R.O.; Hart, P.E.; Stork, D.G. *Pattern Classification*, 2nd ed.; John Wiley & Sons: Hoboken, NJ, USA, 2001.

38. Hastie, T.; Tibshirani, R.; Friedman, J.H. *The Elements of Statistical Learning: Data Mining, Inference, and Prediction*, 2nd ed.; Springer Series in Statistics; Springer: New York, NY, USA, 2009. [CrossRef]

39. Xu, R.; Wunsch, D. *Clustering*; IEEE Press Series on Computational Intelligence; John Wiley & Sons: Hoboken, NJ, USA; IEEE Press: Hoboken, NJ, USA, 2009; Volume 10.
40. Watt, J.; Borhani, R.; Katsaggelos, A.K. *Machine Learning Refined: Foundations, Algorithms, and Applications*, 2rd ed.; Cambridge University Press: Cambridge, UK, 2020. [CrossRef]
41. Kroese, D.P.; Botev, Z.I.; Taimre, T.; Vaisman, R. *Data Science and Machine Learning: Mathematical and Statistical Methods*; Machine Learning & Pattern Recognition Series; Chapman & Hall/CRC Press: Boca Raton, FL, USA, 2019. [CrossRef]
42. Fawcett, T. An introduction to ROC analysis. *Pattern Recognit. Lett.* **2006**, *27*, 861–874. [CrossRef]
43. Flach, P.A. *Machine Learning. The Art and Science of Algorithms that Make Sense of Data*; Cambridge University Press: Cambridge, UK, 2012.
44. Navarro, J.; Diego, I.M.d.; Fernández-Isabel, A.; Ortega, F. Fusion of GPS and Accelerometer Information for Anomalous Trajectories Detection. In Proceedings of the 2019 the 5th International Conference on E-Society, e-Learning and e-Technologies; Association for Computing Machinery, Vienna, Austria, 10–12 January 2019; ACM: New York, NY, USA, 2019; pp. 52–57. [CrossRef]
45. Aminikhanghahi, S.; Cook, D.J. A survey of methods for time series change point detection. *Knowl. Inf. Syst.* **2017**, *51*, 339–367. [CrossRef] [PubMed]
46. Kluever, B.M.; Howery, L.D.; Breck, S.W.; Bergman, D.L. Predator and heterospecific stimuli alter behaviour in cattle. *Behav. Process.* **2009**, *81*, 85–91. [CrossRef] [PubMed]
47. Kluever, B.M.; Breck, S.W.; Howery, L.D.; Krausman, P.R.; Bergman, D.L. Vigilance in cattle: the influence of predation, social interactions, and environmental factors. *Rangel. Ecol. Manag.* **2008**, *61*, 321–328. [CrossRef]
48. Hancock, J. Studies of grazing behaviour in relation to grassland management I. Variations in grazing habits of dairy cattle. *J. Agric. Sci.* **1954**, *44*, 420–433. [CrossRef]
49. Charlton, G.L.; Rutter, S.M. The behaviour of housed dairy cattle with and without pasture access: A review. *Appl. Anim. Behav. Sci.* **2017**, *192*, 2–9. [CrossRef]
50. Phillips, C. *Cattle Behaviour and Welfare*; John Wiley & Sons: Hoboken, NJ, USA, 2008.
51. de Freslon, I.; Martínez-López, B.; Belkhiria, J.; Strappini, A.; Monti, G. Use of social network analysis to improve the understanding of social behaviour in dairy cattle and its impact on disease transmission. *Appl. Anim. Behav. Sci.* **2019**, *213*, 47–54. [CrossRef]

MDPI

St. Alban-Anlage 66

4052 Basel

Switzerland

Tel. +41 61 683 77 34

Fax +41 61 302 89 18

www.mdpi.com

Entropy Editorial Office

E-mail: entropy@mdpi.com

www.mdpi.com/journal/entropy